Reading Kant's Geography

Edited by
Stuart Elden
and
Eduardo Mendieta

SUNY
PRESS

Published by State University of New York Press, Albany

Printed in the United States of America

For information, contact State University of New York Press, Albany, NY
www.sunypress.edu

Production by Eileen Meehan
Marketing by Michael Campochiaro

Library of Congress Cataloging-in-Publication Data

Reading Kant's Geography / edited by Stuart Elden and Eduardo Mendieta.
p. cm. — (SUNY series in contemporary Continental philosophy)
Includes bibliographical references and index.
ISBN 978-1-4384-3605-0 (hardcover : alk. paper)
ISBN 978-1-4384-3604-3 (pbk. : alk. paper)
1. Kant, Immanuel, 1724–1804. Physische Geographie. 2. Geography—Philosophy. I. Elden, Stuart, 1971– II. Mendieta, Eduardo.

B945.D44R45 2011
910'.02—dc22 2010032057

10 9 8 7 6 5 4 3 2 1

Reading Kant's Geography

SUNY series in Contemporary Continental Philosophy

Dennis J. Schmidt, editor

Contents

Acknowledgments

This project began in a conversation between the editors over a beer in Stony Brook, Long Island, in September 2006. It led to sustained discussions over the next few years. A Leverhulme research fellowship allowed Stuart a period of time in New York in the autumn of 2007 working on this project; and a British Academy conference grant enabled a workshop to be held at Durham University in January 2008, which was also supported by the Department of Geography, and ably assisted by Kathy Wood. An earlier workshop was held at Stony Brook Manhattan in November 2007, supported by the Department of Philosophy, Stony Brook University. We would like to thank Domenica Tafuro for logistical support in organizing that workshop. Some of the ideas in the introduction were discussed at a seminar in the Department of Social and Economic Geography, University of Uppsala, Sweden. Stuart is grateful to Gunnar Olsson and Christian Abrahamsson for the invitation. Many of the ideas in the introduction are presented in longer form in "Reassessing Kant's *Geography*."[1]

Above all, we would like to thank the participants in those workshops, where most of the contributors to this volume presented their ideas in an environment of intellectual solidarity. Olaf Reinhardt's translations of Stark's contributions are gratefully acknowledged. A special note of thanks is due to Sam Butler, who not only undertook the translation of Marcuzzi's essay, but also helped with the final assembling of the manuscript, and to Caitlin Woolsey for her work on the index. We also want to thank Dennis Schmidt, Jane Bunker, and their colleagues at SUNY Press for their enthusiastic and unqualified support of this project.

Note

1. Stuart Elden, "Reassessing Kant's Geography," *Journal of Historical Geography* 35:1 (2009): 3–25.

1

Reintroducing Kant's Geography

Stuart Elden

> One can take the classification of organic and living beings further. Not only does the vegetable kingdom exist for the sake of the animal kingdom (and its increase and diversification) but humans, as rational beings, exist for the sake of others of a different species (race). The latter stand at a higher level of humanity, either simultaneously (as, for instance, Americans and Europeans) or sequentially. If our earth-globe [*Erglob*] (having once had been dissolved into chaos, but now being organized and regenerating) were to bring forth, by revolutions of the earth differently organized creatures, which, in turn, gave place to others after their destruction, organic nature could be conceived in terms of a sequence of different world epochs, reproducing themselves in different forms, and our earth as an organically formed body—not one formed merely mechanically.[1]

At the University of Königsberg Immanuel Kant lectured on a variety of topics, including both philosophical and non-philosophical topics.[2] The lecture courses were often well attended, they were widely discussed outside of the classroom, and we have many of them in the form of student transcripts, as well as from some of Kant's own manuscripts and lecture notes. Kant's lectures on logic, metaphysics, and ethics are well known parts of his complete works, and are invaluable sources of knowledge and understanding about his work, its substance, coherence and development. In addition to these subjects, however, Kant also gave courses on anthropology and physical geography. Geography was usually offered in the summer semester while anthropology was given in the winter, and geography was offered forty-nine times over a forty-year period from 1756–96—more frequently than any of his other topics other than logic and metaphysics.[3] Spanning his entire teaching career they thus serve as archeological registers of Kant's work chronicling accretions and shifts in thought.

While the anthropology course was worked up in a book by Kant himself, appearing as *Anthropology from a Pragmatic Point of View* a few years before his death, the lectures on geography had a rather different fate.[4] Indeed, in 1798 Kant thought that a version of them was "scarcely possible" at his own advanced age, for the manuscript he used to lecture from was one he believed only he could read.[5] An unauthorized edition edited by Gottfried Vollmer appeared in 1801.[6] Kant then entrusted the task to Theodor Rink, who made use of student transcripts to produce an edition which appeared the next year.[7] While both editions had a wide circulation at the time, it is the Rink edition that has been seen as the official one, being reprinted in the Akademie Edition of *Kants gesammelte Schriften*. In the early part of the twentieth century, Erich Adickes attempted to get a new version produced but his suggestion was declined. The first full English translation of Kant's *Physical Geography*, a translation based on the Rink edition, is due to appear in the *Cambridge Edition of the Works of Immanuel Kant in Translation* in the near future, in the volume on *Natural Science*—over two hundred years since Kant's death.

Why did Kant lecture on geography? Although it became one of his most popular and best attended courses, and this initially for a *Privatdozent* who existed on student payments, this does not explain things sufficiently. Wilson is valuable in tracking the changing objectives for the geography lectures, suggesting that initially they were "purely scientific, that is, to make a more certain knowledge of believable travel accounts, and to make this into a legitimate academic course of study."[8] But the popularity of the course meant that Kant could begin to suggest that their aim could be "to civilise young students to become 'citizens of the world.' "[9] Zammito has similarly shown how the lectures are related to the *Anthropology* in providing knowledge, but stresses this is for a philosophical purpose.[10] As Louden notes, therefore, their aim was more than merely scholastic, but rather:

> The anthropology and physical geography lectures are thus not primarily intended as further contributions to Kant's critical, transcendental philosophy program . . . [which] was not his only concern. A major portion of Kant's teaching activity was devoted to trying to enlighten his students more about the people and world around them in order that they might live (pragmatically as well as morally) better lives.[11]

For Louden, anthropology and geography are thus "intersecting halves of a larger whole."[12] The problematic link between Kant's views on geography and anthropology, and, especially, on race and his cosmopolitan ethics are highlighted below, but the point here is somewhat different. This is that Kant sees these lectures as providing a broad knowledge of the world as a foundation to the more general studies of his students, and that together the physical geography and pragmatic anthropology give an empirical grounding for his thought more generally. In a postscript to his 1775 article "*Von den verschiedenen Racen der Menschen* [On the Different Races of Human Beings]" Kant suggested that the two lecture courses together were *Weltkenntnis*.[13] This would usually translate as "world-knowledge," but Wilson has suggested the felicitous "cosmopolitan knowledge," with "cosmopolitan philosophy" for the related *Weltwissenschaften*.[14] This knowledge of the world, for Kant, was integral to the moral and political life of the citizen. Both geography and anthropology were taught by Kant because of their "pragmatic" dimension, the way in which this knowledge can guide us in our moral and practical life.

This world-knowledge, this cosmology, is essential to his other writings. Kant suggests that physical geography is about the world as an "object of external sense"; and anthropology as an "object of inner sense."[15] Wilson is therefore clear that the lectures on anthropology must be seen as philosophy:

> Kant explicitly argues that the anthropology is a type of cosmopolitan philosophy. It is not a scholastic philosophy, and it is not critical philosophy, but it is a type of philosophy . . .
>
> The twofold field of physical geography and anthropology are viewed cosmologically and pragmatically. In other words, Kant considered these two disciplines, in the way he taught them, to be philosophy, and philosophy that was useful for the world.[16]

These lectures were to serve as a propaedeutic for "practical reason," and are a "history of the contemporary condition of the earth or geography, in the widest sense."[17] This, for Kant, is the preliminary exercise in the knowledge of the world."[18] Knowledge of the world is thus of both "the human being and nature."[19] Physical geography studies nature, anthropology the human, but the latter outweighs the former, since "nature exists for the sake of the human being. The human being is the end of nature."[20] Nonetheless, the twofold field of *Weltkenntis* needs to be treated cosmologically.[21] Anthropology and geography are thus crucial to Kant's entire enterprise, through his career and across the so-called pre-critical and critical periods.

The ways these two aspects of Kant's work have been treated has differed dramatically. The *Anthropology* has been available in English for several years, with translations in 1974 and 1978 and an entirely new recent translation by Robert Louden. Meanwhile, a complete and reliable translation of the *Geographie* is still forthcoming (although parts of it have been available in English since the late 1960s).[22] Kant's work on anthropology has been discussed by figures of the stature of Martin Heidegger and Michel Foucault.[23] The *Geography*, in contrast, has not received anything like the same amount of attention. It generally merits an entry in dictionaries of Kant's work, but these tend to be pretty brief.[24] There is no explicit discussion of these lectures in *A Companion to Kant*,[25] and a recent edited book on *Kant and the Sciences* makes only a tangential reference to geography.[26] Robert Hanna's comprehensive study, *Kant, Science, and Human Nature*, makes only two passing references to geography and offers no sustained engagement.[27] Many other accounts on related topics are similar. The same neglect can be found in works concentrating on Kant's theories of space.[28]

In part to remedy this glaring neglect, this collection discusses Kant's work on geography—predominantly focusing on the text, or texts, of the *Physische Geographie*, but not confined to that work. It is the product of conversations—initially between the editors, but then broadening to a wide range of contributors across different disciplines. The initial conversations between Eduardo and myself were illuminating, since the pairing of our intellectual interests—a political theorist who now works in a geography department and a spatially minded philosopher with an expertise on race—seemed to cover many of the key bases. Yet we quickly realized that the richness and complexity of Kant's thoughts required a project of many more hands. The texts and contexts, the content, and the relation of the ideas to Kant's work as a whole and to a range of other issues provide an extensive set of positions from which to approach this work.

Accordingly, this book seeks to provide a range of essays discussing, contextualizing, and criticizing Immanuel Kant's work on geography. It brings together scholars of geography, philosophy, and related disciplines to allow a broad discussion of the importance of Kant's text for philosophical and geographical work, both historically and in the contemporary context. At the moment when the wider English language audience will have access to Kant's work on geography in translation, this book will offer a range of ways of interpreting that text, but also criticizing it. Unlike most other topics in Kant scholarship, we are able to build on little preceding work, although the pioneering studies of J. A. May and Erich Adickes are given appropriate references throughout.[29]

Contexts

The book is structured around a number of themes. We first situate Kant's work in its context, before discussing a range of textual issues concerning translation, the edition, and its conceptual claims concerning the relation between history and geography. The other sections are thematic, looking at three key themes: the relation between geography and anthropology; the question of the relation of geography to Kant's critical thought as a whole or what is here called the "geography of reason"; and the issues of gender, race, history and geography.

Kant understood geography in a very broad sense, including much of what we would today understand as human geography under his title of physical geography. Kant was an innovator in geography, if for no other reason than that he was one of the very first to lecture on it as an explicit topic, before it was common to have chairs in geography in Germany. While others lectured on it in a way that was more akin to travel writing, Kant attempted to systematise the subject, synthesizing insights from a range of different sources. Indeed, his outline for the course was unique, and he had to ask for special dispensation from the Minister of Education in order to give a course for which no textbook could be found.[30] The more accurately "physical geography" elements include descriptions of the earth and its terrain; earthquakes and the nature of electricity; climate, the atmosphere. and temperature; and rivers and water. There are also extensive discussions of flora, fauna, and minerals. The final part of the lectures comprised a series of descriptions of particular regions and places in the world. There are many issues and questions in the history of ideas that need to be understood. This book therefore opens with two remarkable chapters, from a historical geographer and a physical geographer, locating and interrogating Kant's position within the history of geographical thought. Michael Church discusses the way Kant's ideas fit into discussions of physical geographical knowledge as a whole; Charles Withers offers perspectives on Kant within the history of the discipline.

Church contests the standard view that the geography of Kant's period was largely practical, and that Kant's role was important in terms of the codification and ordering of knowledge. Rather, Church shows that Kant was one of a number of compilers of knowledge, and that his work appeared just as the focus shifted to more field-based science, partly based on the studies being undertaken in the new world. The relative neglect of Kant's geographical work is, he suggests, closely bound up with this development in the history of the discipline. Withers seeks to show how Kant was not, as geographer, the *sui generis* figure that he is sometimes seen to be in other fields, but rather one whose geographical work links to a wide range of debates. To make this comparative study, Withers claims require an understanding of geography in relation to the Enlightenment thought more generally, which he provides in his essay. As the tensions between these two essays illustrate, the history of ideas is not a straightforward story. Kant's interlocutors, his inspirations and sources, and, in turn, those he influenced are legion. Whether or not that influence bears any relation to his geography is open to question, and one of the key concerns of these essays.

Texts

Criticism of Kant's *Geography* must not merely be for its content, but also for its form. As Werner Stark shows in his contribution, the Rink edition is hopelessly corrupt, and extremely problematic as the basis for any careful study of Kant.

There are thus serious philological difficulties relating to reading the lectures that go beyond their inaccessibility in English. Based on his painstaking archival work compiling all known student transcripts for the Akademie edition, Stark demonstrates how Rink worked, as well as highlighting some of the key issues that must be understood as a basis to a thorough hermeneutic. There are many textual and linguistic issues for us to grapple with as we approach Kant's text. As a supplement to Stark's essay we include his discussion of one of the most notorious passages of the Rink edition, concerning race, showing how this philological approach complicates our understanding of what Kant said and wrote.

The process of translating Kant into English is fraught with difficulty, and we are therefore pleased to be able to include an essay by the translator of Kant's *Physical Geography* for the Cambridge Edition, Olaf Reinhardt. In this essay, Reinhardt looks at the technical issues concerning translation, noting that for specialized texts it is often a case of translating from one foreign language to another. Yet while the technical language of geography is important, it is incumbent on any philosophical translation that it retain a consistency with other texts by the same author, for which the Cambridge Edition has set high standards.

The final chapter of this section is by one of the translators of the French edition of the lectures, Max Marcuzzi. In this essay, specially translated for this volume by Samuel Butler, Marcuzzi discusses the relation between history and geography in Kant's lectures, focusing especially on the introduction to the *Physical Geography.* In doing so, Marcuzzi opens up a number of key themes in that text that relate to Kant's wider concerns, including the understandings of space and time; the relation between science and philosophy; race and breeding; and the relation between geography and anthropology. It thus acts as an effective bridge into the concerns of the second half of this book.

Towards a Cosmopolitan Education: Geography and Anthropology

Kant saw a particular relation between his geography and anthropology lectures, and indeed the anthropology lectures were initially part of the geography course. Even when he split them apart, Kant continued to see them as closely related. He believed that physical geography and pragmatic anthropology together provided *Weltkenntnis*, knowledge of the world, an empirical grounding for his thought. This knowledge of the world, for Kant, was integral to the moral and political life of the citizen. Both geography and anthropology were taught by Kant because of their "pragmatic" dimension, that is to say, they way in which this knowledge can guide us in our moral and practical life. This world-knowledge, or cosmology, furthermore, is essential to his other writings, suggesting that physical geography is about the world as an "object of external sense"; and anthropology as an "object of inner sense."[31] Knowledge of the world is thus of both "the human being and nature,"[32] and anthropology and geography are thus "intersecting halves of a larger whole."[33]

In 1765–66 Kant offered a detailed discussion of how his work had developed over the past decade:

> I have gradually expanded this scheme, and now I propose, by condensing that part of the subject which is concerned with the physical features of the earth, to gain the time necessary for extending my course of lectures to include the other parts of the subject, which are of even greater utility. This discipline will therefore be a *physical, moral* and *political geography.* It will contain, first of all, a specification of the remarkable features of *nature* in three realms. The specification will, however, be limited to those features, among the innumerably many which could be chosen, which particularly satisfy the general desire for knowledge, either because of their rarity or the effect which they can exercise on states by means of trade and industry. This part of the subject, which also contains a treatment of the natural relationship which holds between all the lands and seas in the world, and the reason for their connection, is the essential foundation of all history. Without this foundation, history is scarcely distinguishable from fairy-tales.
>
> The *second* part of the subject considers *human beings,* throughout the world, from the point of view of the variety of their natural properties and the differences in that feature of the human which is moral in character. The consideration of these things is at once very important and also highly stimulating as well. Unless these matters are considered, general judgments about man would scarcely be possible. The comparison of human beings with each other, and the comparison of the human today with the moral state of the human in earlier times, furnishes us with a comprehensive map of the human species. *Finally,* there will be a consideration of what can be regarded as a product of the reciprocal interaction of the two previously mentioned forces, namely, the condition of the *states* and nations throughout the world. The subject will not be considered so much from the point of view of the way in which the condition of states depends on accidental causes, such as the deeds and fates of individuals, for example, the sequence of governments, conquests and intrigues between states. The condition of states will rather be considered in relation to what is more constant and which contains the more remote ground of those accidental causes, namely, the situation of their countries, the nature of their products, customs, industry, trade and population.[34]

Kant therefore sets out a range of distinctions—the physical, moral, and political geography is alluded to here, but the actual analysis is rather more complicated. By the mid 1770s he offers a range of possibilities:

- Physical geography: the foundation or ground for other types of geography as well as history—a general study or outline of nature;
- Mathematical geography: concerned with the measure of the shape, size and motion of the earth, and its situation in the solar system;
- Moral geography: the relation between moral codes and customs and regions, a kind of spatial differentiation;
- Political geography: the relation of political systems and political laws to physical features of geography, part of the reason why these are only nominally universal;
- Commercial [*Handlungs*] geography: concerned with the geographical elements of trade in surplus products;
- Theological geography: concerned with theological attitudes and principles and their relation to physical features of the landscape; again a form of spatial differentiation.[35]

Physical geography is "the physical description of the earth [and] is the first part of knowledge of the world."[36] Indeed, in his essay on "The Conflict of the Faculties," Kant divides the philosophy faculty into two parts—the one that deals with "pure rational knowledge" and one that deals with "*historical knowledge*." The former contains metaphysics of nature and morals, along with pure philosophy and mathematics; the latter includes history, geography, philology, the humanities, and the empirical knowledge of the natural sciences.[37] These therefore produce some *philosophical difficulties*, particularly concerning how we should see these lectures in relation to Kant's work as a whole.

In his *Logic*, Kant suggests that there are four fundamental questions.

1. "What can I know?"
2. "What ought I to do?"
3. "What may I hope?"
4. "What is the human being?"

Kant suggests that "*Metaphysics* answers the first question, *morals* the second, *religion* the third, and *anthropology* the fourth. Fundamentally, however, we could reckon all of this to anthropology, because the first three questions refer to the last one."[38] Just as these other realms of thought rest on the fundamental question, namely anthropology; so too do Kant's reflections on the material world rest on the understanding of geography. For Kant, knowledge of the world is not pragmatic merely when it is "extensive knowledge of *things* in the world, for example, animals, plants and minerals from various lands and climates—but only when it contains knowledge of the human being as a *citizen of the world*."[39] In other words, pragmatic anthropology is the relation of the human to the world, and is thus what Robert Louden calls "impure ethics."[40] Although much of the material in the geography lectures would also come under that remit—parts of the *Anthropology* derived from earlier lectures on geography[41]—we might make a case that much of the rest of the geography is "impure physics," in other words the empirical detail that inhabits the categories of abstract thought.[42]

Two of the most important scholars of the anthropology—Robert Louden and Holly Wilson—offer a range of perspectives on the relation between these two texts. Louden focuses on the role of the human in the physical geography, but more broadly in terms of the relation of the human to nature. Louden's claim is that attempts to draw a strict demarcation between geography and anthropology fail, and that Kant's analyses of what we would today call "human geography" are inadequate in the *Physical Geography*, and, especially when they concern race, deeply problematic. Wilson takes a rather different approach, concentrating on how Kant saw these lectures as educational, and underlines their "pragmatic" purpose. One of their aims, Wilson contends, is to show how natural events can be explained without reference to the will of God, stressing a significant emphasis on the question of causality in the text. In the final essay of this section, David Morris shows how Kant's work on nature generally, and the question of the organism specifically, relates to his work on reason and questions of teleology. In doing so, Morris begins to open up the questions of the next part of the book.

Kant's Geography of Reason: Reason and Its Spatiality

These next essays are concerned with showing how Kant's thought as a whole was concerned with geographical questions. While in part an attempt to provide a new key to unlocking Kant's thought, more generally it seeks to deepen and enrich already existing analyses. Jeff Malpas and Karsten Thiel show how Kant can be understood as pursuing a "geography of reason"; and Onora O'Neill thinks about the way spatial and political categories function in Kant's thought more generally. Jeff Edwards shows how Kant's work as a whole offers a particular political geography, given its concern with the question of land and its acquisition. The problems of community, and the relation of groups to the places they inhabit, are key concerns in Kant's treatment of private law. These essays thus begin a valuable project for understanding Kant's geography beyond his *Geography*. Kant's position within the discipline of geography is too often reduced to a caricature, again more often cited than read. We believe a careful study of the *Geography* needs to take place alongside a reexamination of the role of geography and space in his critical philosophy generally and his work on the philosophy of history and political history more specifically. Together they raise a range of political and philosophical questions.

Gender, Race, History, and Geography

One of the key political issues in Kant's work over recent years has been his understanding of race. Race, for instance, is only very briefly discussed in the *Anthropology*,[43] but at much greater length in the *Geography*. The latter text includes discussions of what is called "moral geography" concerning the "customs and characters" of different peoples,[44] and some extensive discussion of race. In a key essay, David Harvey related the lectures to the interest in Kant's cosmopolitanism,[45] suggesting this renders this particular concept deeply problematic. Harvey notes that many Kantians want to dismiss the work on geography as " 'irrelevant,' 'not to be taken seriously' or [suggest that] it 'lacks interest,' "[46] in much the same way that Benno Erdmann described the *Anthropology* as the "laborious compilation of a seventy-four year old man as he stood on the threshold of decrepitude."[47] For Harvey "the content of Kant's *Geography* is nothing short of an intellectual and political embarrassment."[48] One of the aims of this book was to create a space for an encounter between Harvey and philosophers who were taking the *Geography* seriously. Following Harvey's updated account of these claims, we publish a response by Edward S. Casey, originally delivered at the Stony Brook Manhattan workshop.

There follow three essays that pick up, in different ways, on these issues. Robert Bernasconi offers a detailed analysis of how the geography lectures fit into his pioneering analysis of Kant's racial thought.[49] The essay explicitly engages those, like Pauline Kleingeld,[50] who have attempted to show that Kant had "second thoughts" on racial questions, particularly in the 1790s. Bernasconi gives Kleingeld due credit for shifting the terrain of the Kant and race debate onto the terrain of these lectures, but contests, point by point, her analysis and the claims of others such as Sankar Muthu and Peter Fenves. Walter Mignolo's essay provides a helpful development of these claims, offering a reading against Kant, trying to show how his work is both part of the problem of a colonial mode of thought but also, paradoxically offers resources for thinking against it. His reading is particularly insightful in looking at the structure of Kant's geography lectures, especially in their treatment of parts of the world.

The final chapter is by this volume's co-editor, Eduardo Mendieta, and serves both as a contribution in its own right as well as an epilogue to the volume as a whole. While again treating questions of race, Mendieta also forces us to think about Kant's attitude to women; the relation of sexual difference to racial difference, and both to the distinction between history and geography. The chapter thus provides a recapitulation of many of the key themes of this volume, as well as opening a range of questions for how work on this aspect of Kant's work might develop in the future. These essays, then, provide the basis upon which we might rise to Harvey's challenge: to examine the relation of Kant to contemporary cosmopolitan thought. In this sense, in this section the book moves from being a contribution to Kant scholarship to one that speaks to wider political and social concerns, especially concerning post-colonial or de-colonial thought.

Conclusion

In summary, we attempt to show that Kant's work on geography is not simply a minor concern, but a key topic that needs to be taken much more seriously by scholars of his thought generally. While much of the detail of his lectures may be outdated and therefore of merely historical interest, Kant's way of structuring geographical knowledge and its relation to his thought as a whole is potentially of enduring importance. This importance lies both in the way he understands geography as a counterbalance to history, and in terms of the organization of knowledge. All perceived things are located in logical classifications such as those of Linnaeus; and in space and time. Logic deals with the first; physics with space and time, and of these, geography deals with space—history with time.[51] Geography therefore allows us access to the ordering and categorizing of the world. Indeed, Kant distinguishes geography as the description of the whole world from topography as the description of single places and chorography as that of regions. Orography and hydrograpy—the description of mountains and water—are also mentioned as divisions.[52]

The forthcoming translation of these lectures provides an opportune moment to take stock of their historical importance and contemporary relevance. Yet the availability of the material should come with a warning. Those reading that translation need to be aware of at least two key things: how corrupt the version being translated is, and an awareness of the debates on and around the lectures in recent years. This book provides the basis for understanding just such issues. To read Kant's work on geography is an inherently interdisciplinary venture, which encompasses both human and physical geography and philosophy, but also German studies, anthropology and race studies; and this why the volume of essays include contributions from a range of disciplines. The issues raised by these texts of Kant's are textual and linguistic, philological and hermeneutic, philosophical and political, even as we consider their relation to geography and the wider history of ideas. Only a multidisciplinary, and multihanded, approach can do justice to their complexity. This book seeks to rise to that challenge.

Notes

1. Immanuel Kant, *Opus Postumum*, ed. Eckart Förster, trans. Eckart Förster and Michael Rosen (Cambridge: Cambridge University Press, 1993), 66–7; translation slightly modified. See Peter D. Fenves, *Late Kant: Towards Another Law of the Earth* (New York: Routledge, 2003).

2. For an excellent resource on "Kant in the Classroom," see Steve Naragon, "Kant in the Classroom," Manchester College, http://www.manchester.edu/kant/Home/index.htm.

3. These figures differ in various accounts, but the most reliable source is Robert Louden, *Kant's Impure Ethics: From Rational Beings to Human Beings* (Oxford: Oxford University Press, 2000), 5.

4. Immanuel Kant, *Anthropology from a Pragmatic Point of View*, ed. and trans. Robert B. Louden (Cambridge: Cambridge University Press, 2006). These lectures also exist in different forms. Volume 25 of the Akademie Edition of *Kants gesammelte Schriften* (hereafter AK, followed by volume and page number) is devoted to variant drafts; selections from which are forthcoming in the *Cambridge Edition* as *Lectures on Anthropology*, ed. Allen W. Wood and Robert B. Louden (Cambridge: Cambridge University Press).

5. Kant, *Anthropology from a Pragmatic Point of View*, 6n.

6. *Immanuel Kants Physische Geographie*, 4 vols., ed. Gottfried Vollmer (Mainz: Vollmer, 1801–05).

7. *Immanuel Kants Physische Geographie*, ed. Friedrich Theodor Rink (Königsberg, 1802), here cited according to AK 9.

8. Holly L. Wilson, *Kant's Pragmatic Anthropology: Its Origin, Meaning and Critical Significance* (Albany: SUNY Press, 2006), 9.

9. Wilson, *Kant's Pragmatic Anthropology*, 8.

10. John H. Zammito, *Kant, Herder, and the Birth of Anthropology* (Chicago: University of Chicago Press, 2002), 285–6. See also Chenxi Tang, *The Geographic Imagination of Modernity: Geography, Literature, and Philosophy in German Romanticisim* (Stanford: Stanford University Press, 2008), 98–123.

11. Louden, *Kant's Impure Ethics*, 65; see Wilson, *Kant's Pragmatic Anthropology*, 20.

12. Louden, *Kant's Impure Ethics*, 95.

13. *Von den verschiedenen Racen der Menschen*, AK 2: 443.

14. See Wilson, *Kant's Pragmatic Anthropology*, 20, 113–5.

15. *Physical Geography*, AK 9: 156–7; see *Anthropology from a Pragmatic Point of View*, ed. and trans. Louden.

16. Wilson, *Kant's Pragmatic Anthropology*, 5, 115.

17. *Nachricht von der Einrichtung seiner Vorlesungen in dem Winterhalbenjahre von 1765–1766*, AK 2: 312.

18. *Von den verschiedenen Racen der Menschen*, AK 2: 443. See also the remarks in §70 of the Education lectures, cited by Joseph A. May, *Kant's Concept of Geography and its Relation to Recent Geographical Thought* (Toronto: University of Toronto Press, 1970), 132.

19. *Physical Geography*, AK 9: 158.

20. *Anthropologie Friedländer*, AK 25: 470; *Anthropologie Pillau*, AK 25: 733; cited in Louden, *Kant's Impure Ethics*, 95.

21. *Von den verschiedenen Racen der Menschen*, AK 2: 443.

22. Roland L. Bolin, "Immanuel Kant's *Physical Geography*," (MA Thesis, University of Indiana, 1968).

23. See Martin Heidegger, *Kant and the Problem of Metaphysics*, trans. Richard Taft (Bloomington: Indiana University Press, 1997). Foucault translated the *Anthropology* as *Anthropologie du point de vue pragmatique* (Paris: Vrin, 1964). This edition includes only a brief "Notice historique," which is all that was published of Foucault's secondary thesis on the *Anthropology*. See Michel Foucault, "Introduction à l'Anthropologie de Kant," (Thèse complémentaire pour le doctorat dès letters, 1961).

24. See Howard Caygill, *A Kant Dictionary* (Oxford: Blackwell, 1995), 214–5; Helmut Holzhey and Vilem Mudroch, *Historical Dictionary of Kant and Kantianism* (Lanham: Scarecrow Press, 2005), 131–2.

25. Graham Bird, ed. *A Companion to Kant* (Oxford: Blackwell, 2006).

26. Eric Watkins, ed. *Kant and the Sciences* (Oxford: Oxford University Press, 2001).

27. Robert Hanna, *Kant, Science, and Human Nature* (Oxford: Oxford University Press, 2006), 56–7, 268–70.

28. See, for example, Christopher Browne Garnett Jr., *The Kantian Philosophy of Space* (New York: Columbia University Press, 1939); Arthur Melnick, *Space, Time, and Thought in Kant* (Dordrecht: Kluwer, 1989); Henry Sidgwick, "Kant's 'Exposition' of Space and Time," in *Immanuel Kant: Critical Assessments*, 4 vols., ed. Ruth F. Chadwick and Clive Cazeaux (London: Routledge, 1992), 1: 89–102; and Michael D. Newman, "The Unity of Time and Space, and its Role in Kant's Doctrine of A Priori Synthesis," in *Immanuel Kant: Critical Assessments*, 2: 185–200.

29. Earlier discussions include Richard Hartshorne, "The Concept of Geography as a Science of Space, from Kant and Humboldt to Hettner," *Annals of the Association of American Geographers* 48:2 (Jun 1958): 97–108; May, *Kant's Concept of Geography and its Relation to Recent Geographical Thought*; Paul Richards, "Kant's Geography and Mental Maps," *Transactions of the Institute of British Geographers* 61:1 (Mar 1974): 1–16; and D. N. Livingstone and R. T. Harrison, "Immanuel Kant, Subjectivism and Human Geography: A Preliminary Investigation," *Transactions of the Institute of British Geographers* 6:3 (1981): 359–74. May's is by far the most thorough account in English. There are also brief discussions in survey histories of the discipline, such as Richard Hartshorne, "The Nature of Geography: A Critical Survey of Current Thought in the Light of the Past," *Annals of the Association of American Geographers* 29:3 (Sep 1939): 173–412; 29:4 (Dec 1939): 413–658 (reprinted in book form in Lancaster: Annals of the Association of American Geographers, 1939); David N. Livingstone, *The Geographical Tradition: Episodes in the History of a Contested Enterprise* (Oxford: Blackwell, 1992), 113–7.

30. Louden, *Kant's Impure Ethics*, 184–5 n. 6, drawing on Karl Vorländer, *Immanuel Kants Leben* (Leipzig: Felix Meiner, 1911), 41–3.

31. *Physical Geography*, AK 9: 156–7.

32. *Physical Geography*, AK 9: 158.

33. Louden, *Kant's Impure Ethics*, 95.

34. *M. Immanuel Kant's Announcement of the Programme of his Lectures for the Winter Semester 1765–1766*, in *Theoretical Philosophy 1755–1770*, ed. and trans. David Walford (Cambridge: Cambridge University Press, 1992), 298–9; AK 2: 312–3, translation modified.

35. *Physical Geography*, AK 9: 160–1, 164–5. Rink changes the order of their first presentation, adding "literary geography" (161), which is not in the 1774 transcript or in the fuller elaboration. He also adds detail to the elaboration, usually based on material later delivered in the lectures themselves. Some of the changes—"civil society" for "society" or the replacement of "*Handlungs Geographie*" with "*merkantilische Geographie*"—are more interpretative. May claims that "his concept of the limits and scope of geography is inevitably much broader than any contemporary concept can reasonably be" (*Kant's Concept of Geography*, 153).

36. *Physical Geography*, AK 9: 157. As Brian Jacobs ("Kantian Character and the Science of Humanity," in *Essays on Kant's Anthropology*, ed. Brian Jacobs and Patrick Kain (Cambridge: Cambridge University Press, 2003), 132 n. 49) notes, in Kant's *Groundwork for the Metaphysic of Morals*, the same role is played by physics.

37. Immanuel Kant, "The Conflict of the Faculties," in *Religion and Rational Theology*, ed. and trans. Allen W. Wood and George Di Giovanni (Cambridge: Cambridge University Press, 1996), 256.

38. Immanuel Kant, *Lectures on Logic*, ed and trans. J. Michael Young (Cambridge: Cambridge University Press, 1992), 538. The first three questions appear in *Critique of Pure Reason*, trans. Paul Guyer and Allen W. Wood (Cambridge: Cambridge University Press, 1998), A805/B853.

39. Kant, *Anthropology from a Pragmatic Point of View*, 4.

40. Louden, *Kant's Impure Ethics*.

41. Benno Erdmann, *Reflexionen Kants zur Anthropologie* (Leipzig: Fues, 1882), 48; cited and discussed in Louden, *Kant's Impure Ethics*, 62–3.

42. See Immanuel Kant, "Metaphysical Foundations of Natural Science," trans. Michael Friedman, in *Theoretical Philosophy after 1781*, ed. Henry Allison and Peter Heath (Cambridge: Cambridge University Press, 2002), 181–270.

43. Kant, *Anthropology from a Pragmatic Point of View*, 223–4. However, note that Kant suggests in that work (199) that the observations on the relation of physiognomy to race "belong more to physical geography than pragmatic anthropology."

44. *Physical Geography*, AK 9: 164.

45. David Harvey, "Cosmopolitanism and the Banality of Geographical Evils," *Public Culture* 12:2 (2000): 529–64; and "Geographical Knowledges/Political Powers," *Proceedings of the British Academy* 122 (2004): 87–115.

46. Harvey, "Cosmopolitanism and the Banality of Geographical Evils": 532.

47. Erdmann, *Reflexionen Kants zur Anthropologie*, 37; cited in Robert B. Louden, "The Second Part of Morals," in *Essays on Kant's Anthropology*, ed. Jacobs and Kain, 60.

48. Harvey, "Cosmopolitanism and the Banality of Geographical Evils": 532.

49. Robert Bernasconi, "Who Invented the Concept of Race? Kant's Role in the Enlightenment Construction of Race," in *Race*, ed. Bernasconi (London: Blackwell, 2001), 11–36; and "Will the Real Kant Please Stand Up: The Challenge of Enlighten-

ment Racism to the Study of the History of Philosophy," *Radical Philosophy* 117 (Jan/Feb 2003): 13–22. The literature on these questions is now extensive, and is discussed in Bernasconi's essay in this volume. See also David Farrell Krell, "The Bodies of Black Folk: From Kant and Hegel to Du Bois and Baldwin," *boundary 2* 27:3 (2000): 103–34; and more generally *The Tragic Absolute: German Idealism and the Languishing of God* (Bloomington: Indiana University Press, 2005).

50. Pauline Kleingeld, "Kant's Second Thoughts on Race," *The Philosophical Quarterly* 57:229 (Oct 2007): 573–92.

51. *Physical Geography*, AK 9: 159–60, 162. For a criticism of this division, see Fred K. Schaefer, "Exceptionalism in Geography: A Methodological Examination," *Annals of the Association of American Geographers* 43:3 (Sep 1953): 226–49.

52. *Physical Geography*, AK 9: 159.

I

The Invention of Geography

Kant and His Times

2

Immanuel Kant and the Emergence of Modern Geography[1]

Michael Church

Introduction

At the turn of the twentieth century, the German geographer Alfred Hettner—the most significant student of geographical method in his day—discovered the geographical work of Immanuel Kant.[2] He found in Kant's characterization of geography a salutary reinforcement of his own view of the discipline as the chorological description of Earth's surface in its full complexity. Accordingly, he approvingly noted Kant's position in his 1927 major treatise on geographical method.[3] The reference was picked up by the American methodologist Richard Hartshorne and cited as important support for his view of geography as a chorological analysis of Earth space,[4] a view that was widely and controversially argued in the mid-twentieth century. These circumstances raise the question "what was the influence of Immanuel Kant's geography on modern conceptions of the subject?" It is apparent that both Hettner and Hartshorne developed their views on the basis of their own reading of contemporary trends in the subject,[5] independently of Kant's work. They appropriated Kant's view as helpful reinforcement for their views. But were those views, and those of other geographers, molded by a passage of Kant's views through the succeeding generations of geographical practitioners? In this essay, I will attempt to consider this question by examining something of the context out of which twentieth-century geography grew.

A particularly important question within this context is what influence Kant's views may have had on the immediately succeeding generation of geographers. This question has centered largely on the relation between Kant and the work of Alexander von Humboldt, a dominant geographical scholar of that generation. Accordingly, I will give some attention to that matter. Doing so raises a further question about how geography was made in the time of Kant and thereafter, for Humboldt was not obviously a "geographer," despite being appropriated as such by all subsequent generations of geographers. Geography in the eighteenth and early nineteenth centuries was largely a practical discipline, a subject made and remade by explorers, surveyors, traders, and travelers. The systematic gathering of geographical information was, in the main, dictated by the needs of statecraft and commerce. Most of this work recognizably became "geography"—one may say "formal geography"—only when compiled into summary academic textbooks. It is easy to think of geography as being defined by those compilations. In this essay I adopt an alternative view that geography is made by those who gather the facts and make the primary interpretations in the field, for it is those facts and interpretations that eventually define the focus and evolution of the discipline.

During Kant's time and immediately thereafter, there were epochal changes in the way information about Earth was systematically collected and analyzed, changes that did much to redefine formal geography. The question of Kant's influence on the subsequent development of geography is essentially tied up with that fact.

At the Beginning

In 1755 and 1756 Immanuel Kant presented the theses required to qualify as *Magister* in the university in his home city of Königsberg and, accordingly, became a *Privatdozent* of the university. He immediately announced his lectures in physical geography, notes for which have come down to us as *Physische Geographie*.[6] In the ensuing forty years, he reportedly gave the course of lectures forty-six times, more frequently than any course except logic and metaphysics.[7] In those forty years, a paradigmatic revolution occurred in the perception and practice of the sciences of the earth, including geography. The changes initiated in this period eventually defined the modern subject of geography. Kant and his lectures were both a harbinger and a casualty of that revolution.

Knowledge about Earth ca. 1750, and for some decades thereafter, was pursued in the intellectual circles of Europe under four broad rubrics.[8] The senior pursuit in many respects was the study of mineralogy, which involved the classification of earth materials, but also of fossils—what we recognize today as palaeontology. It was a science (though that appellation was not yet in vogue) of specimens collected in the field and examined in the study. Collections were made both for scientific and for broader cultural reasons. While fossils were well-known to be the remains of animals and plants, their implications for earth history and for biological evolution were not yet understood. Mineralogy as a systematic pursuit was equivalent, for inorganic nature, to botany and to zoology as practiced in the day, largely preoccupied with collecting, naming, and classifying distinctive entities. Accordingly, it attracted the attention of Carl von Linné (Linnaeus). Together, these subjects formed three recognized "kingdoms"of nature.

Mineralogy was concerned with the fabric of earth materials. It did not encompass the larger features of the landscape, which could not be collected and which did not easily lend themselves to Linnaean schemes of classification. The description of the landscape was the province of geography—specifically of "physical geography." This was a subject built on fieldwork and on mapmaking, a science of spatial distribution and spatial relations—a descriptive science. It was largely driven by utilitarian needs. Like mineralogy, it encompassed a wider range of phenomena than today, extending then to peoples and cultural features. There was no definitive distinction made between human geography and physical geography, but the systematic observation of society was relatively little developed and physical geography *sensu stricto* dominated the discipline.

Geognosy (today, we would say "geology") encompassed the third dimension of earth space. It was concerned with the structure of the rocks beneath Earth's surface; again, a subject of field activity. It was a science born of the mining industry and—in a world of learned enquiry that largely revolved around Paris—it is noteworthy, though not surprising, that it was most highly developed in Germany, where mining in the Hercynian rocks of central Europe was an important economic activity. The structural order of the rocks was observed and reconstructed in sequence by means of section diagrams, a descriptive exercise designed to aid the finding of ore-bearing rocks. Sequence carried the first hint of a history. But in the eighteenth century that hint did not extend, within geognosy, to any systematic appreciation of earth history.

History was the province of the last major division of "earth science," earth physics. This subject (not equivalent to modern geophysics) was concerned with seeking regularities amongst the phenomena of the descriptive earth sciences with a view to determining their causes. This was mainly speculative work and, in a world constrained by a biblical time scale, it largely was restricted to considering the proximate causes of local features. In the late eighteenth century, for example, the origin of valleys was a persistently troublesome topic.[9] But questions of origins immediately gave rise to a confrontation with time and to recognition that Earth must have a history. These considerations led to attempts to develop a comprehensive Earth history—to develop a "theory of the Earth" that could encompass origins and development to the point that the present-day appearances could be explained.

The first three of these subjects of learned enquiry were recognized to fall within the realm of natural history—the description of nature. (To appreciate this appellation it is necessary to realize that, in the eighteenth century, "history" did not necessarily incorporate an extended temporal dimension.) These subjects were the province of naturalists. The fourth, "earth physics," fell within the province of natural philosophy.[10] Natural philosophy—the science of nature—encompassed all of what we would regard as the causal science of external nature and, in addition, mathematics. Natural philosophy was the province of philosophers. The science of basic causes, including causes for the appearances of nature, was regarded as part of metaphysics. In the eighteenth century both philosophy and metaphysics encompassed far more than they do today. These divisions notwithstanding, it was not unusual for the attentions of an individual savant to move from one to another amongst these subjects, in particular from one or another branch of natural history to natural philosophy.

The circumstances sketched above go far toward explaining Kant's posture in relation to natural history and natural philosophy. As a philosopher, physics (including what we would recognize as physics today) fell within his purview, as did mathematics. Kant's long-standing interest in the foundations of Newtonian mechanics—not the obvious province of a modern metaphysician—is thereby rationalized, and it becomes clear why he was interested in cosmology[11] (*cosmogeny* in his day) and explanation of earthly natural phenomena.[12] (In the eighteenth century, the history of the universe and the history of Earth's landscape were regarded as topics much more closely related than they are today, in part because, together, they constituted the science that treats physical space and, in part, because the very contracted age of the universe that was accepted at the time gave rise to the conviction that the two topics must be closely related.)

These interests notwithstanding, Kant's *Physische Geographie* is, in the main, an eighteenth century academic description of Earth's surface and the phenomena found upon it. It is an organized compendium of facts (as then perceived) about Earth's surface, and about the animals, plants, and peoples found upon it. Such compendia were the standard fare of formal "geography" in those days. They were serially compiled from travelers' reports, from navigators' maps, and from earlier compendia, and they were always prefaced—as is the *Physische Geographie*—by a discussion of the figure of the earth and its place in the cosmos.[13] Such compilations arose at the time of the Reformation and the release of northern European scholarship from the conventions of mediaeval Catholicism. The archetype was the *Geographia generalis* of Bernhard Varenius (1622–50),[14] once a student in Königsberg. As a set of elementary lectures intended for incoming students, it is possible that Kant considered natural history to be more approachable than abstract philosophy and, certainly, he would have regarded it as an essential preparation for studies in natural philosophy.

Kant was, then, a "formalizer" of geography. The *Physische Geographie* represents a more fundamental aspect, however, of Kant's thought. His theory of knowledge distinguished "inner knowledge" or pure reason and "outer knowledge" or empirically grounded reason.[15] Empirical knowledge he divided according to the dimensions of our normal perception—into spatial knowledge and temporal knowledge. The former he regarded as the province of geography, the latter, history. From this division he drew the claim that geography is purely a study of space and spatial variation, with no significant temporal dimension. In this he was conforming with the conventional division of knowledge about Earth recognized at the time, as described above.[16] But he realized that this division could lead to difficulties for he acknowledged that, whilst geography provides the ground for the enactment of history, history furnishes the explanation for geography. Consequently, whilst Kant heavily criticized "history of nature" (that is, theory of the Earth) as fragmentary, speculative, and inexact—because of the fragmentary nature of historical knowledge itself—he realized that a properly drawn history of nature is the *explanans* of geography.[17] In his own *General natural history* he made a serious attempt to limit his speculations to ones that were consistent with Newtonian physical principles, as then understood. He thereby demonstrated the inextricable connection of natural history with natural philosophy, even though his definition of geography seemed to deny it. Kant also emphasized knowledge as a systemic construct. To make sense, the objects of our reasoning or of our perception must fit into a larger, coherent framework (he used the concept of a house to illustrate the idea; before commencing the piece-by-piece construction, it is necessary to have an idea of the whole edifice). Accordingly, a concept of the entire world would be the framework for more local geographies. Furthermore, geography itself would be the systemic framework within which we contextualize our everyday knowledge of the world. Finally, he emphasized the provisional nature of empirical knowledge—that it consists merely of appearances rather than absolute data—hence any historically based explanation must remain provisional.

The circumstances described above place the *Physische Geographie* squarely in the mainstream of learned discourse about the world in the mid-eighteenth century. They also make clear that it was by no means a novel or exceptional exposition.[18] But Kant's emphasis on the fundamental position of geography in the edifice of knowledge and the importance of geographical knowledge as the framework to make everyday sense of the world around us were important departures (though the latter was, in a sense, amply prefigured by the practical importance accorded, at the time, to geographical knowledge). So was his attempt to avoid speculative content, particularly of a providential kind, even though many of the contemporary "facts" that he reported were indeed fanciful. So far as we know, Kant maintained the outline and general format of his lecture course right to the end. But the approach to the production of geography and, with it, all of earth science changed dramatically through the years during which Kant delivered his lectures.

Sea Change

In the years after 1756, the quality of geographical exploration and the character of enquiry into earth history changed dramatically. Critical developments occurred in both marine and terrestrial exploration.[19] At sea, navigators such as Bougainville, Cook, Lapérouse, and Vancouver turned marine exploration from a matter of serendipitous adventuring into scientific exploration. We may examine James Cook as a representative of these pivotal figures. Between 1768 and 1780 he made three voyages into the Pacific Ocean for which he systematically recruited to his crews scientifically trained officers, naturalists, astronomers, and artists (to visually record their discoveries). Cook himself was a largely self-taught surveyor-mathematician and a superb navigator. His first major command was a voyage to the South Pacific ostensibly under the scientific sponsorship of the Royal Society to observe a transit of Venus. Barely beneath the surface, of course, lay British imperial ambitions and instructions from the Admiralty concerning the charting and claiming of newfound lands. The expedition was a huge success, not least because of the presence of Joseph Banks, a naturalist who later became president of the Royal Society and a dominating influence in British scientific exploration. Through Banks's efforts, Linnaeus's notion of systematic botanical and zoological collections became a major theme in European expeditionary science.[20] There was another prominent reason for Cook's success. He carried with him John Harrison's chronometers: timepieces of such accuracy that they finally solved the long-standing problem of determining longitude. Cook produced maps of unprecedented accuracy with the aid of the most modern instruments.

Cook's subsequent voyages were no less successful scientifically (though Cook was killed on the third voyage). On the second, he carried with him the German geographers Johann Reinhold and Johann Georg Forster, father and son. The Forsters' published accounts of the voyage cemented the success of this new, systematically empirical style of exploration, bringing its effectiveness forcefully to attention in Prussia. Their work was of major importance in establishing a new tradition of disciplined, first-hand observation as the basis for an authoritative geography. The younger Forster, in particular, urged the necessity of focused observation in pursuit of systematic objectives, as opposed to the mere fact gathering of earlier explorations—a view that might have come directly from Kant.[21] On these voyages, major attention was paid to anthropology, to the study of the peoples of the Pacific islands that they visited. Cook himself proved to be a competent anthropological observer and reporter. The major lessons planted in the learned circles of Europe were that empirical observation yielded progress in problems that had remained immune to speculative enquiry, that systematic (especially instrumental) measurement yielded observations of superior reliability, and that there were fascinating things to be learned about societies living in exotic environments.

On land, change in the style of geographical study was signaled by Nicolas Desmarest, an almost exact contemporary of Kant (1725–1815), when, in 1757, he wrote, for Diderot's *Encyclopédie*, that sound physical geography must be based on systematic, first-hand observation. He subsequently engaged in a detailed field study of the volcanic province of the Auvergne and by 1771 had arrived—on the basis of field work—at startling conclusions concerning successive epochs of vulcanism and valley formation by running water.[22] He had successfully connected earth history with the geography of the contemporary landscape. At the same time, Peter Simon Pallas was conducting extensive exploratory travels into Siberia,[23] while the Genevan naturalist-geographer Horace Benedict de Saussure was engaged in an exhaustive field survey of the western Alps,[24] culminating in the first ascent of Mont Blanc in 1787. Saussure carried (or, rather, his mountain guides carried) barometers, thermometers, hygrometers, an "electrometer," and geological sampling equipment to the summit. Measurement was his central purpose. Saussure, like Desmarest, firmly believed that the observed facts of physical geography (and, presumably geognosy)—not traditional accounts and not philosophical speculation—formed the appropriate basis for geography and for understanding the origin of contemporary landscapes.[25]

By 1800, the notion of the emerging earth sciences, including geography, as empirical subjects to be pursued toward systematic ends in the field by instrumental measurement and by expert, first-hand observation was becoming well-established. The most prominent exponent of the new method in the generation that succeeded Kant was Alexander von Humboldt. The new perspective was formally consecrated by Georges Cuvier—arguably the most influential natural historian of his day—in his report on the state of the natural sciences for Napoleon Bonaparte.[26] He emphasized that physical geography and geognosy were true field sciences that engaged the efforts of the best observers—meaning the Saussures, Desmorests, Humboldts, and the like—to replace the conjectures of earlier periods. The orientation of Cuvier's report was strongly influenced by the *Agenda* (1796) of Saussure, a sketch for a history of Earth designed to explain contemporary geography that would be based firmly on observations in the field. The significance of this report is its initiation of a further development from emphasis on systematic observation, exemplified by Cook and the Forsters, toward emphasis on purposefully systemic observation.

As an academic philosopher dependent for his living upon the fees for his lectures, Kant had no opportunity to travel and record new observations nor, it must be said, had he either the constitution or inclination to do so. Apart from reading the field explorers' reports, he was isolated from the new style. Kant's *Kabinetgeographie* was, in comparison with this, anachronistic as a method for the creation of geography. But this is not to say that Immanuel Kant had no influence on developments in physical geography. Two aspects of his thinking were critical to the development of geography and, indeed, of all the natural sciences in those years. First, and most important, Kant's theory of knowledge led him directly to the claim that no empirical evidence for the existence of God could possibly be found. It follows directly that scientific knowledge of the world around us must entirely consist of the provisional knowledge gained by human experience. No providential direction (which would be empirical evidence of God) could be adduced. Kant, the most influential philosopher of the German Enlightenment, provided the definitive intellectual blow against a biblically-based geography and earth history (not that they have ever entirely gone away) and against teleologically biased geographies. Practically, Kant was following the opinion of savants from at least the time of Bartholomäus Keckermann,[27] but Kant's word possessed an authority that no serious scholar could ignore.

Second, Kant insisted on geography as the study of Earth space: it was to him a discipline that described the present condition of Earth's surface and the distribution of the organisms that lived upon it, including the external and material aspects of humans and human societies. It was conceived to be a science with no significant temporal dimension, though he pondered at length the reciprocal relations of geography and history. He furthermore implied a sense of space that is particular, that is defined by the relations of the various objects and activities situated in the space. He adopted the classical concept of the *chorography* of earth space, meaning the description of space according to perceived differences from place to place.[28] This has remained a recurring *leitmotif* of academic and practical geography ever since. It is easy to suppose that Kant's insistence upon a system of nature was a major guiding concept as well, although the substantive development of the earth sciences in the last years of the eighteenth century were forcefully bringing home that lesson quite independently.

As a natural philosopher, Kant was also a student of science. He knew well the currents of contemporary natural philosophy. He recognized that Earth must be far more ancient than the Genesis account would have it. But in this he was following authoritative opinion, in particular that of the Comte de Buffon.[29] No serious savant, after the mid-eighteenth century, believed in the literal interpretation of the biblical account. Perhaps more remarkably, Kant was well aware of the notion of the evolution of species. In the introduction to the *Physische Geographie* we find the following:

> If, for example, one were to consider how the various breeds of dogs descended from one line, and what changes have befallen them through all time as a result of differences in country, climate, reproduction, etc., then this would constitute a natural history of dogs.[30]

This long before Darwin, indeed, before Lamarck's major statements. Again, the idea was in the air, but Kant's matter-of-fact discussion of it must have constituted no small contribution to the growth of its credibility just at the time when the evidence of fossil organisms was first beginning to be organized into evolutionary sequences, especially through the efforts of Cuvier.

Kant's eventual division of *Geographie* and *Anthropologie* was also a harbinger of the future development of a more narrowly delimited physical geography and the emergence of a more specialized earth science. Accordingly, it gave a hint of the growth of academic tension between those geographers who would see the subject as primarily a systematic (and "scientific") study of phenomena of Earth's surface—particularly the phenomena of physical geography—and those who would see it as a study, largely regional and chorological in character, of human occupance of Earth. This tension dominated nineteenth-century German geography and, inasmuch as modern academic geography began there, it persists almost everywhere today. The *Physische Geographie* contains significant hints of some of the directions that nineteenth-century geography would take.

The Humboldt Enigma

Between the fifteenth and nineteenth centuries, geography in Europe was a very important discipline. It provided the existential basis for the organization of states and for their sustenance by systems of land taxation, and it furnished the intellectual underpinnings for European exploration, mercantile expansion, and imperialism. It was of critical importance for military operations. Hence it was largely an applied discipline, prosecuted by practical men focused on survey and mapping. Academic geography was in large measure a sideshow that fed on the reports of these practitioners. Geography had only a tenuous basis, predicated on the personal interests of individual scholars, in the universities of the period. With the sea change of the late eighteenth century, geography needed intellectual champions who could continue to develop the formal discipline.

The most prominent such person in Europe—at least in the view of later commentators—was Alexander von Humboldt (1769–1859), second son of an aristocratic Prussian family. Along with Karl Ritter, professor at Berlin and the only prominent professor of geography in the world of his time, Humboldt carried forward the tradition of intellectual geography in the immediately post-Kantian period. Humboldt was more widely renowned in their time—indeed, regarded as one of the most eminent naturalist-scholars of the period and certainly the leading scientist-statesman. He was taught briefly (1785) by Markus Herz, a major student of Kant and, in 1788, set out to read Kant's works.[31] However, his older brother, Wilhelm, who made a close study of Kant's philosophy, complained that Alexander had little taste for metaphysics.[32] So far as is known, Humboldt never met Kant. He was trained in economics and science at Göttingen (where he was taught by Blumenbach) and then in mineralogy at Freiberg under Abraham Gottlob Werner (1749–1817), the leading mineralogist of his time and one of the principal disputants in the controversies of the day about the origins of the rock strata. In 1790 Humboldt traveled along the

Rhine and to England with Georg Forster.[33] This trip, in the company of a renowned companion of Cook (and translator into German of Bougainville's journals), during which he was introduced to a number of members of the English scientific elite, including Banks, appears to have made a great impression on Humboldt. He must inevitably have absorbed much of Forster's opinions on scientific observation in the course of their travel along the Rhine, and the trip certainly raised his enthusiasm for exploration.[34] He subsequently became an assistant inspector of mines in Franconia and, in the 1790s, the Humboldt brothers became part of the intellectual circle in Jena that included Goethe, Schiller, Herder, and Fichte. Herder was one of Kant's early students; Fichte one of his last acolytes, but both developed their philosophies beyond Kant and ultimately raised his ire.[35] From Herder, in particular, Humboldt received not only reinforcement of the importance of experience in forming a view of nature, but also of the need for generalization from the facts of experience.[36] From the group he absorbed elements of their *Naturphilosophie*, and that must have included comparisons with Kant's views. Herder and the Jena group also held (with Kant and with Forster) decidedly liberal social views, views that became an important part of Humboldt's outlook. But Alexander had made up his mind to be a scientific traveler in the tradition of his friend Forster and he possessed the wealth necessary to carry out his plan. In 1799, he set out with Aimé Bonpland for the New World, where he spent five years traversing most of northern South America and Central America. His reports of their findings made him a European scientific celebrity. Upon his return in 1804 he settled in Paris to write up his results.

Humboldt's first report, *Essai sur la géographie des plantes* (1805),[37] began to develop an idea that he had first considered in the 1790s and became fully developed in *Kosmos*. This was his idea of a *physique du monde* (universal natural science[38]), complemented by his ambition to write an *histoire physique du globe* (natural history of the earth). In his *Relation historique du voyage* (1814) Humboldt variously expressed his project as *physique du monde*, *théorie de la terre* or *géographie physique*. Humboldt's maturing scientific outlook and its origins are largely tied up in this conflation of terms. "Theory of the earth" appears to connect Humboldt's vision with the old, speculative theories of the earth that were part of the natural philosophy of earlier centuries, and that had been dismissed by Kant and convincingly demolished by Cuvier and his associates.[39] However, It may be argued, in the light of Humboldt's interests, that *physique du monde* represents a view ahead to something more like modern geophysics—that is a science of Earth firmly underpinned by observation and experiment. Humboldt's idea was considerably more broad than the modern science, though, for his invocation of *géographie physique* makes clear that he intended the geography of plants, animals and of humans to be part of the study (and was the usual German construction of physical geography at the time). For Humboldt, "theory of the earth" is more apt, I judge, to have signified a description of the earth supplemented by appropriate explanatory generalizations—perhaps following Herder—about the observed phenomena. His relation of plant zones to temperature and/or elevation in the cordilleras of the Americas is an apt example.

Humboldt's geography is evidently based upon three major principles: (1) that measurements are of paramount importance, and that measurements must be made of as many qualities of the environment as possible; but (2) that an holistic and aesthetic sensibility must be present in one's summary appreciation of and report on landscape; and (3) that geography is strictly about the compilation and synthesis of the present facts of the landscape—it is neither an historical nor causally interpretive science. It is widely reported that Humboldt's views were strongly influenced by Kant.[40] What is the evidence?

It has been claimed that Humboldt's infatuation with measurement—he carried a remarkable range of instruments with him on his American travels—derived from his respect for "French Laplacian science" and, in particular, the chemistry of Lavoiser, Berthollet, and others.[41] He indeed made substantial enquiries into the chemistry and chemical physiology of plants and other organisms in the 1790s and no doubt learned a great deal about the importance of exact measurement in the course of that work. But he confessed that he wished, as a traveler-scientist, to emulate Cook and Saussure (and, no doubt, Forster) rather than the laboratory-bound Laplacians. It seems at least equally likely that his respect for field measurements derived, following his illustrious predecessors, from his perception of the importance of measurements to improve the quality of field observations and, certainly, from his experience as a mines inspector. They also usefully supported his aim to understand the systematic geography of extensive regions and, in particular, the relations among various distributions, for which purpose comparative measurements appeared to be essential. At any rate, the appreciation of measurement in geography (and in science in general) was rapidly growing at the end of the eighteenth century and had little to do with Kant, apart from its congruence with the general climate of rationalism that Kant had played a significant role in establishing.

Humboldt's aesthetic sense, insofar as his scientific writings were concerned, expressed itself as a conviction that it was necessary to appreciate the character of the entire landscape—to appreciate the system of nature (a term he emphasized)—as well as the component parts. In later life, he discussed the necessity to appreciate the "homogeneity" of the system of nature, by which I infer that he meant the way in which various observed phenomena relate to each other. This idea might very well have derived from Kant's doctrine of the system of knowledge, and from Kant's specific application of it to geography. In his *Observations on the Feeling of the Beautiful and Sublime* (1764), Kant presented a basis to develop a "geographical aesthetic," but there is no evidence that this work influenced Humboldt. It seems much more likely that his aesthetic sense derived from his direct and substantial exposure to the *Naturphilosophie* of the Jena crowd—which espoused similar principles—and, indeed, to the entire drift of *fin de siècle* German philosophy toward romanticism, than directly from Kant.[42]

The American geographer Richard Hartshorne has made the closest examination of the third and most critical issue.[43] The earliest documentary suggestion of some connection between Humboldt and Kant—based on the congruence of their views about what geography is—appears in Humboldt's *Florae Fribergensis specimen plantas cryptogamicas praesertim subterraneas exhibens* (published in Berlin by H. A. Rotmann in 1793). In a long footnote (translated by Hartshorne) Humboldt clearly laid out his view of geography as the description of the present state of Earth's surface. He specifically contrasted this viewpoint with the study of earth history. These are views expressed by Kant, certainly (most forcefully in his 1788 rejoinder to Forster[44] but most fundamentally in the introduction to the *Physische Geographie*[45]), but they were also the views of Werner, who subscribed to the division of knowledge about the Earth described early in this paper.[46] They were common views in the late eighteenth century, views supported by the *encyclopédistes*. Humboldt is much more likely to have learned this viewpoint from his teacher, Werner—whom he acknowledged—than from unpublished (in 1793) students' notes of Kant's physical geography. That is not to say that he might not have assumed it from other of Kant's writings, for it ultimately derives from Kant's theory of knowledge rather than specifically from the *Physische Geographie*. In various forms, Humboldt repeated these views in later writings, including a virtually identical statement in *Kosmos*. Hartshorne's comparison of texts led him to believe that he had read the *Physische Geographie* and paraphrased statements from it concerning the character of geography.[47] But, while Humboldt uses some specific terminology that also appears in Rink's text, the overall texts are not similar and Humboldt never acknowledged Kant's geography as a source.[48]

Why would he fail to do so? Granted that conventions about citation were much more cavalierly observed (or not) at the turn of the nineteenth century, Humboldt was relatively conscientious.[49] But he was also a member of a Prussian aristocratic family that would have seen its primary *métier* in the service of the state. Indeed, Alexander served on occasion as a plenipotentiary and ambassador for the Prussian court while Wilhelm became a high state official—a Minister of Education who effected reforms that launched the"German century" in science and scholarship. Kant, on the other hand, in his declining years was decidedly *non grata* with the Prussian state. In 1794 he was formally censured by King Frederick William II (ostensibly for his allegedly anti-religious philosophical teachings, though as likely, in fact, for his libertarian political writings). Association with Kant became a risky affair for those ambitious for state advance. In 1797, the king died and was replace by Frederick William III. More liberal than his father, he was nevertheless not a decisive man. The stigma of former censure was not entirely dissipated. Even at his death in 1804, some of Kant's Königsberg acquaintances declined to attend his funeral.[50] In this light, it is rather less surprising that Humboldt would perhaps have been discreet about any intellectual connection he might have had with the great man. The issue might not have touched Alexander deeply, but there was his brother's position to consider as well as the more general issue of the standing of his family. It would appear to be impossible, today, to conclusively establish a direct connection between the technical writings of Kant and those of Humboldt.

There is, withal, no doubt that Humboldt's insistence upon geography as a descriptive science of the current state of Earth (in all its aspects, as emerges in the *Kosmos* project) is consistent with Kant's views. To Humboldt, instrumental observations were data descriptive of the quality of Earth space and of the phenomena of Earth (primarily of Earth's surface). And all the data of geography were empirical. If Humboldt's views derive from Kant's teachings, they much more likely arise from the *Metaphysical Foundations of Natural Science* (1786)—published just before the time when Humboldt was attempting to read Kant—than directly from Kant's geographical teachings. It is reasonable to suppose that Kant's published and widely known theory of knowledge—extensively analyzed by the Jena circle—would be the principal source of any influence on Humboldt. But we have no direct evidence.

Humboldt's treatment of geographical space is not inconsistent with Kant's views, but he strongly emphasized systematic analysis. Humboldt was concerned to map and to understand the patterns of variation in geographical phenomena from place to place. He thought he could achieve this by plotting his measurements in cartographical space and interpolating lines to represent loci of equal quantity or equal intensity. It is widely claimed that he invented isolines[51] and he notably produced the first map of isotherms—lines of equal mean temperature. His ultimate aim was to find some measure of vegetation structure or variation and to map quantitatively the variation of vegetation over the surface of the Earth. (Humboldt was interested in plant distribution from the outset of his career and viewed plants as the most significant neglected phenomenon in geography. Moreover, the influence of Linnaeus and of Banks had made plants an important focus of attention for European scientific explorers.) Humboldt was promoting an exercise in structural inference amongst observations, and an enquiry into the general structure of earth space. Abstracting general patterns from specific data is a quintessentially scientific activity, a part of "general" or "systematic" geography—part of the search for covering laws or universal patterns. It is consistent with Kant's conviction that empirical generalizations could be entertained by way of explanation of geographical appearances. A similar sort of explanatory generalization in the historical realm was underway in Paris as Cuvier's colleague, Alexandre Brogniart, worked out the Tertiary history of the Paris basin during the years of Humboldt's residence there.

The possibility for this sort of development is clearly appreciated in Kant's theory of knowledge and is more or less reflected in the order of topics in the *Physische Geographie*. In one of the earliest prospectuses for his course Kant wrote

> I present this material first in the natural order of classes, and proceed later to set forth, in a geographical manner of instruction, through referring to all countries of the earth, the particular inclinations of men that derive from the regions in which they live. . . .[52]

Humboldt wrote

> I wished to make known the countries I had visited, and to collect noteworthy facts pertaining to a science that is as yet scarcely sketched out that has been rather vaguely named Earth physics, theory of the Earth, or physical geography. Of these two purposes, the latter seemed to me the more important.[53]

It seems clear that both Kant and Humboldt understood both characterizations of geographical space (as structural continuum and as differentiated regions) that have dominated the modern discipline. Humboldt's systematic perspective has dominated modern physical geography.

The Sequel

By the 1790s, the winds of philosophical change were blowing strongly through the German realm. Current debate was moving beyond Kant—indeed, beyond the entire Enlightenment worldview—even before he died. And debate was decisively reoriented by the facts of the French Revolution and the succeeding Napoleonic period. The *Physische Geographie* was not published until after 1800, in two editions that were immediately controversial because, plainly, neither was a genuine product of Kant's hand. Conflated from notes recorded at various times throughout Kant's career, and with substantial editorial additions, it is not clear that either publication represents Kant's entire view of the subject at any stage.[54] The notes appear to have had little immediate impact. In fact, Kant's views on geography appear to have lain neglected throughout the nineteenth century.

Kant retained a pervasive influence in German culture throughout the nineteenth century. Following the Napoleonic debacle, Wilhelm von Humboldt established new school curricula in Prussia that included study of Kant's philosophy in the secondary level. But the focus of study was a summary of his great philosophical works. The rapidly emerging empirical sciences, meanwhile, increasingly distanced themselves from philosophy, in part because—as

Kant had clearly seen—philosophy could not provide a secure foundation for scientific results as absolute knowledge. One early nineteenth-century Kantian philosopher, Jakob Friedrich Fries,[55] saw this problem clearly and attempted to remedy it by elaborating the role of empirical regularities—the aspect that had given Kant most difficulty, but that had most interested Humboldt—within the framework of Kant's theory of knowledge.[56] His work did not become as well known as perhaps it should have; certainly, it did not immediately impinge upon the development of geography. Instead, the methodological issues that were to preoccupy formal geographical work for the duration of the century—indeed, for a century and a half—arose from within the discipline, and the directions they took were strongly influenced by both substantive and formal developments within the empirical sciences.[57]

Developments within geography in the early nineteenth century were dominated by Humboldt and Karl Ritter. Intellectual Europe was excited by the reports that Humboldt made of his travels in the Americas and by the rich sample material he returned. His residency in Paris during the years immediately after his return brought his publications—mostly issued initially in French—immediately to the attention of the leading earth scientists. His results were widely quoted in the geographical textbooks of the day but, beyond statements such as the one initiated in 1793 and his definition of geography as a descriptive science of space, he made no general methodological statements.

In 1820 Karl Ritter was appointed Professor of Geography in Berlin—the leading German university and the only university anywhere to appoint an active professor of geography in the first half of the nineteenth century. Ritter was to a considerable degree concerned with the description of regional space and driven significantly by teleological views; in many respects he was the intellectual heir of Anton Friedrich Büsching.[58] Trained in history, he was also drawn toward a historically based explanation of his geography—a predilection that paralleled the emerging empirical study of Earth history but which appeared to contradict the Kantian definition of geography. Accordingly, it was Ritter who firmly established the term *Erdkunde* in preference to *Erdbeschreibung* as the appellation for a scientific geography. But while his major work, *Erdkunde: allgemeine vergleichende Geographie* (. . . "general comparative geography") was primarily a work of regional description with emphasis on human societies, he was well aware of the possibility for systematic geographies to be constructed (after Humboldt's return to Berlin in 1827, he and Ritter became good acquaintances). This was well revealed by his 1830s controversy with Julius Fröbel over the proper conduct of geographic enquiry.[59] Fröbel criticized Ritter by way of advocating a Humboldtian systematic physical geography as the proper basis for a scientific geography. In his discussion, Fröbel conflated statements of both Kant and Humboldt about geography as a systematic description of Earth space, but his references were not taken up by others, nor were Ritter's rejoinders—the closest he came to something like a Kantian statement—widely noted in comparison with his influential substantive work.

Instead, subsequent generations of geographers read methodological postures into these two founding figures of modern geography from their substantive work. As implied by the Fröbel-Ritter debate, Humboldt was made the champion of systematic geography and Ritter was made the advocate of geography as a regional, chorological study. Both of the putative protagonists, it seems fair to state, knew better. And a reading of Kant, in the introduction to the *Physische Geographie*, would certainly have provided a correction.

An important tangential issue that was to have a strong influence over nineteenth-century debates about the nature of geography arose out of the same late eighteenth-century ferment about the history of Earth that formed a backdrop to Kant's (limited) speculations on Earth history. In the years around the turn of the nineteenth-century Earth science (read geology) came to be defined in something approaching its modern form. Initiated by the efforts of Cuvier and his colleagues in Paris as an attempt to construct a description of Earth's physical substrate (at topographic rather than global scale) and then its proximate history, it threatened to absorb what had heretofore been regarded as the most significant part of physical geography.[60] This development was given decisive impetus by the emergence, after 1810, of a dominant British school of geologists who possessed no historical connections with formal geography. Nor did any of these "new men" have strong philosophical orientations. Physiography (in the modern sense; in the early nineteenth century the dominant part of physical geography) was appropriated by the emerging Earth science and has remained sometimes bitterly contested ground ever since. This circumstance reinforced the view of many geographers that the discipline should follow a regional/chorological orientation rather than a systematic one.

In 1859 both Humboldt and Ritter died and Charles Darwin published *The Origin of Species*. In that work, Darwin defined a vitalistic implication of "deep time" to match the palaeontological implication elaborated by Cuvier and his collaborators and the geophysical implication publicized by Charles Lyell. Thus was forged at the empirical level the reciprocal relations of geography and Earth history that had troubled Kant in his exploration of geographical and historical knowledge. After Darwin, there could no longer be any reasonable pretence that Earth history and contemporary physical geography (*sensu stricto*) could be pursued in isolation from each other. At a stroke, the strict definitions of geography espoused by both Kant and Humboldt were demolished. But it required some decades for these implications to be fully appreciated.

In the larger context, geography continued through the first two-thirds of the nineteenth century in its traditional informal course as a practical subject associated with exploration and mapping. For the Europeans this increasingly meant serving national projects of colonial expansion. In the emerging American states, it meant internal exploration and disenfranchisement of native peoples. Regional inventory and interpretation were emphasized in official and diplomatic reports. To the extent that this activity contributed to the definition of geography, it reinforced the comparative regional method espoused by Ritter.

After 1870, the practical importance of geography began to be acknowledged in newly unified Germany by the establishment of university chairs. This development was followed near the end of the century by the foundation of geography departments in France, America, and Great Britain but the intellectual leadership remained firmly in Germany until after the turn of the twentieth century. The first generation of the new academic geographers were not themselves, of course, trained in academic geography. Most were trained in history and philosophy, which did not prevent some amongst them from taking up an attitude—possibly as a reaction to the more teleological aspect of Ritter's writings, and certainly within the context of nineteenth century materialist/mechanistic conceptions of science—that a properly "scientific"geography must be based on measurement and would largely, perforce, be a physical geography. Prominent among exponents of this view—partisans of Humboldt and Fröbel—were Oskar Peschel, who emphasized physiography and found himself under strong attack from the geologists, and Georg Gerland, who attempted to define geography as what we would recognize as early modern geophysics.[61] The views of both were strongly contested by geographers more concerned with the study of human societies, most importantly Friedrich Ratzel, whose monumental *Anthropogeographie* (1882–1891) decidedly tilted the subject in that direction. Ratzel, under strong Neo-Darwinian influence, also cemented the essential association between geography and history that had begun to take form in Ritter's work.[62]

These skirmishes signal a more fundamental reason for methodological controversy surrounding the issue of systematic geographies, physical or human, versus regional study. With the emergence of the modern systematic sciences, both physical and social, more and more of what Kant and his contemporaries had recognized as the substance of systematic geography was being abstracted into newly defined disciplines (political economy, sociology . . .). The possibility for a distinctive geographical discipline appeared more and more to center upon regional method. Friedrich Marthe[63] reintroduced the term "chorology" ("chorography") in 1877 in the course of advocating regional study as the means to achieve a distinctively geographical explanation on the basis of a causal (historical) analysis of events. Ferdinand von Richtofen emphasized a similar theme in an 1883 inaugural address at Leipzig. Some of the new professors were trained by Benno Erdmann,[64] a prominent Kant scholar of the time, who urged a chorological view of earth space upon his students. Erdmann's view no doubt derived from his reading of Kant, but this influence is not explicitly evident in the substantive work of his students.

What is most surprising through this development, which contests two aspects of Kant's thinking about geographical description—aspects that he appears to have regarded as perfectly complementary—is how almost no appeal to Kant's actual teaching was made. Commentators have attempted to read Kant's influence into the debates, but it appears to be a difficult exercise made somewhat plausible only by the fact that Kant himself largely—in the matters of empirical knowledge and, specifically, spatio-temporal knowledge—enunciated principles of deeper ancestry and persistent presence in European thought. The founding fathers of the late nineteenth-century geographic debate were clearly seen to be Humboldt and Ritter by virtue of their nuanced positions being made something of a caricature. The real substance of the debate was concerned rather with developments in the sociology of knowledge that were occurring as the first generation of academic geographers struggled to define their field within the framework of increasingly specialized nineteenth-century sciences. And neither Humboldt nor Ritter provides us with a signposted path back to the ideas of Kant.

Resurrection and Denouement

Richard Hartshorne reports no significant reference to Kant's geography before the end of the nineteenth century. He credits Friedrich Hahn, professor of geography at Königsberg in 1904, with resurrecting Kant's views on geography on the centenary of his illustrious predecessor's death. Both he and his student, Willy Kaminski, who wrote his dissertation on Kant's concept of physical geography,[65] recognized that the significant aspect of those views lay not in his long obsolete substantive teaching, but in his characterization of geography in relation to the structure of human knowledge. The dissertation came to the attention of Alfred Hettner, the leading German exponent of geographic methodology at the turn of the twentieth century (and also trained in philosophy), who first mentioned Kant's characterization of geography in a footnote to a 1905 paper[66] expounding his own views and then wrote at paragraph length in his authoritative 1927 book.[67] In 1911 and thereafter, Erich Adickes finally produced an authoritative commentary on the *Physische Geographie.*[68]

Hettner, searching for a means to ensure the continuation of a unitary geographical discipline, conscious of the rapidly developing Earth sciences and, now, the social sciences, chose—like his immediate predecessors—to emphasize the chorological aspect of Kant's concept of space. Such an orientation produces an organization of knowledge about the world that is distinctively geographical and that admits all of the phenomena traditionally studied by geographers. Regional geography was declared to constitute the intellectual core of the discipline. With American geography facing similar, perhaps more severe challenges than those facing the Europeans (because physiography/geomorphology was more completely absorbed into the Earth sciences), these views were taken up by Hartshorne, prompted by Hettner's reference, and used to make a general methodological statement for geography[69] that described the field as an endeavor that is primarily concerned with (empirical) differentiation of earth space—that is, a chorological science that describes the heterogeneity of space. This was a statement of the dominant view of geography early in the twentieth century, a view that had been strongly developed and intellectually analyzed during the waning years of the nineteenth century with the establishment of the first academic departments of geography.

There may be some connection between the origins of the chorological perspective and the concept of Earth space propounded by Kant. But the view might equally be ascribed to a thread of thought among geographical scholars that can be traced back via Ritter to Büsching and to the geographical encyclopaedists of the seventeenth and eighteenth centuries. Hettner apparently constructed his arguments before he recognized the relevance of Kant's thinking. Kant's authority was wheeled out by Hettner and Hartshorne more by way of reinforcement for the view than as its source.

Withal, Kant's theory of knowledge allows more than Hettner or Hartshorne adopted. He recognized rational knowledge derived by pure reason, and he recognized empirical knowledge derived by experience and observation. A priori knowledge gained by reason is certain knowledge; empirical knowledge remains contingent and provisional, and it is always referenced within the dimensions of time and space (hence, history and geography) through human cognition. Between these two categories of knowledge, Kant appears to have recognized a kind of theoretical knowledge derived by inference from empirical knowledge. We might regard this kind of knowledge initially as observed empirical regularities, but we may reformulate these regularities in terms of physical "laws" expressed through the specification of particular boundary and forcing conditions (e.g., the "law" of breaking waves—one of Kant's examples).[70] The question whether such empirically derived laws had any relevance in a properly constructed geography lay just below the surface of the methodological debate at the turn of the twentieth century. Kant remarked in this respect

> This twofold interest manifests itself also among students of nature in the diversity of their ways of thinking: those who are more especially speculative are, we may almost say, hostile to heterogeneity, and are always on the watch for the unity of the genus; those on the other hand, who are more especially empirical, are constantly endeavoring to differentiate nature in such manifold fashion as almost to extinguish the hope of ever being able to determine its appearance in accordance with universal principles.[71]

The chorological view of geography was strongly challenged in the mid-twentieth century[72] with the claim that the search for spatial regularities—for empirical laws (constituting a systematic or "nomothetic" approach)—is the more scientific and more appropriate goal for geography. It is Peschel's and Gerland's argument all over. Kant's remark neatly skewers this entire chorologic-nomothetic debate.

Kant recognized that a properly drawn picture of external nature might focus on the individual perceptions that differentiate and give character to earth space, while at the same time recognizing the repeated patterns that give rise to generalizations about the structure of earth space, and to empirical principles (as advanced by Humboldt). He further realized that a properly drawn history of nature is the *explanans* of the current geography, though it is possible to know the history only fragmentarily so that explanation is not assured and (he claimed) no first causes might be discerned. These perceptions anticipate the developments, centered in Paris, that were going on at the end of the eighteenth century, fuelled by the increasingly detailed and purposeful reports of scientifically qualified voyagers and travelers. Kant appears to have opposed the view that geography is mere description. He described geography as "a system of knowledge"—no system, no practical value. We would say "no system, no science."

Kant appears to have effectively foreseen the intellectual development of views of Earth that were going on around him but, as a sedentary philosopher, his own geography was purely pedagogical compilation. He could do little to contribute directly to those developments. The question remains, did his philosophy substantially influence those developments? It appears probable that, in a general sense, it did but, like many threads in intellectual history, it may now be impossible to close the issue.

Notes

1. The essay has benefited greatly by comments by C. W. J. Withers, O. Slaymaker, and the editors, even though they would disassociate themselves from some of the views it contains. I wish to thank the editors for having the courage to invite a philosophical outsider—a geomorphologist with an interest in the history of earth science—to participate in the project. I am sure that I have learned much more than anyone will learn from reading my paper.

2. Through reading Willy Kaminski, *Über Immanuel Kants Schriften zur physischen Geographie. Ein Beitrag zur Methodik der Erdkunde* (PhD diss., Königsberg, 1905). The thesis was prompted by the centenary of Kant's death and supervised by Friedrich Hahn, professor of geography in Königsberg. The history is recounted by Richard Hartshorne, "The concept of geography as a science of space, from Kant and Humboldt to Hettner," *Annals of the Association of American Geographers* 48 (1958): 97–108.

3. Alfred Hettner, *Die Geographie, ihre Geschichte, ihr Wesen und ihre Methoden* (Breslau: F. Hirt, 1927), quoted in Hartshorne, "The concept of geography as a science of space."

4. Richard Hartshorne, *The Nature of Geography* (Lancaster: American Association of Geographers, 1939).

5. The history of Hettner's and Hartshorne's discovery of Kant's views is recounted in Hartshorne, "The concept of geography as a science of space."

6. The notes were not published by Kant. Some 27 student editions are known, of which 17 are extant (Werner Stark, unpublished notes), and published editions were edited by Gottfried Vollmer (1801) and by Friedrich Theodor Rink (1802). The latter is said to be an authorized edition and has formed the basis for modern editions to date, but it evidently is constructed from more than one set of notes and contains many editorial interpolations. We do not have a complete, homogeneous edition of the *Physische Geographie.*

7. According to Robert B. Louden, *Kant's Impure Ethics: From Rational Beings to Human Beings* (Oxford: Oxford University Press, 2000), 5. Joseph A. May, *Kant's Concept of Geography and Its Relation to Recent Geographical Thought*, (Toronto: University of Toronto Research Publications, 1970), 4, claims 48 times.

8. In this and the immediately following paragraphs I mainly follow Martin J. S. Rudwick, *Bursting the Limits of Time* (Chicago: University of Chicago Press, 2005).

9. See Barbara A. Kennedy, "The Trouble with Valleys," in *Process and Form in Geomorphology*, ed. David Ross Stoddart (London: Routledge, 1997), 60–73.

10. The concepts of natural history and natural philosophy were codified in the *Encyclopédie* of Denis Diderot (vol, 1, 1749), though they are in fact much older.

11. Cf. *General Natural History and Theory of the Heavens, or an Essay on the Constitution and Mechanical Origin of the Whole Universe, treated in Accordance with Newtonian Principles* (1755). Kant's preliminary title is perhaps more revealing: "Cosmogony or an Attempt to Derive the Origin of the Cosmos, the Formation of the Heavenly Bodies and the Causes of their Motions from the General laws of the Motion of Matter According to Newton's Theory." The book was in large measure prompted by the work of Buffon.

12. For example, his works on the theory of winds: "New remarks about the explanation of the theory of winds" (1756); "Whether the westerly winds in our environs are so humid because they blow over a large ocean" (1757), both published as part of the announcement (prospectus) of his lectures. Kant wrote on other geophysical phenomena, including earthquakes (the great Lisbon earthquake occurred in 1755), fireballs, volcanoes, and the craters of the moon (which he took to be volcanic edifices).

13. This tradition continued into the late twentieth century in physical geography textbooks, largely as a preface to the discussion of global climate.

14. A German edition of the *Geographia Generalis* was published in 1755, at just the time when Kant began his lectures. The earliest departure of a Geographia from a strictly scriptural basis appears to be that of Philipp Melancthon (1497–1560), professor of philosophy at Wittenberg and Martin Luther's colleague. His *Cosmographia* remained in print until 1628. See David Livingstone, *The Geographical Tradition* (Oxford: Blackwell, 1992), on Reformation geography.

15. The outlines of this division of knowledge are clearly presented in the *Encyclopédie* and go back at least as far as Descartes. Kant's construction of knowledge was by no means original.

16. May, *Kant's Concept of Geography*, 51–2, points out that J. M. Franz, a mid-eighteenth century geographer, specifically contrasted geography and history and Hartshorne, "The concept of geography as a science of space" states that Franz's is the earliest such explicit statement that he could find. Kant does not acknowledge Franz, but that is not surprising given the scholarly conventions of the day.

17. Some Kant scholars assert that Kant heavily criticized the concept of natural history, but in fact "natural history," as understood in the eighteenth century (and still informally used today) would have been, in the main, congruent with Kant's concept of geography. It is possible that some confusion arises over different meanings of *Naturgeschichte* and "natural history" between the German and English-speaking worlds. While it is less relevant here, it should be noted that Kant classified mathematics with pure reason, so that it would not, in his view, be part of natural philosophy.

18. Which raises the question why Kant chose to write his own notes when the practice in German universities of the eighteenth century was for lecturers to follow a published text. Kant acquired special permission from the Prussian Ministry of Education to lecture without such a text. In respect of the *Physische Geographie*, the reason may simply be that knowledge about the world was continually changing, so that the perceived facts needed to be updated regularly. This would have been a relatively easy enterprise in a seaport city, particularly as Kant enjoyed the social company of international merchant traders based in Königsberg (see Manfred Kuehn, *Kant: A Biography* (Cambridge: Cambridge University Press, 2001) on Kant's relations with Joseph Green and Robert Motherby, said to have been his closest friends). Furthermore, Kuehn also reports that the literature of travel figured prominently in Kant's leisure reading.

19. The following account is, of necessity, reduced to highly selected sketches of the changes that occurred in the latter eighteenth century. An authoritative account, but from a decidedly francophile perspective and with a focus on the earth sciences more narrowly defined, is provided by Rudwick, *Bursting the Limits of Time.*

20. Shortly before setting out with Cook, Banks worked with Daniel Solander, a student of Linnaeus, who then accompanied Banks and Cook on the first voyage.

21. But we do not know. Georg Forster was 18 years of age when he set out with Cook in 1772 and only 25 when he published *A voyage round the world* in 1777. But Kant's views may have been transmitted to him by his father. The view could as easily have arisen with Banks, who was well aware of contemporary scientific developments.

22. Desmarest's first major publication on this theme was "Mémoire sur l'origine et la nature du basalte à grandes colonnes polygones, déterminées par l'histoire naturelle de cette pierre, observée en Auvergne," *Histoire de l'Académie Royale des Sciences* 1771 (Paris; published in 1774). Kant believed in the efficacy of running water to modify the surface of the earth by erosion and sediment transport. Assertion of this position is given in "On the question whether the Earth is aging from a physical point of view" (1754). The key passage, translated by S. G. Martin (1924), is quoted by May, *Kant's Concept of Geography*, 87–8.

23. *Reisen durch verschiedene Provinzen des Russischen Reichs*, 4 vols. (St. Petersburg: Akademie der Wissenschaften, 1771–76).

24. *Voyages dans les Alpes, précédés d'un essai sur l'histoire naturelle des environs de Genève*, 4 vols. (Neuchâtel: Fauche-Borel, 1779–96).

25. The views of Desmarest and Saussure were strongly reinforced by Déodat de Dolomieu, Professor at the Muséum d'Histoire Naturelle in Paris and one of the most senior French natural historians at the close of the eighteenth century.

26. Georges Cuvier, *Rapport historique sur les progress des sciences naturelles depuis 1789, et sur leur état actuel* (Paris: Imprimerie Impériale, 1810). Cuvier was First Secretary of the Institut de France—the reformed Académie Royale—senior researcher at the Muséum d'Histoire Naturelle and Professor at the Collège de France.

27. Whose *Systema Geographicum* was published in 1611. The work notably distinguished "general" (or systematic) geography from "special" (or regional) geography.

28. In particular, in the introduction to the *Physische Geographie*, he states that "there is chorography, that is description of regions and their peculiar features" (trans. Reinhardt [Oxford: Oxford University Press, forthcoming]). Kant did not, however, deny the importance of the systematic analysis of geographical phenomena—indeed, his text is organized into successive sections that analyze phenomena first systematically, and then regionally.

29. Buffon suggested that Earth was 168,000 years old and privately estimated half a million years. In the *General Natural History* Kant posited a universe that required millions of years to form and has no evident end—all terrifyingly modern, apart from the error of 10^3x in the estimate of the age of the earth—an error later repeated by Darwin and even exceeded by leading physicists. Later in Kant's career, a figure with similar views who was in closer proximity to Kant than Buffon, though much less renowned, was Johann Friedrich Blumenbach (1752–1840), professor of medicine at Göttingen. Blumenbach made special studies of fossils and was one of the earlier naturalists to seriously treat the possibility of extinctions. This led to the recognition among scholars of the time of a pre-Adamite ("prehistorical") world—a world much older than Genesis would have it. Blumenbach published his ideas in his *Naturgeschichte* (1779 and later editions), but only after 1788, very late in Kant's career, did he fully articulate them.

30. Trans. Reinhardt, AK 9: 8. Kant is discussing the distinction of geography from history, but the example gives a clear indication of his view on the possibility for the evolution of species. Arthur O. Lovejoy, "Kant and Evolution," in *Forerunners of Darwin*, ed. Bentley Glass, Owsei Temkin, and William L. Strauss, Jr. (Baltimore: Johns Hopkins University Press, 1959), 173–206, and Ernst Mayr, *The Growth of Biological Thought* (Cambridge: Harvard University Press, 1982) both deny that Kant had a concept of evolutionary development of organisms. One may agree that he had no idea of general evolution.

31. According to a letter written by Humboldt, 27 February, 1788, cited by Charles Minguet in *Alexandre de Humboldt, historien et géographe de l'Amérique espagnole 1799–1804* (Paris: François Maspero, 1969), 37.

32. In a letter of July, 1789, quoted by Minguet, *Alexandre de Humboldt*, 40.

33. Forster had recently engaged in a controversy with Kant concerning the origins of the human race—or races, as were perceived at the time—which turned in part on the nature and limits of geographical description, in particular on the admissibility of historical perspectives into the discipline. In his response, Kant emphatically emphasized the distinctiveness of historical and geographical perspectives.

34. Humboldt's book, *Mineralogische Beobachtungen über einige Basalte am Rhein* (Braunschweig: Schulbuchhandlung, 1790) was dedicated to Forster, but it defended Werner's mistaken Neptunist views on the origin of basalt. Forster knew the truth of the matter. Aaron Sachs, *The Humboldt Current* (New York: Viking, 2006) quotes Humboldt as claiming (in 1799) that it was Forster "to whom, for the most part, I owe what little knowledge I possess" (53).

35. The relationships of Kant with his students are discussed by Kuehn, *Kant: A Biography*.

36. See Margarita Bowen, *Empiricism and Geographical Thought: From Francis Bacon to Alexander von Humboldt* (Cambridge: Cambridge University Press, 1981), 217. Cf. Herder's *Ideas on the philosophy and history of man* (1784). Herder also espoused a teleological concept of *Erdgeschichte* from which Humboldt disassociated himself.

37. The German edition (1807), *Ideen zu einer Geographie der Pflanzen, nebst einem Naturegemälde der Tropenländer* was, significantly, dedicated to Goethe. The narrative in this paragraph is drawn from Bowen, *Empiricism and Geographical Thought.*

38. The present translations are my own. The words are variously translated. The variations significantly inflect the meaning and, by extension, Humboldt's inspiration and intentions.

39. Bowen makes this point about "natural history" (which has been used as the translation of *physique du monde*) and ignores the historical implication of "theory of the earth." I think she has ignored the customary meaning, in English, of the former term and the history of the latter. Humboldt's strictly spatial construction of physical geography must also be weighed in these interpretations, else he must be accused of inconsistency. The usual German words for "earth history," used by both Kant and Humboldt, are *Erdgeschichte* or *Naturgeschichte*.

40. See, for example, Livingstone, *The Geographical Tradition*, 134; George Tatham "Geography in the Nineteenth Century," in *Geography in the Twentieth Century*, 3rd ed., ed. Griffith Taylor (New York: Methuen, 1957), 28–69; Bowen, *Empiricism and Geographical Thought*; Sachs, *The Humboldt Current.*

41. See, for example, Michael Dettelbach, "The face of Nature: Precise Measurement, Mapping, and Sensibility in the Work of Alexander von Humboldt," *Studies in the History and Philosophy of Biology and Biomedical Science* 30 (1999): 473–504.

42. Tatham, "Geography in the Nineteenth Century" reprints a quotation from a letter of Humboldt that confirms this viewpoint: "I was impressed with the conviction of the powerful influence that had been asserted upon me by the society I enjoyed at Jena, of how, through association with Goethe, my views of nature had been elevated, and I had, as it were, become endowed with new perceptive faculties." Goethe, of course, knew Kant's philosophy well and was particularly influenced by Kant's arguments about teleological and aesthetic judgment as expounded in the *Kritik der Urteilskraft* (Critique of Judgment), published in 1790, shortly before Humboldt's interaction with the group began. Humboldt, however, scorned teleological arguments in relation to his science and expressed his aesthetic sense relatively narrowly.

43. Hartshorne, "The concept of geography as a science of space."

44. "Über den Gebrauch teleologischer Prinzipien in der Philosophie" (On the use of teleological principles in philosophy), which appeared in *Der Teutsche Merkur* (Jan/Feb 1788). See Kuehn, *Kant: A Biography*, 343–4.

45. Where, according to Erich Adickes (cited by May, *Kant's Concept of Geography*, 72), his full definition appeared as early as 1775. May reviews the history of Kant's concept of geography in some detail. Phillip R. Sloan, "Kant on the History of Nature: The Ambiguous Heritage of the Critical Philosophy for Natural History," *Studies in History and Philosophy of Biological and Biomedical Sciences* 37 (2006): 644, emphasizes the cogency of a connection between Humboldt's definitions and Kant's views as printed in the 1788 essay. It was indeed in that year that Humboldt set out to read Kant. His subsequent association with Forster may also have directed his attention to this piece.

46. Hartshorne's translation is to some degree bedeviled by subtle changes in the meaning of words that were occurring even in the period when Humboldt was writing, and by the fact that the original text was written in Latin, but presumably with Germanic inflections on the meaning of words. The quotation opens as follows: "Geognosy ["Geognosia, Erdkunde," both words given in the original] studies animate and inanimate nature . . . It is divided into three parts: solid rock geography [Geographia oryctologica, quam simplicitur Geognosiam dicunt], which Werner has industriously studied; zoological geography [Geographia zoologica] . . . and the geography of plants [Geographia phytologica]. . . ." The English words are Hartshorne's; the Latin is given in his footnotes. Humboldt appears here to be equating geognosy with all three classical "kingdoms of nature" and appears to associate "geography" with each of these studies. Hartshorne notes that "geognosy" later came to be identified with geology *sensu stricto*, but he also ascribes the term to Werner. Now Werner, whom Humboldt names in his text, defined "oryctognosie" as the study of minerals and "geognosy" as the study of rock formations (i.e., geology), which Hartshorne acknowledges in the paragraph following his translation. Humboldt's statement is correct insofar as each of the three kingdoms will have a geographical component associated with its study, but the Latin word "Geognosia" appears to leave some room for confusion. Humboldt appears here to conflate "geognosy" with "Erdkunde": the latter later came to mean "geography" and the entire question whether the study of Earth's surface—particularly what we today call "physiography"—is properly a geological or geographical study became, later in the nineteenth century, a matter of sometimes bitter academic dispute that continues today. The entire quotation does emphasize that Werner was a prominent source of Humboldt's ideas. The last part of the quotation clearly gives Humboldt's position regarding geography: "Thus zoological history, the history of plants, and the history of rocks, which tell only the past state of the earth, are to be clearly distinguished from geography." Humboldt used the terms *Historia telluris* (*Erdgeschichte*) to signify "Earth history." Bowen, *Empiricism and Geographical Thought* gives a slightly varied translation.

47. Hartshorne, "The concept of geography as a science of space" records a controversial acknowledgment by Humboldt of Kant; controversial because reported in an edition of an 1827 lecture printed almost a century later: in *Kosmos,* the same phrase is used without acknowledgment. Hartshorne also collates several passages of Kant and Humboldt concerning the "system of nature," but it is not at all clear that Humboldt's text is derivative.

48. On the contrary, there is a long footnote in *Kosmos* 1, in which Humboldt lauds the *Geographia generalis* of Varenius as the exemplar of a general and comparative geography from an earlier time. It is, of course, possible that he chose not to use the

Physische Geographie as his exemplar simply because he was aware that the published editions were by no means due only to Kant.

49. Humboldt does cite Kant in *Kosmos*, in which, where he refers to him as "the distinguished philosopher" and "one of the few philosophers who have escaped the imputation of impiety." Humboldt bases his discussion of the great Lisbon earthquake of 1755 in large measure on Kant's account and credits Kant as having defined "with rare sagacity the limits of physical explanations in his celebrated essay *On the theory and structure of the heavens*" (i.e., the *General Natural History*, quoted from E. C. Otté, trans. (New York: Harper, 1870), 210), but he also mounts an attack on Kant's obsolescent cosmological ideas, contained therein.

50. Kuehn, *Kant: A Biography*, 3.

51. In fact, the first map to employ isolines appears to be an isogonic map produced by Edmund Halley in 1701. (An isogon passes through points of equal magnetic declination.) Both this map and Humboldt's more celebrated isotherm map are exceedingly crude because based on very few actual observations. By noting the systematic variation of temperature with altitude, Humboldt did invent the principle of reducing observations to a common basis.

52. From *Entwurf und Ankündigung eines Collegii der physischen Geographie* (1757), quoted by May, *Kant's Concept of Geography*, 66.

53. In *Voyages aux regions équinoxiales du nouveau continent* (Paris: Librarie grecque-latine-allemande, 1816), vol. 1. Adapted from a translation by John Leighly, "Methodological controversy in nineteenth century German geography," *Annals of the Association of American Geographers* 28 (1938): 241.

54. Although it is considered that the Introduction to the notes, dated to ca. 1775, represents his mature methodological views.

55. 1771–1843. He briefly intersected the Jena circle during his early training, in particular listening to Fichte's lectures and reading Schelling. See Frederick Gregory, "Extending Kant: The Origins and Nature of Jakob Friedrich Fries's Philosophy of Science," in *The Kantian Legacy in Nineteenth-Century Science*, ed. Michael Friedman and Alfred Nordmann (Cambridge: MIT Press, 2006), 81–100.

56. See Helmut Pulte, "Kant, Fries and the Expanding Universe of Science," in Friedman and Nordmann, ed., *The Kantian Legacy in Nineteenth-Century Science*, 101–21.

57. Kant's work did remain relevant for some scientists in nineteenth-century Germany, notably the mathematical physicists who were naturally interested in the foundations of their knowledge. The most famous example is that of Hermann von Helmholtz, whose interest in the foundations of geometry and in spatial perception naturally led him to Kant's views on space and perception. Reciprocally, the focus of attention of the Neo-Kantian Marburg school of the 1870s was mathematical physics.

58. Anton Friedrich Büsching (1724–93) was Professor of philosophy in Göttingen. His *Neue Erdbeschreibung*, published in 11 volumes beginning in 1754, was a "special" (i.e., regional) geography with providential overtones. Büsching was exactly Kant's contemporary and the first publication of his work occurred just two years before Kant's first lectures. His work has been cited as one of Kant's major sources. Kant would have been quite comfortable with the regional character of the work, which conformed with his sense of space as relatively differentiated, whilst eschewing the providential aspect.

59. The Ritter-Fröbel controversy is reviewed by Leighly, "Methodological controversy in nineteenth century German geography." Hartshorne, "The concept of geography

as a science of space" unaccountably regards Fröbel's intervention as obscure, though his 1939 account (in *The Nature of Geography*) is fair.

60. Humboldt's role in this development has not been deeply explored despite the existence of a large corpus of work on "Humboldtian science," but his orientation toward a systematic "geophysical" geography must have been regarded as sympathetic toward the emergence of a more narrowly defined Earth science, despite his own decidedly catholic view. Spending part of the 1790s and the years 1804–1827 in Paris, he knew Cuvier and his colleagues very well and pursued discussions with them; reciprocally, the French were the first recipients of the largest corpus of his primary scientific work.

61. Gerland's *Beiträge zur Geophyzik*, established in 1887, became an important geophysical journal. Gerland himself actually published widely in geography. His concept of geophysical geography was attacked by Hermann Wagner, a methodological commentator of the period, because Gerland, following Humboldt, retained plant and animal biogeography within his concept (curiously enough, aspects of ecology are being reintegrated into Earth-surface geophysics today). Gerland later wrote a commentary on Kant's geographical work: "Immanuel Kant, seine geographischen und anthropologischen Arbeiten," *Kant Studien* 19 (1905): 1–43 and 417–57, but this was after the "rediscovery" of Kant's geographical work by Hahn and Kaminski (see below).

62. Ratzel remains a highly controversial figure in the history of geography because of his politico-geographical ideas. See Livingston, *The Geographical Tradition*, 196ff on Ratzel's ideas and influence. Ratzel was also a member of the "Leipzig Circle," an informal association of scholars active between about 1890 and 1914 that attempted to erect a history of German intellectual accomplishments as a reinforcement of Wilhelmian nationalism. In this group, the Humboldt brothers were judged to be figures of the first importance and *Kosmos* was regarded as a scientific synthesis of major cultural importance. See Nicolaas A. Rupke, *Alexander von Humboldt: A Metabiography* (Frankfurt: Peter Lang, 2005).

63. Friedrich Marthe, "Begriff, Ziele und Methode der Geographie," *Zeitschriftder Gesellschaft für Erdkunde* (1877), 422–67. Marthe was professor at the Royal Military Academy in Berlin.

64. Erdmann obtained his PhD at Berlin in 1873 and pursued his career at Berlin, Kiel and Breslau. He remained active until ca. 1920.

65. Kaminski, *Über Immanuel Kants Schriften zur physischen Geographie.*

66. Alfred Hettner, "Das System der Wissenschaften," *Preussisches Jahrbücher* 122 (1905): 251–77.

67. Hettner, *Die Geographie, ihre Geschichte, ihr Wesen und ihre Methoden.*

68. Erich Adickes, *Untersuchungen zur Kants Physische Geographie* (Tubingen: Mohr, 1911).

69. Hartshorne, *The Nature of Geography.*

70. Arguments about the epistemological status of such empirical "laws" go back to Buffon and to David Hume. See Sloan, "Kant on the History of Nature." Kant's final views are expounded in *Kritik der Urteilskraft* (*Critique of Judgment*) of 1790: see Angela Breitenbach, "Mechanical Explanation of Nature and Its Limits in Kant's *Critique of Judgment*," *Studies in History and Philosophy of Biological and Biomedical Sciences* 37 (2006): 694–711. It was the ambiguous nature of Kant's conception of empirical regularities that was taken up by Fries in the attempt to save the philosophical foundations of empirical science in the early years of the nineteenth century.

71. *Critique of Pure Reason*, trans. Norman Kemp Smith (London: Macmillan, 1950), 683.

72. Fred K. Schaefer "Exceptionalism in Geography: A Methodological Examination," *Annals of the Association of American Geographers* 43 (1953): 232–5; rejoinder by Richard Hartshorne " 'Exceptionalism in Geography' Re-examined," *Annals of the Association of American Geographers* 45 (1955): 218–24. Immediately after this debate the entire issue largely died away. A reason as important as any, I expect, was that the simple growth of the discipline produced an assuredly pluralistic outlook. But it is true that regional geography has largely remained in eclipse for half a century now, even though geographical study remains firmly place-based.

3

Kant's *Geography* in Comparative Perspective

Charles W. J. Withers

"Immanuel Kant is the outstanding example in Western thought of a professional philosopher concerned with geography."[1] He was far from being the only one. This essay considers both claims through an examination of Kant's geography in comparative perspective.

My title permits of several interpretive possibilities. One way of reading Kant's geographical writings and his lectures on the subject, which latter formed the basis to his *Physical Geography*, would be to do so in relation to his better-known philosophical work. Another would be to read Kant's work in relation to geography as a whole, either in relation to later forms of geography, to "modern geography" however understood, or in relation to that longer-run tradition of geographical writing and "definition" in which Kant may be placed from the second half of the eighteenth century. Further, a selective comparative historiography of Kant's geographical work would likewise be instructive since it is clear that, where his geography has been read by geographers and others, Kant has been read differently at different times and so has been interpreted to serve different intellectual ends. We might then consider the interpretive communities that have engaged with Kant's geography, including those who in their time did listen to his lectures on the subject or who read his later published work on the subject. Perhaps most evidently, Kant's geography teaching, and his views of the intentions behind his *Physical Geography*, might be examined in relation to other contemporary geographical work. In that sense, Kant's lectures in geography might be understood in relation to other expressions of Enlightenment geography as taught and as written and to what Enlightenment geographers and other Enlightenment philosophers concerned with geography—where they can be identified with certainty—actually did.

These possibilities are of course connected. They are, moreover, reflected in extant work to one degree or another. In his 1970 work, May dismisses Kant's credentials as a geographer, noting that "Kant was not a geographer at all and made no pretence of being one," yet May's remains the only substantive treatment by a geographer of Kant's geography in its wider intellectual context.[2] Since then, others have paid variable attention to his geography or work in the earth sciences. In 1974, Richards—for whom, *contra* May, "Immanuel Kant was an enthusiastic geographer"—took Kant's interests in spatial relations and perceptual experiences to be a precursor to geography's interests in mental maps and cognitive cartography.[3] In doing so, however, Richards subscribes to those presentist interpretations in which Kant has been read, as a regional geographer by Hartshorne and others in the 1930s for example, and by those in the 1960s who saw in his attention to the question of "spatial relations" an intellectual genealogy for geography's then turn to spatial science. In his claims for geography's intellectual basis to rest in the science of areal differentiation, Hartshorne saw in Kant's attention to geography as chorological science a means to defend his own views through reference to a particular intellectual genealogy for the subject.[4]

The connections between Kant's geography and his philosophy are treated seriously, albeit briefly, by Margarita Bowen in her *Empiricism and Geographical Thought* (1981). Yet Bowen tends to treat Kant's work and the history of geography in general in progressivist tones, seeing Kant as a necessary and straightforwardly empiricist contributor to geography's "modernity" en route to its later apotheosis in the form of Humboldtian science. As Bowen put it, "his [Kant's] works evidently provided a fruitful source of inspiration for Humboldt," and, further, "In particular, the unformed study of *physical geography*, on which Kant himself had offered introductory lectures without attempting a serious publication during his years of vigour, came to be chosen by Humboldt as a focus for his own contribution."[5] In one modern assessment of Kant's legacy, there is no suggestion that he was a significant influence upon nineteenth-century German geography.[6] Büttner has considered Kant's geography in relation to a longer-run physico-theological tradition, attributing to Kant the "emancipation of geography" from a position of intellectual servitude to theological dogma.[7] Historians of science and others have discussed Kant's thoughts on earthquakes, volcanicity, and the Earth's age—what we might take as part of his interest in Earth processes, on seeing the 1755 Lisbon earthquake as an influence upon Kant's previously teleological orientation and of a shift towards causal explanations in his writings.[8]

By contrast, there is nothing on Kant's geography in a discussion of his work on the exact sciences.[9] Kant's modern biographer pays virtually no attention to his geography.[10] Where brief attention has been paid by one historian of geography to the connections between the cognitive content of Kant's geography and his concerns to lay out particular domains for the various sciences, a further leading geographer has dismissed the content of Kant's *Physical Geography* as "nothing short of an intellectual and political embarrassment."[11]

Mindful, then, of these comparative possibilities and recognising that in terms of contextual background Kant's geography, if discussed at all as part of his overall work has been more written about than understood, this essay focuses on Kant's geography lectures in relation to other university geographical teaching in the late eighteenth century and upon his *Physical Geography* in relation to other textual geographies of the time. Even this relatively limited focus presents problems as to its remit. It demands attention to the nature and cognitive content of Kant's geography in relation to what was understood and taught as geography in the later eighteenth century, and, more specifically, as "physical geography" by others elsewhere. And it demands consideration of the claims made by various practitioners as to geography's intellectual range and instructional purpose. Even then, we cannot take such content in isolation, either from what else Kant wrote or without recognising that his lectures in geography, given as they were over forty years and changed over time as Kant incorporated new material. In that respect, his ideas have been mediated for us through his students' lecture notes (of which there are several surviving versions). Kant did not publish a geography book nor base his lectures upon any single text. There are, in effect, different "original" editions of the printed material if one takes account of Vollmer's limited compilation of 1801 and Rink's fuller but amended work of 1802.[12]

It is possible, nevertheless, to consider the nature of Kant's geographical teaching and the claims made for geography in his *Physical Geography*. Indeed, addressing the content and the purpose of Kant's geographical work—its "what" and its "why"—requires that something be said about its "how" by examining the world of geographical letters in which he, with others, was placed. As Mayhew has shown of John Pinkerton's *Modern Geography* of 1802, the citational practices of late Enlightenment geographical authors can illuminate the cosmopolitan networks in which they worked and the sources upon whom, as sedentary scholars and university teachers, they were dependent for much of their geography.[13] Using the references within the surviving student lecture notes, it is possible to follow this methodology to see who Kant drew upon, in what context and, at least in outline, to address "when" questions by noting changes over time in his citational geography.

Central to what follows, then, is a concern that Kant's geography be understood in relation to what else was going on in the world of Enlightenment geography. Kant was not the only professional philosopher concerned with geography in the later eighteenth century. He was far from alone in teaching the subject or writing about the range and instructional value of geography. Knowing how far Kant's view of the nature of geography as articulated in his teaching of it in Königsberg over forty years was shared by practitioners and audiences elsewhere depends upon knowing this wider context. In addressing these issues, I consider evidence of geography's textual culture in England,

Scotland, France, and Germany and examine the geography curriculum in the universities of Aberdeen, Glasgow, Göttingen, and Geneva. In particular, and in regard to its substantive content, its textual production, the claims made by Kant as to its nature and the fact that it was given as a set of lectures, Kant's *Physical Geography* is here considered in relation to what is known of the practices of late Enlightenment geographical textual compilation and of its place in the teaching programmes of the philosophers Thomas Reid and Horace-Bénédict de Saussure, and among some less evidently "geographical" personnel such as Colin Maclaurin, Condorcet, and Nicolas Desmarest.

Interpretations of Kant's *Physical Geography* in comparative perspective depend, then, upon our recognising differences in what geography was and how it worked in the Enlightenment. But these are matters of greater significance, too, for in such an approach, we may see the Enlightenment itself as geographically different, as having certain centers of geographical endeavour just as it had centers of philosophical enquiry and of political economy.[14] In those terms, the focus of interest is less about Kant's geography alone than it is to do with knowing where to place Kant on the map of Enlightenment geographical knowledge.

Kant's Geographical Writings and His *Physical Geography*

Kant's lectures on geography—what became his *Physical Geography*, first in limited printed form in 1801 and more fully from 1802—were first given in Königsberg in 1757. They were the result of his compilation of others' work and were not the result of his exploring, traveling, or observing for himself the things he wrote about.[15] They focused on physical geography and were given at first-year level although outside visitors and others did attend them. Kant considered geography "the propaedeutic for knowledge of the world,"[16] a term explored below.

In addressing Kant's concept of geography, May asked five questions: What are the origins of Kant's ideas on geography, or are they original? What development did Kant's ideas on geography undergo in his own thinking? What influence did Kant's ideas on geography have? What does Kant mean? To what extent can Kant be regarded as an adequate foundation for modern geography?[17] His focus was with answers to the last two questions.

Here, different questions are pertinent, questions which allow Kant's geographical work to be read comparatively: How does Kant's concept of geography relate to other work of the time, particularly to the teaching of the subject? Are his claims on the relationship of geography to history, on the subdivisions of geography and on the overall utility of geography apparent in others' views of the subject as taught? Was "physical geography" a common term, commonly understood?

Kant's earliest statement of his concept of geography appears in 1757.[18] Knowledge of the Earth is possible, he notes, through mathematical study, attention to "political doctrine" (forms of government and commercial systems and so on) and physical geography. Later, in the Introduction to his *Physical Geography*, Kant noted a sixfold distinction (including physical geography) between mathematical geography, "in which are treated the shape, size and motion of the earth, as well as its relation to the solar system"; moral geography, "the customs and characters of people according to the different regions"; political geography, which dealt with the connections between civil society, universal laws and a country's natural setting; mercantile geography, which centred on trade and resource exchange; and theological geography, in which comparative differences in theological principles were explained, again in relation to natural setting—what Kant termed "fundamental changes according to differences of soil." Physical geography drew these divisions together: "Physical Geography is thus a general outline of nature, . . . it is not only the ground of history but also the basis for all other possible geographies." Kant incorporated the study of man within his view of physical geography—"Physical geography considers only the natural condition of the earth and what is contained on it: seas, continents, mountains, rivers, the atmosphere, man, animals, plants and minerals."[19]

Kant's use of the term "physical geography" was thus broad, embracing as it did the relationship between natural setting and patterns of human occupance, not just the features of the physical earth. And it was intended to be useful. As Kant put it:

> I present this material first in the natural order of classes, and proceed later to set forth, in a geographical manner of instruction, through referring to all countries of the earth, the particular inclinations of men which derive from the regions in which they live, the variety of their prejudices and ways of thinking, in so far as all this serves to acquaint man better with himself, including a brief comprehension of his art, commerce and science, and a report on the products of the above mentioned lands in their proper places, climate, etc., in a word, everything that belongs to a physical consideration of the earth.[20]

As Kant made clear in the second major outline of his course, in 1765–1766, geography was considered by him a sum of knowledge that was agreeable and easily understood.[21] Kant presented physical geography as useful in teaching morality, as an instrument for political geography, a summary of natural history and the key to theoretical physics and to theology and because, overall, it brought "ennobling satisfaction."[22]

In the Introduction to the *Physical Geography*, Kant took geography with anthropology to "together constitute knowledge of the world." "We are taught knowledge of human beings by anthropology. We owe our knowledge of nature to physical geography; that is, to a description of the earth." Physical geography may have been concerned with description of the earth's surface features, but such a concern was the key to other forms of knowledge. "The physical description of the earth is thus the first part of knowledge of the world. It belongs to an idea (*Idee*) which is called the propædeutic to understanding our knowledge of the world."

Kant saw geography as a propaedeutic in relation to science and to life: as an essential part of a child's education, part of a general framework for knowledge, and as an introduction to the other, more theoretical, natural sciences. Geography was also an end-product of knowledge, even an intellectual synthesis, if less clearly a "discipline."

In here addressing geography's concern with the description of the Earth's surface, Kant is delimiting his concept of geography in relation to the long-run Classical definition of geography as concerned, in terms of scale, with the whole Earth (in contrast to cosmography and to chorography), and in terms of method to geography's concern with description.[23] Geography's descriptive focus on the objects of nature meant that it dealt with the "system of nature." Physical geography was a concrete and largely empirical system of nature in contrast, for example, to physics, a more abstract and theoretical system of nature.

In Kant's concept of geography, geography and anthropology were both useful empirical sciences, part of that knowledge of the world "which serves to procure the pragmatic for all the acquired sciences and aptitudes, . . . useful not only for school but also for life."[24] Anthropology was concerned more with what Kant termed man's "inner sense," questions of the soul or self, in contrast to geography's attention to "outer sense," the world as mediated through sensate empiricism. The distinction between inner and outer sense is problematic and features more widely in Kant's writings than in his concept of geography, notably in his attention to mental or moral character—and thus to "moral geography"—to race and to the distinctions between pragmatic anthropology and physiological anthropology.[25] Kant's lectures on anthropology, which grew out of his lectures in physical geography, were first given in 1772–1773.[26]

History, likewise an empirical discipline for Kant, was in his conception more clearly distinguished from geography than was anthropology. In the Introduction to his *Physical Geography*, Kant states clearly that geography is essentially concerned with spatial relations, history with temporal relations. Various expressions of this distinction were articulated: "History is description according to time; description according to space is geography"; "History concerns events which, under the aspect of time, have occurred one after the other. Geography concerns appearances under the aspect of space which occur simultaneously"; "History is a narrative, but geography is a description." But they are not without contradiction: thus, "The history of nature contains the manifold qualities of geography, namely how things were in different epochs, but not how things are at the present time because that would be a description of nature."[27]

The intellectual affiliation between history and geography and the importance of phrases like "Geography therefore is the substratum" [to history] ought not, I suggest, to be read, as in part they have been by May, in terms of an ontological priority for geography, and certainly not read in terms of any strict separation between history and geography. The connections are better read in terms of that commonplace in Classical humanist tradition, namely of geography—strictly, chorography—as one of the "eyes" of history, by convention the left eye, with chronology (history) the right eye. It ought also to be seen in terms of a more specifically geographical methodology, between "ancient geography," which took as its focus scriptural claims and accounts of "ancient authorities," and "modern geography" which, in being based on the latest observations and reports, was always in a state of becoming even as it was being aspired to and advertised in the titles to geography's books (as in Pinkerton's *Modern Geography* of 1802). Ancient geography and modern geography were methodological labels in the Enlightenment *and* part of longer-run epistemological debates, not just particular classificatory consequences of Kant's concern to see geography as the science of spatial relations read in historical context.[28]

In sum, Kant conceived of geography as a science confined to descriptive study of natural phenomena on or near the Earth's surface, in which conception man was included inasmuch as he was part and a product of nature. His lecture notes emphasise physical geography, were brought together and published under that title, were reliant upon evidence from others' texts and observations, and incorporated other divisions—mathematical, moral, political, mercantile, and theological, geography. Geography was a system of nature, and, less clearly, a history of nature in that geography's concern with things in space extended to things in time via "ancient geography" and "ancient history." Geography's concern with spatial relations and history's with temporal relations are important, but not absolute, distinctions. Geography in Kant's view embraced some of the concerns of anthropology, notably around human character and customary practice. Kant took geography to be a propaedeutic in several senses: as a basis to education, as an introduction to the more theoretical systematic sciences, as an ordering principle for knowledge, and as an end in itself.

Kant's Geographical Work in Comparative Perspective

These facts have important implications for interpretation of Kant's geographical lectures and for consideration of the Introduction to his *Physical Geography* as a mature expression of his concept of geography in relation to his lectures. Knowing that his lectures survive in various student "editions" and that Kant drew from others' work without ever writing a geography text of his own does not demand that we suppose a false and essentialist premise of logical internal consistency that ought to obtain for Kant's geographical writings for them to be comparable to others'. Such a notion is clearly untenable, for Kant or anyone else: authors' works are not always logically consistent, authors' ideas change over time, "bodies" of written work are fluid things, supplemented by authors and readers and open to different interpretation by different communities. Different comparative possibilities have their own legitimating logic. Yet since what is being compared varied in its material production and mode of delivery—as individual lectures, a geographical curriculum, as a later conspectus incorporating statements concerning geography's intellectual classification and utility (but not a textbook)—we do need to be attentive to the material forms of geography's production and communication as well as to its cognitive content. Similarly, we need to be attentive, where we can be, to the social and institutional context in which geography was being taught, written, read, listened to, and debated. Kant's geography has a textual space then and now as, for his audiences, it once occupied a dynamic speech space and site of enquiry.[29]

For no other reason than the claim made by May that "Many of Kant's ideas on geography can be found, in embryonic or undeveloped form in the works of J. M. Franz, a geographer active in the period from 1740 to 1770,"[30] let me turn first to Franz and thence to geography teaching in Göttingen. From Göttingen, I move to Geneva and to geography lecturing in the universities of Enlightenment Britain and elsewhere before turning to Kant's text.

Johann Michael Franz was involved in the establishment in 1746 of the Cosmographical Society of Nuremburg, and in developing there that synthesis of mathematics, practical mapping, survey, and navigation known as mathematical cosmography and in plans, not all fully realised, for a "Staatsgeographus" (state geographer) and corresponding classes for geography. From 1755, Franz was professor of geography at Göttingen. His teaching there was mainly mathematical in orientation. Geography was also given there by Johann Christoph Gatterer, professor of history, who emphasized geography's place as the basis to historical understanding and used his sixty-six map *Atlas Complet des Révolutions* to teach comparative history and ancient geography.

Geography was also taught in Göttingen by Anton Friedrich Büsching, professor of philosophy there between 1754 and 1761, and author of *Neue Erdbeschreibung*, which appeared in eleven volumes between 1754 and 1792 (and which was available in a six-volume English translation from 1762). In his teaching, Büsching took a wide-ranging view of geography "as a well-grounded account of the natural and civil state of the known earth surface." Büsching's geography, unlike Kant's, was in part based on his own travels. Yet, like Kant's and many others', it was dependent upon up-to-date information from credible sources for some of its content: "Its sources" [Büsching noted] "must not be other geographies, but good descriptions of particular lands and places, and specific investigations." Unlike Kant, Büsching offered no wider philosophical reading of geography's place in systems of knowledge other than through a conventional prefatory appeal entitled "Of the Utility of Geography," which stressed the benefits of geography to the statesman, and through his primary division of the subject into natural and civil aspects, which, in respect of the first, was then ordered in relation to mathematical and physical geography.[31] In his discussion of "Of the Natural State of the Earth," Büsching considered the Earth's atmosphere, mineralogy, and waters and paid attention to man within a discussion of the animal kingdom.

Bowen has observed that Kant's "lectures on geography at Königsberg appear to have owed a considerable amount to Büsching's example, especially in their commencement in 1756."[32] This claim is borne out in comparison of the respective ordering and treatment of the earth's atmosphere, its waters or hydrography, and features such as mountains, deserts, and valleys in the work of both men. It is less clear in respect of Büsching's treatment of man. Where Kant saw latitude and climate as determining influences and in those terms equated geography with social capacity, Büsching takes a more measured view of the climate's determining influence than, for example, either Kant or Montesquieu:

> Men also differ greatly in their make, size, method of living, and in their manners. But as to the difference in their intellectual faculties, we are not to look for that in their nature, or climate, but in the greater or less opportunity they have of improving and exercising their mental powers. An inhabitant of *Greenland* or *Lapland*, a *Moor* or a *Hottentot*, is in his way as intelligent as one among the more civilized nations; and if the former had the same opportunities of improving his understanding and regulating his passions as the latter enjoys, he would not be at all inferior to him.[33]

Kant certainly owes more and is closer in content to Büsching than to Franz, not least in the propædeutic value attached to geography and in recognizing geography's place with respect to history for the reason, noted above, that it was a commonplace understanding. Kant drew upon Büsching in preparing his lectures.[34] Büsching's published work is cited in the notes to Kant's lectures; Franz's is not.

Kant's attention to the physical realm in his geography lectures is echoed in the lectures on geography given in Geneva in 1775 by the philosopher Horace-Bénédict de Saussure. This course seems to have been given only once and what we know of it is from the lecture notes of a Swiss theological student, Jacques-Louis Peschier. For Saussure, geography was divided into three different parts, astronomical, physical and political of which division, he notes, "The last part is almost the only one to be commonly taught, and it deals with the states and customs of mankind, without paying attention to the actual Earth." If geography "were to be taught precisely," continued Saussure, astronomical geography would be taught first, since "other aspects cannot be well understood without such prior consideration of the Earth in general relative to the sun." This cosmographical perspective was not something he would begin with because, noted Saussure, "this requires a certain knowledge of mathematical principles which you have not yet attained," [and so] "it will be sufficient to render the understanding of physical geography easier by presenting here certain introductory points concerning its fundamental principles."[35]

Saussure's physical geography lectures were ordered around six parts: the Earth's surface in general; landmasses; waters; the atmosphere enveloping the globe; the changes or variations which have occurred on our globe; and the products of the land and sea. The study of man as part of nature and mankind's relationship to the physical environment was not considered although man was briefly considered as an agent of change through agricultural modification. Like Kant's, Saussure's lectures were about description of the physical earth, much less with its theoretical interpretation, and drew on his own observations in the Jura and the Alps and from a trip to Italy in 1772–1773. Description of the physical earth was consistent with what Kant called "description of nature" rather than with what he saw as a "history of nature," which involved the transformations nature has gone through in different times (a focus more apparent in Buffon's *Théorie de la Terre*, for example). Unlike Kant, Saussure's geographical teaching had no place for consideration of moral geography.

In the forty-year period over which Kant gave his geography lectures in Königsberg, two university professors in England gave lectures on geography: Nathaniel Bliss and Thomas Hornsby of Hart Hall, Oxford, respectively Savilian professors of geometry and of astronomy. Both lectured on mathematical geography. In Scotland there were five, all in Aberdeen: Thomas Gordon, professor of humanity at King's College, who lectured on ancient geography; and, in Marischal College, John Stewart, Thomas Blackwell Junior, James Beattie, and Robert Hamilton, who variously taught mathematical geography, ancient geography and geography's connections with civil history. Extending the survey to include near contemporaries such as the "Common Sense" philosopher, Thomas Reid, who lectured on "the elements of geography" in King's College in 1752–1753, or the Newtonian mathematician Colin Maclaurin, professor of mathematics at Edinburgh who taught geography as part of his classes in 1741, or James Millar, professor of mathematics at Glasgow, who gave a 43-lecture program on geography from 1802 until 1821, and the mathematical geography classes in Trinity College Dublin, and the comparative picture becomes even richer.[36]

There is not the space here to document the geographical "curriculum" of each of these figures, but elements can be noted inasmuch as they bear upon Kant's geographical teaching.[37] In most British university geographical teaching in the eighteenth century, including within the wider British world of Trinity College Dublin and the American colleges, attention focused upon mathematical geography—testing Newtonian propositions, instilling mathematical principles through geographical examples, or discussing the Earth's dimensions and its measurement—on geography's commonplace connections with history, or upon a mix of the two. Physical geography received little systematic attention. Elements comparable to Kant's do feature. The course of lectures given by Thomas Reid was in four parts: a general regional geography of the globe; the main lands and seas named; a list of notable features associated with each country and major city; and an account of the principal trade winds. In the first and last of these, it is possible to see connections with Kant's topographical descriptions and his limited engagement with mercantile geography. Kant did concern himself with winds in his geography lectures, in terms of trade winds ("*Paßatwinden*") and commercial routes and in regard to winds' "character" in respect of dampness and dryness and, thus, their explanatory role in matters of climate and health. Reid, like Kant, emphasised geography's propaedeutic value in claiming that geography was an essential part of that philosophy "which may qualify Men for the more useful and important Offices of Society."[38] Only in Millar's geographical lectures in Glasgow and then for the period after Kant's death in 1802 does physical geography form a major element in a British university geography curriculum: Millar organised his course around the Earth as connected with the heavenly bodies; the Earth's main physical properties;and the Earth in connection with the inhabitants. There is no internal evidence from the versions of Millar's course extant in student dictates that he drew upon Kant's 1802 published work, but there was common ground in the attention to waters and mineralogy, less so in the attention to man.

Used with caution, the variant students' "editions" of his geography course show that Kant drew widely upon others' natural philosophy and geographical texts.[39] Kant looked to several of the late seventeenth-century physico-theologists such as John Woodward, Thomas Burnet, and William Whiston in framing his course. Classical authors—Aristotle, Strabo, for example—are cited in demonstrating geography's scholarly longevity and the pedigree of its mathematical concerns. The work of Bernhard Varenius, notably his *Geographia Generalis* (1650), is cited in giving attention to "modern" developments in geography's content and method. More significantly, numerous eighteenth-century authors are cited in support of particular points—Buffon on natural history, La Condamine and Maupertuis over geodetic investigations, Gmelin and Peter Simon Pallas for their work on Siberia and the Russias. In lectures attributed to Barth, dating from c.1782, we find the first mention of James Cook and the Forsters, which suggests that Kant revised his content to reflect new findings from the Pacific voyages.[40] In that sense, we ought not to make too sharp a distinction between the human geographical content of his anthropological lectures, which Kant began from 1772, and his geography lectures since it is clear that, in the latter context, Kant continued to consider questions of human variability as it was being revealed through the voyages of Enlightenment navigators even as he expanded upon them more fully in his anthropological teaching.[41]

Kant drew, as did Büsching, upon the work of the Swiss geographer Samuel Engel whose 1765 *Mémoires et Observations Géographiques et Critiques* influenced contemporary views on polar exploration with his claims (erroneous) about an empty and ice-free ocean in the northernmost latitudes: Kant mentions Engel by name, refers to "the hypothesis" and the presence of open seas east of Spitsbergen as if to support it, in section 37 of his *Physical Geography*. Kant also drew upon others' texts in order to structure his lectures, in part upon Buffon's *Théorie de la Terre* (1749) and, notably, upon the 1766 work *Physick Beskrifning Ofver Jordklotet* of the Swedish chemist and mineralogist, Torbern Olof Bergman, via a 1769 German edition published as *Physicalische Beschreibung der Erdkugel.* Bergman also looked to Burnet, Whiston, and Woodward. Saussure drew differently upon Buffon and Bergman. Saussure's six-part division of physical geography is virtually identical to that of Bergman, and Saussure echoes Bergman's attention to observation and facts and disavowal of theory. Buffon's work, by contrast, was conspicuously theoretical beginning with disquisitions about the Earth's formation and age before turning to a variety of "proofs" in support of his own speculations. Saussure was critical of Buffon because of this theoretical focus and because he thought the Frenchman's work to contain too many generalisations without sufficient empirical substantiation.[42]

We know that Kant used the German edition of Bergman's work: the Swede is included in the lists of those whose works he drew upon (although he is not cited in Rink's footnotes to Kant's lectures). Saussure and Kant would both have found the Swede's emphasis upon empirical encounter and observations consistent with their own views. Yet we must be mindful that physical geography was then a common feature of geography's subdivision and that established classificatory and rhetorical practices, which ordered geography in relation to cosmography and physical geography in relation to other subdivisions, were commonplace. The fact that different people ordered and presented their geography in particular ways—in written form, in lectures, as textbooks—reinforces the point about being attentive to place, personal intention and to textual conventions in Enlightenment geography's publishing as well as to its cognitive content.[43]

In Paris, for example, the world of geography teaching and publishing was very different from that in Königsberg. Buffon's multivolume textual and theoretical engagement with the nature of the Earth incorporated the results of new geographical knowledge from numerous *voyageurs-naturalistes*, yet Buffon never taught geography. Later in the century, geography varied in the French capital between Condorcet's programmatic vision for a science of synthesis and of military application in the Ecole Militaire and from 1795, in the National Institute (where geography featured in the Class of Moral and Political Sciences), and as an instructional aid to children and the administrative basis for a new political vision of France altogether in the writings of Edme Mentelle. Based in his workshop in the Quai d'horloge on the Ile de la Cité, the leading geographer and mapmaker Jean-Baptiste Bourgignon d'Anville was producing a new cartography of and for the Enlightenment and corresponding with numerous others throughout Europe while doing so. Yet he never left Paris. D'Anville in France's capital—for many, the capital of the Enlightenment—was, like Kant in Königsberg, a sedentary scholar, both men the epitome of a *géographe de cabinet*.[44]

D'Anville was a contributor to the many geographical essays in the *Encyclopédie* of Diderot and d'Alembert, a work that was itself a geographical metaphor for new systems of knowledge—d'Alembert likened it to a world map—and which was seen by many as the Enlightenment's manifesto. The principal entry on geography in the *Encyclopédie* appears in volume VII of that work, published in November 1757. The entry was written jointly by d'Anville, the mineralogist and chemist Nicolas Desmarest, and the mapmaker Didier Robert de Vaugondy. Geography was considered under three period headings: ancient, of the middle ages, and modern, in which last context it was divided into six subdivisions—natural, historical, civil and political, sacred, ecclesiastical, and physical. Aside from the obvious and varied place of the human world within the realm of geography, the section on physical geography is noteworthy here. This was the work of Nicolas Desmarest. Desmarest was less concerned that "*géographie physique*" should include attention to the earth's external shape and cosmographical place and more that it describe its substance. "More precisely,"

> *Géographie Physique* is a reasoned description of the great phenomena of nature and the study of general results obtained from local and specific observations, combined and associated methodically in different groups and according to a master plan capable of revealing the natural economy of the globe as long as the latter is considered as a barren and uninhabited physical mass.[45]

Desmarest's definition of physical geography is in contrast to Kant's in its lack of attention to man and in its suggestions of a theological, even teleological, argument.

Further textual comparisons are afforded by the work of Philippe Buache in whose 1752 *Essai de Géographie Physique*, geography was divided into three main parts: natural or physical, mathematical, and historical. Buache's *Essai* was known to Kant and used by him in his teaching.[46] Buache's work on the physical geography of drainage basins and upon topographic regions as delimiting the geographical distribution of animal species was an important influence upon the geographical community in Göttingen and, more so, upon Eberhardt Zimmerman, professor of mathematics in Brunswick, whose *Specimen Zoologiae Geographicae Quadrupedum* (1777) drew upon Buache in making associations between "natural" habitat and climate. Zimmermann also wrote a *Political Survey of the Present State of Europe* (1787), effectively a comparative statistical and political geography of late eighteenth-century Europe, and which was, in turn, an influence upon the medical geography of the German Leonhard Ludwig Finke in which concerns for a "general-medical practical geography" were rooted in explanation of the environmental bases to the nature and distribution of disease.[47] Although none of Buache, Zimmerman, or Finke is mentioned within Kant's *Physical Geography*, Kant was aware of the questions with which these and other men were dealing since, in addition to noting the connections between health and winds (see above), he alludes in his lectures to the work of James Lind in this context, although it is not clear whether Kant is referring to Lind's 1753 *Treatise on the Scurvy*, or, as is perhaps more likely, to his *Essay on Diseases Incidental to Europeans in Hot Climates* (1768).

In Conclusion: Kant a Geographer?

Questions of cognitive content *in* geography and intellectual intention *about* geography are always questions *of* geography. I mean by this that to expose late eighteenth-century geography in its university settings to comparative scrutiny is to locate differences in what contemporaries held the subject to be. What geography was held to be depended crucially on who taught it, where it was used, and with what end in view, whether this is so of its associations with Newtonian theory (as in Oxford, Edinburgh and Marischal College, Aberdeen), as part of classical history and commercial utility (as for Thomas Reid in King's College Aberdeen), general mathematics (Glasgow and St Andrews), with astronomy, mapmaking and mathematical cosmography (in Göttingen), or the dynamics of the physical world (as in Glasgow, Paris, Geneva, and Königsberg). Looking at geography as only an instructional discourse directed at university students presents a partial picture. In academies, in the public sphere, and in schools, geography was the subject of considerable attention in numerous ways. In sum, "there was more geography being done in the eighteenth century than there were geographers doing it; and different notions of geography were at work."[48]

Kant "did" geography, but was he a geographer? It depends, of course, on how we define the term and think of his geographical writings. In discussing the textual traditions of late eighteenth-century English geography, Robert Mayhew notes that the term "geographer" did not mean then what it means now:

> Rather, most authors who compiled information about the earth's surface at a global scale were either impoverished Grub Street

> journalists or wealthier scholar historians. Most of the so-called geographers in seventeenth- and eighteenth-century Britain, be they poor or rich, were in fact authors in numerous genres and rarely of great intellectual distinction.[49]

Kant does—and more substantially does not—fit this picture. He was a compiler of others' geographical facts and lectured in geography and was in those two (weak) senses a geographer. He did both things because he considered geography useful in several ways. Kant was no Grub Street hack. Yet his sedentary engagement with geography and his commitment to its teaching for forty years makes him more of a geographer than many who did so describe themselves at this time and more so than many of his professional Enlightenment peers, philosophers, historians, or mathematicians.

Kant's engagement with geography was indicative of the intellectual interests of the age in which he and others worked. Geography was an established intellectual enterprise with understood relations to history, to mathematics (in mathematical geography), and to cosmography (often, in the last context, realised as astronomical geography), and with an emphasis upon description. Quite what was described and at what scale—as a matter, properly, of geography's concern with the whole earth, or of chorography with parts of it—differed in the teaching of different philosophers and others. Kant's divisions within his geography were one expression of a common trend. His incorporation of others' findings was likewise a common authorial conceit. In the later Enlightenment, new knowledge of the terraqueous globe through primary survey, mapmaking, and oceanic voyaging was flooding in, transforming knowledge of the shape and contents of the world as the home of man. Kant's engagement with geography was notable for his attention to moral geography as part of his anthropological teaching and writings—in which regard his discussions with Georg Forster may be characterised as those between a sedentary scholar reliant upon others' accounts and Forster's claims borne of limited but direct observation—and for his view of the subject as a propaedeutic and, in Introduction to his *Physical Geography* rather more than in his lectures, for statements regarding geography as a form of and a basis for knowledge. In this last respect in particular, his is one of the most articulate expositions within the Enlightenment of geography's philosophical reach.

It is in that context in which Kant's geography must be treated comparatively, I believe, rather than to see him as the basis to "modern geography" or, even more particularly (and wrongly), to see his concern for spatial relations to be the forerunner to ontological angst within the geographical community two centuries later. Yet to do an Enlightenment comparative analysis more fully still would involve knowing when exactly Kant drew upon others' findings in amending his teaching, and knowing, perhaps from his correspondence, of any connections between Kant and his fellow philosopher-geographers—in Göttingen, Geneva, or elsewhere. Local matters become important. Since he never left Königsberg to do any geographical investigating for himself, who were his local sources with respect to the faraway: ships' captains, other travellers, booksellers? Can we ever know his patterns of book-buying? Did Kant own a copy of the *Encyclopédie*? Since Kant did not read French, how did he use Buache's 1752 work? Can we know the exact extent of his intellectual debt to Bergman? Kant's geographical writings place him and Königsberg centrally on the map of Enlightenment geographical knowledge. What is also true from this comparative reading of his *Physical Geography* and with his students' mediation of his teaching is that, while he was a significant figure in Königsberg in his geography classes, Kant was at the same time connected through links of varying strength and type to many others on that map and that, through attention to his work in this respect, we can see how one of Europe's leading philosophers engaged with questions posed by the world's changing geography in the Enlightenment.

Notes

1. Joseph A. May, *Kant's Concept of Geography and its Relation to Recent Geographical Thought* (Toronto: University of Toronto Press, 1970), 3.

2. May, *Kant's Concept of Geography*, 12.

3. Paul Richards, "Kant's Geography and Mental Maps," *Transactions of the Institute of British Geographers* 61 (1974): 1–16, quote from page 1.

4. These and other interpretations are discussed in May, *Kant's Concept of Geography*, 8–24. For Hartshorne's attention to geography as the science of areal differentiation, see Richard Hartshorne, "The Nature of Geography," *Annals of the Association of American Geographers* 29 (1939): 173–658. On Hartshorne's views on Kant as a precursor to "modern" spatial science, see Richard Hartshorne, "The Concept of Geography as a Science of Space, from Kant and Humboldt to Hettner," *Annals of the Association of American Geographers* 48 (1958): 97–108, and *Perspective on the Nature of Geography* (Washington: Association of American Geographers, 1959), 22, 25, 67, 109, 180.

5. Margarita Bowen, *Empiricism and Geographical Thought: From Francis Bacon to Alexander von Humboldt* (Cambridge: Cambridge University Press, 1981), 209. Original emphases.

6. See Michael Friedman and Alfred Nordmann, ed., *The Kantian Legacy in Nineteenth-Century Science* (Cambridge MA: MIT Press, 2006).

7. Manfred Büttner, "Kant and the Physico-Theological Consideration of the Geographical Facts," in *Science and Religion: Wissenschaft and Religion*, ed. Anne Baümer and Manfred Büttner, (Bochum: Brockmeyer, 1989), 82–92.

8. Oscar Reinhardt and David Oldroyd, "Kant's Thoughts on the Ageing of the Earth," *Annals of Science* 39 (1982): 349–69; "Kant's Theory of Earthquakes and Volcanic Action," *Annals of Science* 40 (1983): 247–72; "By Analogy with The Heavens: Kant's Theory of The Earth," *Annals of Science* 41 (1984): 203–21. For May, these "early works . . . are not to be construed as works in geography, but as works in speculative

physics of the earth's interior" (May, *Kant's Concept of Geography*, 85). On the view that Kant's explanatory emphases shifted because of the 1755 Lisbon earthquake, see Svend Erik Larsen, "The Lisbon Earthquake and the Scientific Turn in Kant's Philosophy," *European Review* 14 (2006): 359–67.

9. Michael Friedman, *Kant and the Exact Sciences* (Cambridge: Harvard University Press, 1992).

10. Manfred Kuehn, *Kant: A Biography* (Cambridge: Cambridge University Press, 2001) mentions only the "Announcement of a Lecture in Physical Geography," Easter 1757 (135). There is no mention of Kant's geography in Paul Guyer, ed. *The Cambridge Companion to Kant* (Cambridge: Cambridge University Press, 1992).

11. On the connections between Kant's geography and his other concerns, see David Livingstone, *The Geographical Tradition: Episodes in the History of a Contested Enterprise* (Oxford: Blackwell, 1992), 113–7. The dismissive remark is from David Harvey, "Cosmopolitanism and the Banality of Geographical Evils," *Public Culture* 12 (2000): 529–64, quote from page 529. Harvey's earlier engagement with Kant's attention to geography as a chorological science was critical both of the attention Hartshorne and others paid to regional science without it having a basis in science and of Kantian notions being about absolute space rather than about relative space: see David Harvey, *Explanation in Geography* (London: Edward Arnold, 1969), 70–9, 207–12.

12. May, *Kant's Concept of Geography*, 74-5. Erich Adickes, basing his findings on an examination of twenty manuscript copies of Kant's lectures dating between the late 1750s and late 1792, concluded that the later sections in the 1802 Rink edition (from section 53 to the end), dated from before 1760 and were based on Kant's original course notes. The Introduction was also thought to date from this period. Sections 52 and before were thought to be compiled from later manuscript copies of the lectures. It is possible even that Rink wrote some part of the *Physical Geography*, sections 11 and 14 (the first on cosmology, the second on water and hydrography), and may have added the footnotes throughout.

13. Robert Mayhew, "Mapping Science's Imagined Community: Geography as a Republic of Letters, 1600–1800," *British Journal for the History of Science* 38 (2005): 73–92.

14. This concern for the Enlightenment's geographical dimensions is the subject of my *Placing the Enlightenment: Thinking Geographically about the Age of Reason* (Chicago: University of Chicago Press, 2007).

15. Kant's work on earthquakes was published in 1756, that on volcanoes in 1785. It is clear that he drew some material for these from travels in Bohemia and Moravia. It is less clear how much these excursions informed his geography lectures.

16. Quotes relating to Kant's geographical works are here taken from the Akademie Edition of his *Gesammelte Schriften*, published in twenty-two volumes, and here cited as AK, by volume and page. This quote is from AK 7: 120–1.

17. May, *Kant's Concept of Geography*, 14.

18. *Entwurf und Ankundigung eines Collegii der Physichen Geographie*, AK 2: 3–10.

19. *Entwurf und Ankundigung eines Collegii der Physichen Geographie*, AK 2: 3.

20. *Entwurf und Ankundigung eines Collegii der Physichen Geographie*, AK 2: 9.

21. *Nachricht von der Einrichtung seiner Vorlesungen in dem Winterhalbenjahre von 1756–1766*, AK 2: 303–13.

22. May, *Kant's Concept of Geography*, 68.

23. Classical distinctions, based on Ptolemy's writings, separated geography as the study of the Earth as a whole, from cosmography, the earth in relation to other planetary bodies, and chorography, the description of parts or regions of the Earth.

24. *Von den Verschiedenen Racen der Menschen*, AK 2: 443 n.

25. May, *Kant's Concept of Geography*, 107–18.

26. Immanuel Kant, *Anthropology from a Pragmatic Point of View*, ed. Robert B Louden (Cambridge: Cambridge University Press, 2006), 6.

27. May, *Kant's Concept of Geography*, 124.

28. Robert Mayhew, "Geography, Print Culture and the Renaissance: 'The Road Less Travelled By,' " *History of European Ideas* 27 (2001): 349–69.

29. On this, see David Livingstone "Science, Site and Speech: Scientific Knowledge and the Spaces of Rhetoric," *History of the Human Sciences* 20 (2007): 71–98.

30. May, *Kant's Concept of Geography*, 51.

31. Bowen, *Empiricism and Geographical Thought*, 154–7.

32. Bowen, *Empiricism and Geographical Thought*, 159.

33. Anton Friedrich Büsching, *A New System of Geography*, 6 vols. (London: A. Millar, 1792), 1, 48.

34. *Entwurf und Ankündigung eines Collegii der Physischen Geographie*, AK 2: 4.

35. Albert Carozzi and John K. Newman, eds. *Lectures on Physical Geography given in 1775 by Horace-Bénédict de Saussure at the Academy of Geneva* (Geneva: Editions Zoe, 2003), 80.

36. For a fuller account from which these figures are taken, see Charles W. J. Withers and Robert Mayhew, "Rethinking 'Disciplinary' History: Geography in British Universities, c.1580–1887," *Transactions of the Institute of British Geographers* 27 (2002): 11–29.

37. For fuller discussions, see Charles W. J. Withers "Toward a Historical Geography of Enlightenment," in *The Scottish Enlightenment: Essays in Reinterpretation*, ed. Paul Wood (Rochester: University of Rochester Press, 2000), 63–98; Charles W. J. Withers, "Situating Practical Reason: Geography, Geometry and Mapping in the Scottish Enlightenment," in *Science and Medicine in the Scottish Enlightenment*, ed. Charles W. J. Withers and Paul Wood (Edinburgh: Tuckwell Press, 2002), 54–78; Withers and Mayhew, "Rethinking Disciplinary History."

38. Cited in Withers, "Toward a Historical Geography of Enlightenment in Scotland," 70.

39. See the work of Werner Stark, "Informationen zu Kant's Physischer Geographie," Philips-Universität Marburg, http://www.staff.uni-marburg.de/~stark/geograph/geo_start.htm.

40. These names, and the dating of Barth's notes, are taken from the Web listings of Kant's lectures in Stark, "Informationen zu Kant's Physischer Geographie."

41. May, *Kant's Concept of Geography*, 72. See also Todd Hedrick, "Race, Difference, and Anthropology in Kant's Cosmopolitanism," *Journal of the History of Philosophy* 46 (2008): 245–68.

42. Carozzi and Newman, *Lectures on Physical Geography*, 431–46 review Saussure's borrowings from Bergman and Buffon.

43. On this point more generally, see Richard Sher *The Enlightenment and the Book* (Chicago: University of Chicago Press, 2006).

44. For fuller accounts of the nature and history of geography in eighteenth-century France, see Withers, *Placing the Enlightenment*; Anne M. C. Godlewska, *Geography Unbound: French Geographic Science from Cassini to Humboldt* (Chicago: University of Chicago Press, 1999); Sergio Moravia "Philosophie et Géographie à la fin du XVIII siècle," *Studies in Voltaire and the Eighteenth Century* 57 (1967): 937–1001; Numa Broc *La Géographie des philosophes, géographes et voyageurs français au XVIIIe siècle* (Paris: Editions Ophrys, 1974); Michael Heffernan "Edme Mentelle's Geographies and the French Revolution," in *Geography and Revolution*, ed. David Livingstone and Charles W. J. Withers (Chicago: University of Chicago Press, 2005), 273–303.

45. This definition by Desmarest of the nature of physical geography is cited in Carozzi and Newman, *Lectures on Physical Geography*, 447. For a fuller discussion of Desmarest's definition in relation to the nature of geography in the Encyclopédie, see Charles W. J. Withers, "Geography in its Time: Geography and Historical Geography in Diderot and d'Alembert's Encyclopédie," *Journal of Historical Geography* 19 (1993), 255–64.

46. May, *Kant's Concept of Geography*, 15.

47. For a discussion of Zimmerman and Finke, see Withers, *Placing the Enlightenment*, 125, 201, 207–10.

48. This paragraph is based on Charles W. J. Withers, "Eighteenth-Century Geography: Texts, Practices, Sites," *Progress in Human Geography* 30 (2006): 711–29, quote from page 721.

49. Robert Mayhew, *Enlightenment Geography: The Political Languages of British Geography, 1650–1850* (London: Macmillan, 2000), 37–8.

II

From a Lecture Course of Forty Years to a Book Manuscript

Textual Issues

4

Kant's Lectures on "Physical Geography"

A Brief Outline of Its Origins, Transmission, and Development: 1754–1805

Werner Stark
Translation: Olaf Reinhardt

I would like to begin with a provocative thesis by way of introduction. So far, nobody—not even I—could have done in concreto what is the general topic of this volume—that is, to read a certain text that can properly be called "Kant's Physical Geography." The first question to occupy us therefore is: how did this current, somewhat paradoxical situation arise?

We know that Kant did not publish a book with the title *Physical Geography*. Not he himself, but other people produced two books with this title that are completely different in both subject matter and text in the years between 1801 and 1805. Just as indisputable are five facts that seem to be of direct significance here:

1. In his forty years as a university lecturer in Königsberg, Kant announced and gave lectures on Physical Geography with great regularity.
2. In the spring of 1757, the young and little-known Privatdozent Immanuel Kant had an invitation to his lectures printed, namely the eight-page text "Outline and Announcement of a Series of Lectures on Physical Geography."
3. Unlike with his other, more specifically philosophical lectures, Kant did not rely on the printed textbook of another author for this lecture series. Rather, he developed an outline himself,

My thanks to the Fritz Thyssen Stiftung (Köln) for funding my research in the years 2003–2005 and 2006–2007.

and this is sketched in the printed program of 1757. During the whole period of these lectures, Kant retained the practice of not using any printed handbook for his oral delivery.

4. A whole series of manuscripts from the eighteenth century have been handed down to us; these have no immediately obvious relationship to Kant's lectures on Physical Geography.
5. The Academy Edition of the "Physical Geography"[1] published as part of the Kant edition by the then Prussian Academy of Sciences did not overcome the historical problems and editorial challenges associated with its transmission. AK 26 is intended to remedy this finally, and provide a reliable textual basis for international research, so that Kant's *Physical Geography* can actually be read.

Of course, for many years there has been serious research on individual topics of the *Physical Geography* that Kant treated either orally in his lectures or in other publications. With certain reservations we can therefore exchange views and advance research on the Physical Geography. An indispensible precondition for such an exchange is a down-to-earth report on the whole of the transmission of the primary manuscript material.

In what follows, four basic and general questions will be raised and answered that arise from the facts just outlined and that are necessary or relevant for more specialized studies.

I. How and when did the lecture course Kant always announced as the physical geography originate?

II. How was the course transmitted?

III. What is the subject matter of the course?

IV. What do the two historical editions of 1801 and 1805 offer?

The first question relating to the origins of the course can be answered relatively easily and briefly if we cast our eyes on the Königsberg of the first half of the eighteenth century and have access to archival and printed sources that can reveal to us today the small, easy to survey academic world of the Albertina, founded in 1544. Immanuel Kant became an academic member by his matriculation on 24 September 1740. As far as we know, he left his birthplace, Königsberg, and thus the immediate vicinity of the university, in the summer of

1748 for almost six years. We know of no documents or other evidence from this period that would permit us to draw conclusions about wishes, intentions, or plans of the studiosus philosophiae. Nor do we know whether Kant remained away from the city for the whole time or whether he returned for shorter or longer visits to maintain personal relationships.

To all appearances,[2] Kant deliberately returns to Königsberg in the spring of 1754 and very determinedly and energetically begins an academic career: on 12 July 1755, the Albertina confers the academic degree of *doctoris seu magistri philosophiae*[3] on the now 31 year old. In time for the beginning of the winter semester 1755–56, he acquires a general permission to teach in the Faculty of Philosophy by means of a public disputation pro receptione on 29 September. Two obligatory publications in Latin at this time are accompanied by several German language titles, the themes of which can be clearly assigned to physical geography. Kant gives his first lectures on precisely this topic in the summer of 1756. In April of the following year he has the brief program, "Outline and Announcement of a Series of Lectures on Physical Geography," previously mentioned printed. This course of lectures remains a fixed part of the academic teaching of the Königsberg philosopher until the end of his lecturing career in the summer of 1796. It is a success from the beginning, that is, even the Privatdozent Immanuel Kant was able to attract a sufficient number of paying students. The material and the presentation must have been "somehow interesting."

Of course Kant is neither the inventor of physical geography as such nor is he the first lecturer in Königsberg to offer a course on this subject. Karl Rappolt (1702–1753) had already done this in 1750, perhaps in deliberate memory of the former Königsberg student Bernhard Varenius (1622–1650). Varenius had published his Geographia generalis one hundred years earlier in Amsterdam. In the summer of 1750, Rappolt announced a *Geographicam Physicam, Varenio, Woodwardo, Scheuchzero ac Jurino ducibus*. Thus, as the names of the people mentioned in the announcement show, Rappolt knew that the work of Varenius had led to a scientific study of questions of physical geography and that, in this, a leading role had to be assigned to the British scientist John Woodward (1665–1728), particularly for the period around 1700. Even if Kant, as shown above, could hardly have been among Rappolt's audience in the summer of 1750, Rappolt can definitively be shown to have been the only lecturer in the Albertina through whom the studiosus philosophiae Immanuel Kant can have become acquainted with the topics and methods of modern geosciences in the 1740s. Whether, and if so when, Kant read John Woodward's *Essay towards a natural history of the Earth*, published first in English in 1695, then in Latin (1704) or in German translation (1744), or if his knowledge was entirely indirect—via Rappolt, Buffon (1750), or Lulofs (1755)—we do not know.

There is no doubt that the topic area of Physical Geography becomes established in a certain sense as an object of academic teaching in the middle of the eighteenth century. It is only in this way that we can explain the fact that explicit textbooks are published in the German-language area that are dedicated to this subject and not to the "Geography" in general. Similar to the early Kantian program is:

> Georg Wolfgang Kraft: Kurtze Anleitung zur Mathematischen und Natürlichen Geographie, nebst dem Gebrauch der Erd-Kugeln und Land-Charten, zum Nutzen der Russischen studierenden Jugend [295 S.] (St Petersburg: Academie der Wissenschaften 1738; 2te Auflage 1764) Georg Wolfgang Kraft: Short Introduction to mathematical and natural geography as well as to the use of globes and land charts, for the use by the student youth of Russia. [295 pp] (St. Petersburg: Academy of Sciences 1739, 2nd ed. 1764)

What is not clear, however, is whether Kant had intentions other than primarily scientific ones in respect of his course or whether his own interest in Physical Geography is related to other motives. I tend towards the assumption that there was a direct connection to "physico-theology" and to the related question of the relationship between "faith and knowledge." As a result, I count John Ray (1627–1705) and William Derham (1657–1735) among the intellectual forefathers of the course.

The second question in connection with the transmission can also be answered quickly; insofar as one is prepared to make an essential differentiation: two sources of transmission need to be distinguished carefully and fundamentally. On the one hand there is the "concept for the lectures" Kant wrote down early. This outline is directly addressed in the Announcement of spring 1757:

> Everything is to be summarised in written summary essays, which will serve to [facilitate] the revision of this science, which in any case entertains our attention sufficiently by virtue of its pleasant nature.[4]

The end result of these written efforts by the lecturer has been transmitted in the form of the "MS Holstein" (Holstein Manuscript). This consists of a copy made by several people that Kant gave to the still very youthful Count Friedrich Karl Ludwig von Holstein (1757–1816), presumably on the occasion of a privatissimum at the beginning of the 1770s. But work on the content and editing of the text of this manuscript were already completed by the spring of 1759. Some notes in the margins added by Kant himself in the "MS Holstein" date from the time of the privatissimum, namely 1772–73.

In older research, based on Erich Adickes's work on Kant's Physical Geography, the "concept" elaborated in writing is referred to as "Diktat-Text" (dictation text). This name implies the assumption that Kant actually used this text not only pro forma for his teaching into the 1770s or even into the beginning of the 1780s, but literally dictated it. There is, however, nothing to support such a view except for the formulations usual at the time in the official lecture announcements of initially the Privatdozent and later, as of 1770, the professor. I regard it as out of the question that Kant—with the possible exception of the very first years—ever dictated a text to his student audience. Even for the time when Herder sat in his lecture room (1762–1764) such an assumption is contradicted by the surviving handwritten documents.

On the other hand, there is the long "lecturing practice." For this we have to rely almost exclusively on student lecture notes. These, in turn, fall into two classes differentiated chronologically: with the exception of the notes written down by Johann Gottfried Herder in the 1760s, all the student lecture notes go back to lectures given by the Professor of Logic and Metaphysics Immanuel Kant, newly promoted to a full Chair in the spring of 1770. The "lecture notes" of all lectures on Physical Geography given after 1770 can therefore represent nothing that is in principle different to what is the case with the other lectures: namely, the notes do not provide any text that is actually Kant's. Rather, we have here the results of a not entirely clarified process of endeavors and procedures by the student audience as such to render the "spoken word" into a written form. These texts should not be misinterpreted as stenographical reports. As a literary genre of their own, they constitute the only source of information on the structure and subject matter of Kant's teaching over a long period. For the Physical Geography, Table 4-1 can be established.

Overview of Transmission of Texts

Probably only the bottom row is immediately comprehensible: a total of 27 manuscripts have become known; about 100 years ago, Erich Adickes used 22; today we have 17. Of these 5 were unknown to Adickes; 10 have been lost since 1945 along with other materials of the Königsberg libraries. The abbreviations consisting of three letters are derived from the "names" of the manuscripts; thus "Doe" stands for Dönhoff. Adickes on the other hand used no self-explanatory names of this type. In line with the chronology he established, the manuscripts are designated by a single capital letter: (A stands for "Herder" (written in the 1760s), B for the "MS Holstein" (written at the beginning of the 1770s).

A procedure that is similar only in externals can be seen in the right-hand column: the letters **A, B, C, D**, and **X**. The first four stand for concepts or

Table 4-1. Manuscripts

Σ (total)	Date	Adickes (1911/13)	Additional	Lost: (1945: Königsberg)	Σ (extant 2007)	Type (Stark)
6	1757/59	5 (B-Group)	1 (Phi)	2 (C E)	4 (**B** D F, Phi)	A_0/X_1
1	1763	1 (Herder)	—	—	1 (**A**)	A_1
1	1770	—	1 (**Hes**)	—	1 (Hes)	A_2
5	1772/85	5 (G-Group)	—	3 (G K L)	2 (H I)	A_0/X_2
2	1774(?)	1 (Wer [1913])	1 (**Kae**)	—	2 (Wer, Kae)	B_0
5	1776(?)	5 (M N O Z, **Mes**)	—	1 (N)	4 (M O Z, Mes)	B_1
4	1780er	3 (P Q R)	1 (**Doe**)	2 (**Q** R)	2 (Doe, P)	C
3	1790er	2 (S T)	1 (**Doh**)	2 (S **T**)	1 (Doh)	D
27		**22**	5	10	**17**	

for certain structures of the course that—by accident—roughly correspond to the four decades (1756–1796) in which Kant gave the course. **A**: 1757–1772, **B**: 1774–1779, **C**: 1780s, and **D**: 1790s. The **X**, as suggested by mnemonics, stands for mixtures, that is, compilations from different traditions. As distinct from Adickes's method, these letters therefore stand for courses of different content, not necessarily for individual manuscripts. It was only with this procedure, developed in 2004–05, which has a corresponding technical translation into html files, that I succeeded in describing the complicated situation of the transmission as well as in finding a way to represent it. The matter is now obviously simpler than it must appear to anyone who tries to understand it from the basis of the *Untersuchungen zu Kants physischer Geographie (Studies on Kant's* Physical Geography).[5]

The capacity of the typological procedure[6] was demonstrated in June 2007, when it was possible to describe and place chronologically the newly discovered "MS Dönhoff" within a few days: it is the best of the accessible representatives of Type **C**; it will be edited as part of Volume 26 of the Kant Edition by the Berlin-Brandenburg Academy of Sciences. The manuscripts Holstein, Hesse, Kaehler, Messina, Dönhoff, and Dohna provide the indispensable components of this edition. In addition, there is an excerpt from an anonymous manuscript, surviving in a single print from 1833, which seems to go back to the summer of 1791.

In the center column of Table 4-1 are the abbreviations of the five manuscripts not known to Adickes; four of these each represent one type and should therefore to be regarded as editorial copy texts. Of these, Adickes knew no representatives of Type $\mathbf{B}_0$ at the time of his first book, the Studies on Kant's Physical Geography.[7] That is, he can only have become aware of the decisive structural change in the course at the beginning of the 1770s subsequently, that is in 1913, when he published a newly discovered manuscript based on Kant's lectures.[8] In addition, there is the fact that, as editor of Section III "Unpublished Manuscripts" of the Academy Edition, his interest was focused on the "Dictation Text" rather than on the course as such. Although Adickes had access to copies of the Types I have called **C** and **D**, his attention remained limited to the two traditions contained in the Rink edition of 1802.

All things considered, the situation in regard to the sources is clearly more positive and varied today than was to be expected at the beginning of the twentieth century. This situation is one of the astonishing paradoxical phenomena surrounding the Physical Geography of Immanuel Kant.

To summarise, we have before us a transmission to be differentiated fundamentally between:

1. A text written by Kant himself at the beginning of his career before the spring of 1759. Here it is questionable whether,

and if so to what extent, Kant used this text as a basis for a particular semester course.

2. Various student lecture notes of actual lectures held that should not be misunderstood as the "spoken word" of the lecturer.

Earlier in this essay, under **III**, I asked what the subject of the course was. Within the framework of my present brief overview, I would like to give an, as it were, topological kind of answer and dispense with a presentation of the numerous subjects of the Physical Geography from a problem-oriented or developmental perspective. In accordance with the basic division of the manuscripts into two, the following should be differentiated:

The MS 'Holstein' or the Detailed Plan of the Course

The structure of the text shows a reflection of the thirty-seven books "On Nature" by Pliny the Elder. Kant's treatise has a division into three approximately equally weighted parts. The relationships can be summarized as a table as follows (see Table 4-2).

As compared to Pliny, Kant (whether consciously or not) had undertaken a major shift, as indicated in the last two rows above: the geography, including ethnography, occupies the last (the third) position. In addition, the sequence of the ancient world is altered and the New World, "America," is added. Kant treated other given patterns in the same way. This observation is of considerable importance given that around two-thirds of the text of the MS Holstein consists of excerpts from textbooks, travel descriptions, and other works. On this, a third, somewhat simplifying[9] table (Table 4-3) is provided.

Table 4-2. Natural Science

Kant: Physical Geography	Pliny: naturalis historia
Preparatory	Book 1 [**not suitable**]
Part 1: **Physical Geography**	Book 2
Part 2: Natural Science [**0.**›General‹]	Books 3–6
(**1**) Man/(**2**) Animals/(**3**) Plants/(**4**) Minerals	(**1**) Europe, (**2**) Asia, (**3**) Africa
Part 3: Geography (**1**) Asia, (**2**) Africa, (**3**) Europe, (**4**) America	Books 7–37: (**1**) Book 7/(**2**) Books 8–11 (**3**) Books 12–32/(**4**) Books 33–37

Table 4-3. Literary Models (background) in the Ms Holstein

Part 1: Physical Geography	›**Treatise**‹; **Kant 1757**: = > Varenius 1650, Buffon 1750, Lulofs 1755.	**Appendix [last part]:** Varenius 1650		
Part 2: Natural History	**Man** [no direct model]	**Animals** Pontoppidan 1753–54; Halle 1757.	**Plants** [own structure]	**Minerals** Justi 1757
Part 3: Geography	**Asia** Salmon 1732ff. Gmelin 1751f.	**Africa** Ludolf 1684-94; Colbe 1745; Salmon 1748; AHR 1749–1751: vols. 2–5 & 8.	**Europe** Keyßler 1740f.; Büsching 1754.	**America/Polar** AHR 1751–59: vols. 9, 12, 13, 16 & 18; Müller 1758.

If one looks at this arrangement of the whole text, a structural weakness is revealed. The products of nature presented systematically in Part 2 necessarily reappear in the geographical part. After all, the objects from the "three realms of nature" need to be localized somewhere. A close reading reveals—as expected—a mass of repetitions both of subjects and even of wordings. Even if the third part ought to have concentrated more (in modern terminology) on ethnic differences between humans, such repetitions would have occurred, because the human being is already an explicit topic at the beginning of Part 2, and not only in his purely physical differences. Kant evidently accepted the risk of such duplications quite consciously, as it is already present in the Announcement of 1757 ("the products of the land already explained above").[10]

It is with some urgency that, in this situation, we now need to raise the question of the intellectual demands Kant was associating with the course altogether. I would expect that his answer to such a question would have been different for the three Parts and perhaps changing in the course of time. For it is only in the first Part of the text of the MS Holstein that there are statements the linguistic forms, which would allow us to conclude that he had his own theoretical demands. The first part (102–107) leads to his own "attempt at a thorough manner of explaining the ancient history [of the earth]." At the beginning of the second part (114–132) too, we can discern a desire for an individual system of his own in regard to "human beings": the presentation of physical peculiarities in humans culminates in a brief theory on taste. But overwhelmingly Kant saw himself, probably from the early periods of the course, as no more than a reporter of the research of others. If we add the manuscripts of Herder and Hesse, the formal structure of which follow the type established in the Holstein MS, it does not change this diagnosis.

One detail from the third, the geographical Part, seems to me worthy of mention: according to the "Hesse MS," in the summer of 1770, Kant discussed England (243–245) and Central Europe (Germany [245–247], Poland [249]), none of which are considered in the "Holstein MS." If we look at the textbook on which this part of the course was based, the Neue Erdbeschreibung (New description of the Earth) by Anton Friedrich Büsching of 1754,[11] the first two parts of which deal only with Europe, then there is indeed "Great Britain" and "Ireland" toward the end of the second part (1079ff.), but not Central Europe. Evidently a tension exists between the early plan of the course and what Kant actually lectured on.

The Course from the Middle of the 1770s

If we look at the course based on the numerous lecture notes from the middle of the 1770s, a topological consideration will rapidly reveal a significant difference; in the third, the geographical Part, "Europe" disappears. A comparison shows these relationships more clearly in Table 4-4.

Table 4-4. Structural Comparison: Type A and Type B

Type A »Concept 1757/59«	Type B »Course 1775«
§§ only in Part 1: each chapter	§§ throughout the whole text
0. Preparatory/Treatise	0. Prolegomena/Tractatio
Part 1 General: Chapter 1–8 Chapter 9: Shipbuilding and navigation	Section 1: General Part [Elements] Articles 1–4: Water, Land, Air, Geogony [Fire] Article 5: Shipbuilding and navigation
Part 2: [Three realms of nature] A. 1–4: (Man) Animal B. 1–2: Plants C. 1–6: Minerals	Section 2: Specific Part [Products] Articles 1–4: Man, Animals, Plants, Minerals
Part 3: Peculiarities of Nature arranged geographically: Asia, Africa, Europe, America	Article 5: Peoples in three Continents: Asia, Africa, America

Taking the areas of human knowledge treated by Kant in his lectures into consideration, we can rapidly see a convincing explanation for the change: the treatment of the peoples of Europe takes place in the newly established course on anthropology from the winter of 1772–73 onward. It was not called "European ethnology" but under various headings, together with the "Physiognomy" and the difference between man and woman, it forms the second part of the Anthropology, ultimately (1798) called "Characteristics." Indeed, in the middle of the 1770s, the first pages of the "Kaehler MS" and the "Messina MS"—that is of all the representatives of the **B**-Type—contain a detailed exposition of the relationship between "Anthropology" and "Physical Geography." In another manuscript, furthermore, an indication of the altered manner of proceeding is explicitly mentioned. "Pillau" (at the beginning of the 1780s) writes:

> The 4th part of the world, which we inhabit, is Europe, and about this there is no need to mention anything, as one will find everything in Büsching's Description of the Earth, Europe. Yet something will be said about Europe in an appendix.[12]

The last sentence is evidently an addition by the unknown writer of the manuscript, since in his copy, only the wording of the excerpt from Büsching follows, as it had already been written by Kant in the middle of the 1750s, according to the "MS Holstein." There is a similar heading in "Busolt" at the beginning of the section that is part of the geography by its placement but in fact concerns ethnography:

> Concerning the national character and features, the morals, the customs of some people outside of Europe.[13]

Viewed as a whole, the danger of duplication that was built into the first plan has been largely eliminated. The "First Part" is structured explicitly via the scheme of the four "old elements": water, earth, air and fire. This is followed by an "appendix" adapted from Varenius on ship building and shipping, the two large scale technologies of the seventeenth and eighteenth centuries. The "Second Part" comprises the topic areas of sections 2 and 3 of the early plan in five stages.

Although we do not know for certain to what extent Kant covered a particular area of his material in his lectures, a first approach allows us to conjecture that the surviving lecture notes at least resemble the actual process. According to my present topological observations, the quantitative situation can be transferred to a table (see Table 4-5).

Table 4-5 lists the page numbers and in round brackets ( ) the subtotal of the number of pages. The numbers 4–6, indicated by italics contain in part

Table 4-5. The Situation as of the Mid-1770s

	Prolegomena		Phys. Geogr.		Natural History			Geography	No.
Kaehler (1775)	1–48	(48)	49–353	(304)	354–477	(123)	477–530	(53)	1
Messina [**M′**] (1782)	1–16	(16)	17–185	(168)	186–324	(138)	324–369	(45)	2
Dönhoff (1782)	2–6	(4)	6–78	(72)	78–169	(92)	169–196	(27)	3
Volckmann [**P**] (1785)	1–9	(9)	9–66	(57)	66–101	(35)	101–104	(3)	4
Crüger [**L**] (1785)	1–7	(7)	8–120	(112)	120–179	(59)	179–192	(13)	5
Puttlich [**Q**] (1785)	1–7	(7)	7–176	(69)	177–294	(117)	294–372	(78)	6
Ms 1729 [**S**] (1792)	? –?	(?)	? –83	(ca. 80)	83–142	(59)	142–183	(41)	7
Dohna (1792)	1–6	(6)	7–97	(91)	98–206	(109)	207–243	(37)	8
Vigilantius [**T**] (1793)	? –?		? –90	(ca. 80)	90–178	(88)	178–203	(25)	9
Vollmer (1801–1805)		(86)		(1725)		(109)		—	10
Rink (1802)	1–55	(55)	56–312	(257)	1–128	(128)	129–248	(120)	11

text excerpts (Adickess' "Dictation Text") from the early phase of the course. The first column also shows the abbreviations used by Adickes in square brackets [ ] and in **bold type**.[14] The Königsberg manuscripts (numbers 5–7 and 9) have been lost since 1945. The dates also entered in the first column relate to the writing of the manuscript and not necessarily to the lecture course it is based on. The "Messina Manuscript" is autonomous only in the first part (17–185): prolegomena, natural history, and geography are identical with the "Kaehler Manuscript," but for minor copying errors and the like.

Of the manuscripts that have survived in their entirety, only numbers 1, 3, and 8 can be classified as pure, separate "Types"; in these, verifiable percentages can be calculated. As a comparative guide, Table 4-6 adds the values for the original plan (Holstein MS) and the Rink edition, which combines the Holstein and the Kaehler.

The first column shows the year in which it has been established that the course was given. In the case of "Holstein" and "Rink," the number refers to the year the text was edited. Columns 2 and 4 indicate the percentage of the total of each text based on the original pagination; in the fourth column and in round brackets the percentage of the animal kingdom devoted to human beings is also shown. For the years after 1774, five observations emerge:

1. The original treatment in three equally weighted parts comes to a definitive end with the establishment of the anthropology course (1772–73).
2. A clear emphasis on the Prolegomena occurs only in the mid-1770s.
3. If we exclude the special role of the "human being," then the "physical nature of the Earth" and its "products" (Natural History) occupy approximately equal space in Dönhoff and Dohna. The

Table 4-6. Changes in the Weighting of Topics (%)

	Prolegomena	**Phys. Geogr.**	**Natural History (Man)**		**Geography**
Kaehler (1774)	9	58	23	(6)	10
Dönhoff (1782)	2	37	47	(11)	14
Dohna (1792)	2, 5	37, 5	45	(9, 5)	15
Holstein 1757/59)	0, 9	31, 9	33, 3	(5, 3)	33, 9
Rink Edition (1802)	10	46	23	(3, 4)	21

early Kaehler manuscript, on the other hand, places much greater weight on the "physical nature of the Earth."

4. The geographically structured third part with its exposition on people outside Europe always remains clearly less than one fifth of the whole.
5. The quantitative situation of the Rink edition is conspicuous; it is one result of its character as a compilation.

These overviews will suffice for my present purpose. What remains is to answer the fourth question I posed at the outset: what are we to think of the two historical editions of 1801 to 1805?

If I were to give an answer that was as brief as possible, then it would be: we should think nothing of them! And this is exactly what I mean. Both editions lead the reader to believe that reading them would provide one with a reliable insight into Kant's "Physical Geography." But, according to my explanation thus far, this cannot be the case: we, who are born at a later date, must abandon the idea of being able to produce the Physical Geography of Immanuel Kant. There are two decisive reasons for my unambiguous rejection of the idea:

1. Both of the historical editions contain text that goes back to student notes of orally delivered lectures.
2. Neither of the two editions makes any explicit statements about the point in time or even refers to a narrow time frame within which Kant is supposed to have made the statements, let alone to have written them.

If I were to answer less provocatively and therefore in a somewhat longer way, then I would have to proceed in a more differentiated manner and begin with a "public declaration" by Immanuel Kant, one that he signed in Königsberg on the 29 May 1801. I quote the relevant first passage:

> At the last trade fair, the bookdealer Vollmer published a physical geography under my name, based, as he himself says, on student notes; I do not recognize this text as mine, neither in the subject matter nor in form. I have assigned the legitimate publication of my physical geography to Prof. Dr Rinck.[15]

This establishes that Rink is the editor authorized by Kant. Of course, this is a fine legal point which, as is so often the case, does not help us much. For the question is aimed at the quality of the editions or their information in regard to the conceptions, opinions, and statement that Kant put forward in his lectures on Physical Geography. We would like to know both what Kant said or wrote about the individual topic areas of Physical Geography and to find out what conception of the science he had in the second half of the eighteenth century. More precisely, what (and if possible: how) he presented to his students in this regard.

Now, fortunately we have sufficient manuscript material to satisfy our requirements both in terms of Kant's lectures and of the two historical editions. Needless to say, this satisfaction cannot be attained by means of a brief look nor by some short cut; it is possible that complex procedures will have to be applied or even developed to reach our goal. Relating this to the texts printed under the names of Vollmer and Rink of 1801 to 1805, what we need to do is actually simple; all we need is a mere comparison of texts to discover what the facts of these editions are, that is, to discover how they came about.

First of all, in reference to Rink: the main substance of the text was combined from two chronologically distinct manuscripts:

1. Parts of the early (1757/59) plan of the course written by Kant, that is what Adickes called the "Dictation Text" [A-Rink] and

2. A set of student notes from the course of the summer of 1774 [B-Rink].

In addition, we need a large number of additions and changes by the editor to bring it up to date, especially in the first volume of the edition.

Then, in reference to Vollmer: Table 4-5, given above to answer my third question, shows that the edition as such is a torso:

1. It fulfills only one part of Kant's programme as established for the period after 1774.

2. Even in the parts we do have, there is a conspicuous disproportion.

In so far as we can carry out a comparison of particular points with the extant manuscripts, it rapidly appears that the text presented is something other than what it may appear to us today. What is offered here is not a mere "edition" of student manuscripts based on Kant's lectures, but an expedient publishing project, using lecture notes from Kant's course to produce an up-to-date textbook on Physical Geography. It is a project that failed in its aims and is apparently associated with personal tragedies of those involved with its later parts.[16]

In conclusion, I should like to formulate a brief reminder of the procedure adopted by the Academy Edition of Kant's complete writings: this major cooperative enterprise was conceived in the last decade of the nineteenth century.[17]

While for the Sections I: "Works," II: "Correspondence," and III: "Unpublished Manuscripts" there is mostly clarity about the methodological requirements of the edition, this was not given for Section IV: "Lectures." It was only with the work of Erich Adickes[18] that some clear ideas emerged about the problems associated with the origins and transmission of student lecture notes. No solutions were finalized at the time. The fact that Kant's differential practice with each of the disciplines—logic, anthropology, morals, etc.—and that the different situation of the transmission involved different requirements for the editorial work of the evidentiary texts has been one of the fundamental insights associated with the preparation of AK 25. The fact that the procedures developed for Kant's private course on "Anthropology" were insufficient to do justice to the in many ways much more complex demands of the Physical Geography is one of the insights that has gradually matured within the framework of my "Erneute Untersuchungen."[19] One result of this is the development of an Internet-accessible presence on the documentation of the course on Physical Geography, first made available in August 2007. It is hoped that the URL: http://kant.bbaw.de/base.htm will be released simultaneously with the publication of AK 26.1 (i.e., presenting the MS Holstein) of the Kant Edition.

Notes

1. Immanuel Kant, *Physische Geographie*, in the Akademie Ausgabe of the *Gesammelten Schriften* (hereafter AK, with volume number, page and line numbers, as needed), AK 9 (1923).

2. As revealed by the precise dating of its publication; c.f. AK 1, section 1.

3. C.f. the official "Einladungsdiplom" ("Certificate of Invitation"), facsimile in AK 13.

4. Immanuel Kant, *Entwurf und Ankündigung eines Collegii der physischen Geographie nebst dem Anhange einer kurzen Betrachtung über die Frage: Ob die Westwinde in unsern Gegenden darum feucht seien, weil sie über ein großes Meer streichen (Announcement of 1757)*, AK 2: 9, 26–9.

5. Erich Adickes, *Untersuchungen zu Kants physischer Geographie* (Tübingen: J. C. B. Mohr, 1911).

6. This refers to a procedure adapted from that used by biologists: plants (or texts) actually found are considered as models and described and represented as such. Other plants (or texts) can be categorized as belonging to a species or type on the basis of such exemplars. In this way a complex taxonomic system can be constructed. The first characteristics for establishing a type of lecture course are the external structure and the proportions of the text. A more precise description is based on a detailed structure, a characteristic terminology, and ultimately the whole text.

7. Adickes, *Untersuchungen zu Kants physischer Geographie*.

8. Erich Adickes, *Ein neu aufgefundenes Kollegheft nach Kants Vorlesung über physische Geographie* (Tübingen: J. C. B. Mohr, 1913).

9. Articles in periodicals of the period and various other individual sources have not been considered.

10. Kant, *Announcement of 1757*, AK 2: 9, 23.

11. Full German title: Erster Theil, welcher Dänemark, Norwegen, Schweden, das ganze russische Kaisertum, Preussen, Polen, Galizien und Lodomerien, Hungarn und die europäische Türkey, mit den dazu gehörigen und einverleibten Ländern. / Zweyter Theil, welcher Portugal, Spanien, Frankreich, Wälschland und Groß-Britannien enthält. (Hamburg 1754). English: "Part I, which contains Denmark, Norway, Sweden, the entire Russian Empire, Prussia, Poland,Galicia and Lodomeria, Hungary and European Turkey together with the countries belonging and incorporated in it. / Part II which contains Portugal, Spain, France, Italy and Great Britain (Hamburg 1754). On the further publication of the whole work: in 1757–1759 in three part volumes appeared, the third part "containing the German Empire"; the fourth part (1761) presents "the united Netherlands, Helvetia [. . .], Silesia and Glatz; the fifth part in 1768, the "various countries of Asia" (Peter Hoffmann, *Anton Friedrich Büsching (1724–1793). Ein Leben im Zeitalter der Aufklärung* (Berlin: Arno Spitz, 2000), 268f.; cf. also 159).

12. "MS Pillau," 438.

13. "MS Busolt," 305.

14. Adickes, *Untersuchungen zu Kants physischer Geographie.*

15. Immanuel Kant, "Nachricht an das Publicum, die bei Vollmer erschienene unrechtmäßige Ausgabe der physischen Geographie von Im. Kant betreffend," AK 12: 372.

16. Werner Stark, "*Das Manuskript Dönhoff—eine unverhoffte Quelle zu Kants Vorlesungen über Physische Geographie*," in: Kant-Studien, Bd. 100 (2009), S. 107–109.

17. Werner Stark, *Nachforschungen zu Briefen und Handschriften Immanuel Kants* (Berlin: Akademie-Verlag, 1993); Frithjof Rodi, "Dilthey und die Kant—Ausgabe der Preussischen Akademie der Wissenschaften. Einige editions—und lebensgeschichtliche Aspekte," in Dilthey-Jahrbuch für Philosophie und Geschichte der Geisteswissenschaften 10 (1996); Werner Stark, "Die Kant—Ausgabe der Berliner Akademie—Eine Musterausgabe?" in Kant und die Berliner Aufklärung, ed. Dina Emundts (Wiesbaden: L. Reichert, 2000).

18. Adickes, *Untersuchungen zu Kants physischer Geographie, and Ein neu aufgefundenes Kollegheft nach Kants Vorlesung über physische Geographie.*

19. Werner Stark, *Erneute Untersuchungen zu Kant's Vorlesungen über Physische Geographie*, forthcoming. See also Werner Stark, Was wusste Kant über Asien? / Hinweise und Überlegungen zu Kant's Interesse an Fragen der Geographie, in: Materialien der 10. Internationalen Kant Konferenz. Klassische Vernunft und Herausforderungen der modernen Zivilisation 22–24, April 2009, Kaliningrad (Russische Staatliche Kant-Universität. Institut für Kantforschung; Redaktion: V. I. Bryushinkin, Kaliningrad 2010), Bd. 1, S. 101–121, 10 Abbidungen. / Werner Stark, Herder's Kant-Papiere. Eine jurze Klarstellung, in: Herder Jahrbuch / Herder Yearbook, Bd. 10 (Heidelberg: Synchron 2010), S. 13–24.

5

Historical and Philological References on the Question of a Possible Hierarchy of Human "Races," "Peoples," or "Populations" in Immanuel Kant—A Supplement.

Werner Stark
Translation: Olaf Reinhardt

The text "Immanuel Kant's Physical Geography" published by Friedrich Theodor Rink contains a passage consisting of four sentences that has attracted attention in numerous ways. It appears—in modern parlance—that a kind of hierarchy among the four "populations"[1] of humanity differentiated by color is clearly being asserted. This paragraph is to be found at the beginning of Part Two of the lecture course, which is devoted to the three kingdoms of nature (animal, vegetable, mineral). The treatment of the "Animal Kingdom" opens with a separate section on "humans"; here we read:

> [1] In the torrid zones, humans mature more quickly in all aspects than in the temperate zones, but they fail to reach the same degree of perfection [2/3]. Humanity has its highest degree of perfection *in the* ***white*** *race*. The yellow **Indians**[2] have a somewhat lesser talent. The **Negroes** are much lower, and lowest of all is **part of the American races**.[3]

Three statements can be distinguished here, provisionally indicated by inserted square brackets [ ]:

1. For human beings as individuals, a difference determined by the "climate" is stated with reference to the duration of their

"maturation"; in "torrid zones" this process is said to take place more rapidly than in cooler ones; in the (medium) "temperate" zones the process is ultimately better.

2. Four populations distinguishable by their skin color are differentiated; they can be assigned to different territories of the earth.[4]

3. "Humanity," i.e., the species, is said to attain varying degrees of "perfection" in the four populations; the text says the "whites" are preferred.

Even in an interpretation of this kind, what the text states is not conclusive. There is no explanation of exactly what meaning is intended by "perfection," nor any indications of what a connection of the two "perfections" is to consist of: what does the duration of the maturity, i.e., the development of the individual, have to do with the "perfection" of various populations of the species? Only the expression "talent" can perhaps be of use in explaining the alleged situation, for it is only in reference to it that there is any mention of a "more or less" via the comparative ("lesser"); I return to this below. First, a few observations.

If we adopt the perspective of the origin of the text, it rapidly becomes clear that the sentence quoted occurs at the end of a fairly lengthy insert that, in 9: 315, 35, begins with explicit references to Le Vaillant's *Journeys*[5] and to another author. Ever since Erich Adickes's *Studies on Kant's Physical Geography*,[6] we know that, beginning[7] with the "history of the wells and springs," for his edition Rink used a text that Kant himself had written in the time of the beginning of the lectures on the "Physical Geography" between 1757 and 1759.[8] We still have this original (Type A_0) text today in the form of the "MS Holstein," where we read

> Precisely this is reported by Ludolph on many Ethiopian women.// The Moors and other savages between the Tropics can generally run astonishingly well. They, as well as other savages also have greater strength than other civilised peoples, which originates in the free movement that is allowed them in their childhood.[9]

The two slashes (//)mark the point where the Rink edition[10] contains text that clearly belongs to a later time—as can be seen in the two travelers mentioned by name.[11] Of course, this insert cannot be attributed directly to the activities of Rink, rather, the manuscript he used and that has subsequently been lost probably already contained it. The extant "MS Barth" shows many correspondences with the Rink text, especially in the present context.

> In the torrid countries humans mature altogether much earlier, but do not attain the perfection of [those] in the temperate zone. Humanity is in its greatest perfection in the *white races*, the yellow Indians already have a lesser talent, the Negroes are much more degraded and the Americans most of all, and among these, especially the *inhabitants of Tierra del Fuego*.[12]

The plural in the manuscripts "Pillau" and "Barth" softens our judgment; for it is not about "the whites" as such but about different groups (nations) of this population. In addition, the plural harmonizes better grammatically with the other plural forms listed: "Indians," "Negroes," and "American peoples" or "the Americans." Thus we can probably assume that an oversight or a simple printing error has occurred here. The singular "the race" strengthens the impression that a theory of four races, distinguishable by their skin color, is being advanced, while at the same time, one of them is declared to be the "most perfect." Certainly, unlike Rink, the manuscripts Barth and Pillau do not return immediately to the text of the MS Holstein,[13] but continue the exposition after a separate line, probably conceived of as a heading, "On the profound degradation of the Americans." The text continues with a general observation ("Lack of feeling is their character"[14]), adds various illustrative examples and finally returns to the inhabitants on the southern tip of the New World:

> The Fuegans, however, are so stupid that, regardless of the fact that the cold there is so terrible that 2 Englishmen froze to death in the middle of their summer, they have not hit upon the idea of fitting doors to their houses, whereby they could block the access of the penetrating cold to some extent. Instead all their huts are quite open to the air. Nor do they even cover their lower half, so that they constantly shiver with the cold. How far these people are below even the Greenlanders, in whose lands the cold is far greater still.[15]

The literary source of this remark is clear: John Hawkesworth's *An Account of the Voyages Undertaken.*[16] Volumes two and three depict events, discoveries, and experiences on the first of Cook's circumnavigations of the globe. In Book Two, Section Five ("The Passage through the Straits of le Maire and further description of the inhabitants and natural resources of Tierra del Fuego"), we read:

> Viewed as a whole, these people appeared to be not only the most wretched and most helpless, but also the most stupid of all human beings and the dregs of nature. Their life is constant wandering in

> empty deserts, where two of our men froze to death in the middle of summer; their dwellings nothing other than a miserable hut built from poles in which they are exposed not only to the wind but also the rain and the snow; their dress miserable; they went almost naked, without any kind of comfort, not even equipped with that which even the lowest level of art can produce, they were not even acquainted with any means of cooking their food, and yet they were happy.[17]

The preceding Section Four ("Report of a botanical mountain journey undertaken here") describes the circumstances surrounding the deaths of the two members of the group in the night of 16 January. In Section Five, lifestyle, clothing, and dwellings of the inhabitants are described in detail[18]; with Plate 23, the volume also contains a corresponding copper plate depicting this.

Two other prominent contemporary travel descriptions agree with such a negative characterization of the Fuegans. In Louis Antoine de Bougainville's *Journey Around the World*, he writes:

> These savages are small, ugly, thin and have an intolerable odour about them. They go almost naked; their entire clothing consists of mere hides of seal skins that are too small to cover them completely. They make the roofs of their huts of these same hides and the sails of their dug-outs. They also have some pelts of guanaco but in smaller numbers. The women are ugly and the men do not appear to care much for them. [. . .] They seem to be good-hearted, but at the same time they are so simple that one cannot think of any way to thank them for their good-heartedness. We thought them superstitious and suspected that they believe in certain spirits who do evil. Those who can conjure them up are their doctors and priests at the same time. Of all the savages I have seen in my life, the Pecherais live in the worst and rudest way, or what one can call in the state of nature/in the true sense of the word. If a man who is free and his own master, knows neither duties nor occupational business and is happy with what he has because he knows no better, deserves to be pitied, then I would pity these above all, because not only do they lack everything that belongs to the comfort of human life but they also have to battle with the most terrible climate in the world.[19]

Similarly Georg Forster in the second part of his description of James Cook's circumnavigation:

> After much waving, numerous of these people came into the ship, yet they did not show the slightest sign of pleasure and appeared to be without any curiosity. They were short of stature, none over 5 foot 6 inches (English measure) tall, had wide, large heads, broad faces, very flat noses and the cheekbones very prominent under the eyes; the eyes themselves were of a brown colour, [. . .]. Taken together, these features constituted the most complete and eloquent picture of the profound misery in which this unfortunate kind of humans exists.[20]

These quotations from contemporary travel descriptions verify that the final phrase of the paragraph under discussion ("and lowest of all is part of the American races") is to be understood primarily as a reflection of travel literature. This is the description of a situation that Kant can have presented in a lecture in the mid-1770s at the earliest. We can neither presume nor insinuate any pejorative intention. The passage under discussion is certainly distinctly later chronologically than those around it in the framework of the Rink text, or in that of the manuscripts Barth and Pillau. Perhaps it originated as a marginal note by a student who, in the mid-1770s, had access to a copy of the original 1757–1759 text of the MS Holstein and who updated his copy with details from a current lecture course.

This supposition is supported by a further observation. The final phrase of the passage under discussion has an almost verbatim parallel in another part of the Rink edition.

> Altogether the nations of the southern hemisphere are on the lowest *level of humanity*[21] and they have no interest in anything other than the most sensuous pleasures. The savages towards the north, although they live even closer to the Pole, display far more *talents and skill*.[22]

This passage belongs to that part of the Rink text that goes back to student notes from the mid-1770s.[23] In librarian's parlance, Rink has "done a doublet" by having the same text printed in two different places of his book. The doublet is easily explained by the circumstances: (1) Rink used two different manuscripts; and (2) one of these contained material from different times. Since we have two manuscripts with a lecture course of "Type B_0" we can understand fairly exactly how this doublet text came about. In the "MS Werner" (Type B_0) that had already been used by Erich Adickes and is now stored in Marburg, we find:

> In general the nation in the south is in the *lowest of all degrees of humanity*, since nothing interests them more than eating. The

> savages in the north, although they are closer to the Pole, are far more *interesting*.[24]

The same in subject matter but deviating distinctly in the formulation, "Wolter" (Type B_1) writes:

> At the Straits of Magellan, it has been found that the capacity of *humans* to reason has faded away almost totally and that they approach the *animals* to such an extent that, if one were to travel any further, one would scarcely find any humanity. Their constitutions and conceptions are very wretched and limited. The Greenlanders, who are very skilled are completely distinct from them.[25]

Thus it can still be assumed for good reasons (1) that in Kant's course of lectures of Type B, the primary opposition is "animality" and "humanity," and (2) that reference is being made to two populations of humans who have completely differing ways of life despite being in climatically very similar conditions: the Eskimos living in the cold zones of the north on the one hand and the southern Fuegans on the other.

II

As shown at the beginning, the assertion of a hierarchy among the different human populations is tied to the language of the expressions "perfection" and "talent." Since the entire transmission of the course has been digitized, a few terminological observations can be rapidly established:

1. The German-language[26] term "*Menschheit*" appears only twelve times[27]; this is not only conspicuously seldom, but on closer observation it is downright revealing. Considering the typological arrangement of the manuscripts, it turns out that this term belongs to "Type B." It is typical for the lectures of the mid-1770s. The fact that it appears in the compiled "Type X_2" (Barth, Pillau) does not contradict this. Only the occurrence in "Herder" in the middle of the 1760s varies from that chronology. The passage is:

   > Negroes can [. . .] only be got to work by force:—they get pleasure from childish things, glass corals, copper barrel rings: they are the least well-brought up, everything by imitation, nothing through their own abilities: Hume[28] cites

> an example: they must therefore have an essential fault in the main characteristic of humanity.[29]

Kant's contemporaries could have read a passage that corresponds to this content in an almost contemporary, highly successful essay by Kant, *Observations on the Feeling of the Beautiful and Sublime*:

> The Negroes of Africa have by nature no feeling that rises beyond the silly. Mr Hume challenges everybody to produce a single example where an Negro has shown talent and asserts: among the hundreds of thousands of blacks who are abducted from their lands to somewhere else, even though very many are set free, yet there has not ever been found one who has presented something great, be it in art or science or in any other notable area, [. . .][30]

It is not only the tenor in which this statement resembles the passage of the Rink text quoted at the beginning, but also in the terminology; here too "talent" is mentioned. But first to "perfection."

2. The noun "Vollkommenheit" (perfection) is used in all the Types of lecture notes[31] and with 29 occurrences in total it is admittedly not very common. If one focuses only on the occurrences together with "Mensch" (human being) as subject matter, the number is drastically reduced: the "Holstein" (Type A_0) can be left out of consideration because of the quite different context[32]; the notes by Herder seem altogether too cryptic[33] and in "Type B" of the 1770s, it is reported ("MS Kaehler") that the various kinds of grain used by people have "developed to much greater perfection through cultivation." The only remaining occurrence is in the passage that was the subject of our discussion at the beginning. This at least shows that the expression "humanity . . . in its greatest perfection" is extremely conspicuous within the framework of a course on *Physical Geography* by Kant.[34] Without any doubt, the term was used by Kant himself in the context of his lectures in the program of the summer of 1775—more on this below.

3. "*Talent*" is the third expression the use of which needs to be examined to clarify whether and in what sense "a kind of hierarchy

among the different populations of humanity" is being advanced. Somewhat surprisingly, it turns out that the term is first used in the "MS Hesse," which was written in 1770, and five times at that, always in the context of a possible differentiation between various "populations."

> The talents of the mind are not so evident [in the inhabitants of the Antilles] as it is in others, as in the Europeans. So, *e.g.* the Negroes are generally very stupid and additionally in the whole of Africa they have something quite silly and childish about them.[35]

> All Negroes have a quite different character from the whites. In the countless numbers that have been brought from *America to Europe*[36], who have been trained in all sorts of arts and sciences, not a single one has shown any talent at all for arts and sciences, while on the other hand, almost every white has an aptitude for something.[37]

> [The Chinese] lack talents almost as much as the Negroes do. They have not distinguished themselves in the sciences in so many centuries, so that not a single one of them has become famous and one might learn something from them.[38]

> The Spaniards have more talents than they have shown so far, for occasionally a great mind distinguishes itself among them.[39]

> The Germans are the noblest people in respect of their talents and arts, skilled in all kinds of arts and sciences and excellent in some.[40]

In a section of the "MS Pillau," which was written later, there is an observation quite parallel to "Hesse":

> Thus there is a nation [the Chinese], that has no talents for invention or insight into matters of the mind.[41]

Evidently the expression "talent" is used in direct relation to "arts and sciences," which as such, cannot be regarded as naturally innate qualities or characteristics of humans. Apart from this,[42] the expression occurs only in the supplements to "MS Friedländer,"[43] all of which must have been written before 1778.[44] On the other hand, "talent" is absent in the earliest Type (A_0 "MS Holstein") of the course and in all the other Types B, C, and D, which arose after the introduction of the course on anthropology in the winter of 1772–73. Does an explanation of "talent"[45] actually belong in the complex of topics of a course on *Physical Geography*? A quick comparison with the *Anthropology* course will show that this is the place for "talent" in the system: the expression occurs more than 800 times. In the *Anthropology from a Pragmatic Point of View*, first published in 1798, Kant gives the following definition: On the talents in the cognitive faculty.[46]

Not quite so didactically in the first part of the Friedländer manuscript of 1775/76, which is devoted to the individual psyche:

> In the human mind, disposition, talent and genius must be distinguished. The difference between disposition and talent is this. Disposition is a capacity of the mind, but talent is a gift of the mind. Disposition is the quickness to grasp something, talent to produce something. Ease of being educated is disposition, but [of] talent is to invent something, *e.g.* the memory belongs to disposition. The difference between talent and genius is: talent is the degree of power of the mind by which something can be produced if instruction has preceded it. But genius is a talent, which cannot be a product of instruction. Genius does without all instruction, and replaces all art.[47]

In the second part, which is to be seen as differential psychology, there is the following wording:

> In respect of the mind, we can divide the principles of activity into the disposition, talent and temperament. The disposition is the capacity of receptivity to receive certain objects, hence disposition belongs to capacity. Talent is an ability to bring forth products, so it belongs to power. Temperament is the union of both. In respect of the disposition, a person is called tractable, quick to learn, mild; in respect of the disposition he is passive. Disposition is required of the apprentice. Talent of the teacher.[48]

For the present purposes it is irrelevant how the various capacities are delimited and determined; for there can be no doubt that "talent" and "disposition" belong to the capacities of the mind and do not represent a mere physical quality.

III

As the genesis of the text under discussion and the language used in it can now be regarded as having been adequately examined, it is at least very doubtful whether Kant actually did maintain that there is a natural rank order of different human populations in an oral lecture. There is little doubt that the assumption that the "inflammatory" statement quoted at the beginning goes back to a student's lecture note that cannot have been written before the mid-1770s at the earliest. In all probability, the formulation was ultimately made more pointed and intensified in tone by Friedrich Theodor Rink.

There is no doubt that, in his essay *On the Different Races of Human Beings* first[49] published in 1775, Kant maintained that there were natural differences between the varying populations of humans. The question is thus justified as to whether there are consequences associated with this concept that suggest or even require the acceptance of a "kind of hierarchy among the populations of humans." I shall explicitly not attempt to answer this question within the framework of the present essay; I restrict myself to indicating four points to help place this question systematically and historically in Kant's work.

1. From the above, there is a possible *systematic place* in Kant's work where such assertions are put forward or statements made. This is the second part of the lectures on anthropology, which was ultimately called "characteristics." From the winter of 1772–73 until the publication of 1798, this is where the great European peoples are presented and differentiated from one another. Years before (1764) Kant had already treated this topic under the then usual title "On National Characters" in his essay *Observations on the Feeling of the Beautiful and Sublime.*[50] It goes without saying that the people of the three non-European "parts of the world"[51] are also present here, without the use of any concept called "race."

2. If we ask about the *systematic purpose*[52] Kant was pursuing in 1775 with his essay, then the concept of human "races" represents an attempt to distinguish the territories of the two only academic courses Kant developed by himself. As a being within nature, "mankind" is treated in the *Physical Geography*; in so far as he is

differentiated from nature, he is the topic of the *Anthropology*.[53] The lecture program of 1775 culminates in the statement that there are four races to be separated geographically that (a) came about in the far distant past, and that (b) are ultimately based on the climatic conditions of the respective areas.[54]

3. In this essay, Kant himself articulates the "perfection"[55] that is of interest in the present context; in the first printed version of 1775, we find:

    > If one were to ask with which of the present races the first human tribe might have the greatest similarity, then one would presumably have to declare oneself, although without any prejudice on account of the claimed greater perfection of one colour over another, for that of the whites. For the human being, whose descendants were to adapt to all regions, was able to be best suited for this if he was fit for the temperate climate from the very beginning; because this lies in the middle between the extreme limits of the circumstances into which he was to find himself. And it is precisely here that we find the race of the white since the oldest times.[56]

    According to this it is conceivable that a student listener misunderstood an explanation of this position in an oral presentation and thus wrote down the formulation that appears in Rink.[57]

4. In the years following, Kant constantly developed his "concept" further and presented it in his lectures on *Physical Geography* and *Anthropology*. It would go well beyond the framework of what were intended to be some indications, if I were to outline or comment on this development.[58]

The question on a natural rank order among the different human populations thus has a certain, admittedly "narrow" place of its own in Kant's work, at or on the borderline between the *Physical Geography* and the *Anthropology*. If there were conceptual changes or modifications in details within these two areas, this borderline will have been influenced by them. The lecture notes taken by students on *Physical Geography* provide a wellspring flowing constantly for many years that can shed light on the positions Kant held on "humanity" as a whole and in their relationship to other lecture courses (especially those on anthropology) as well as on the relevant publications on the philosophy of history in the 1780s and 1790s.

Notes

1. In order to avoid misunderstandings, I am using a current scientific term.

2. This, and subsequent uses of the word "Indian" refers to inhabitants of India. The "Indians" of America are referred to as "American Indians."

3. *Physical Geography*, Rink edition (1802), vol. 2: 10. Reprinted in the Academy edition of Kant's collected works (hereafter AK, followed by volume, page and line numbers), AK 9: 316, 4–9. Emphasis in the excerpt is mine.

4. C.f. "MS Kaehler," 357–9; correspondingly, "MS Werne," 400–1.

5. François Le Vaillant (1753–1824), *Voyage dans l'intérieur de l'Afrique par le Cap de Bonne Espérance, dans les années 1780–85*, 2 vols. (Paris, 1790); translated several times into German. (Kant read all his sources in German translation. English titles given in [ ] should not be taken to suggest that he read any of them in English.) Cf. Hans Fromm, *Bibliographie deutscher Übersetzungen aus dem Französischen, 1700–1948* [Bibliography of German Translations from the French], 6 vols. (Baden-Baden: Verlag für Kunst und Wissenschaft, 1950–3).

6. Erich Adickes' *Untersuchungen zu Kants physischer Geographie* (Tübingen: J. C. B. Mohr, 1911).

7. This is different from the text printed on the preceding pages: elaborated and modified by Rink, it is based on a student's version of a course of lectures given be Kant in the mid-1770s. The "MS Kaehler" is the primary representative of this type of lecture today.

8. *Physical Geography*, AK 9: 273, 21.

9. "MS Holstein," 123.

10. *Physical Geography*, AK 9: 315, 35–316, 9.

11. That is, Le Vailliant (AK 9: 315, 35–6; see n. 5 above) and Petr Ivanovic Rytschkov (1712–1777, mentioned at AK 9: 316, 1–3). See Petr Ivanovic Rytschkov, *Orenburgische Topographie, oder umständliche Beschreibung des orenburgischen Gouvernements* [*Topography of Orenburg, or, detailed description of the province of Orenburg*], 2 vols., trans. Jacob Rodde (Riga: Hartknoch, 1772). Alternatively, see *Orenburgische Topographie*, trans. Christian Heinrich Hase, in Büsching's *Magazin für die neue Historie*. Vol. 1 appeared in 1771, nos. 5, 457–530; 6, 473-516 and 7, 1–64; vol. 2 appeared in no. 7 (1773), 65–188.

12. "MS Barth," (1782?), 122, emphasis mine. On dating this, see Adickes, *Untersuchungen zu Kants physischer Geographie*, 113ff. Cf. the "MS Pillau," 221ff.

13. In "Barth" this does not occur until two pages later (24) with the heading "On the changes that people *really* undertake on their bodies [*Gestalt*]." Cf. "MS Holstein," 127: "On the changes that people *arbitrarily* undertake on their bodies [*Gestalt*]." Rink has changed the wording to: "On the changes that people make to their bodies themselves" (AK 9: 318,11f.). From the context of the "Holstein" it can be directly deduced that the second formulation emphasized by me above is the one that fits the

facts. The preceding heading is: "The human being considered according to his other innate characteristics over the whole Earth" (121).

14. "MS Barth," 122.

15. "MS Barth," 123f.

16. Full title: John Hawkesworth, *An Account of the Voyages Undertaken by The Order of His Present Majesty, For Making Discoveries in the Southern Hemisphere, And Successively Performed by Commodore Byron, Captain Wallis, Captain Carteret, And Captain Cook, In the Dolphin, the Swallow, and the Endeavour: Drawn up from the Journals which were kept by the several Commanders, and from the Papers of Joseph Banks, Esq.*, 3 vols. (Berlin, 1774). From the *Wöchentlichen Nachrichten* (Berlin), edited by Anton Friedrich, it is clear that by the end of August 1774, all three volumes of the German translation were complete. Cf. *An Account of the Voyages Undertaken*, 272ff.

[*Trans*: I have quoted the title of the English original, which has "present majesty" instead of "Britannic majesty" and "Southern Hemisphere" instead of "South Sea."]

17. Hawkesworth, *An Account of the Voyages Undertaken*, vol. 2, 58f.

18. Hawkesworth, *An Account of the Voyages Undertaken*, vol. 2, 54ff.

19. Louis Antoine de Bougainville, *Reise um die Welt welche mit der Fregatte la Boudeuse in den Jahren 1766, 1767, 1768 und 1769 gemacht worden* (Leipzig, 1772), 126ff.

20. Georg Forster, *A Voyage round the World in His Britannic Majesty's Sloop Resolution, Commanded by Capt. James Cook, during the Years, 1772, 3, 4, and 5* (London, 1777), ch. 9. Translated into German as *Reise um die Welt während der Jahre 1772 bis 1775 in dem von Seiner itztregierenden Großbritannischen Majestät auf Entdeckungen ausgeschickten und durch den Capitain Cook geführten Schiffe "the Resolution" unternommen*, 2 vols. (Berlin: Haude & Spener, 1778–80).

21. This expression is found only in the Rink text; cf. "level of humanity" (AK 9: 267, 24), in a note originating from Rink.

22. *Physical Geography*, AK 9: 230, 16–21, emphases mine. The word "skill" [*Adresse*] seems completely out of place in the present context; it does not occur anywhere else in the text corpus of and relating to Kant's lectures on physical geography; this corpus has been completely digitized and is searchable. Perhaps "*Delicatesse*" ("sensitivity") is intended.

In any event, cf. Joachim Heinrich Campe, *Wörterbuch zur Erklärung und Verdeutschung derunserer Sprache aufgedrungenen fremden Ausdrücke. Ein Ergänzungsband zu Adelung's und Campe's Wörterbüchern. Neue stark vermehrte und durchgängig verbesserte Ausgabe* (Braunschweig, 1813), 88, which recognizes the meaning of the word "*Adresse*" as "*Geschick. Geschichlikeit, Gewandenheit. Behendigkeit.*" This formulation should also refer back to Rink.

23. *Physical Geography*, AK 9: 156, 1–273, 29. Cf. note 7, above.

24. "MS Werner," 136, emphasis mine. Cf. "MS Kaehler," 140.

25. "MS Wolter," 48f. Emphases mine.

26. The similar term "*Menschlichkeit*" (humanity), like its possible Latin-derived counterpart "*Humanität*" (from *humanitas*), does not occur in the corpus of the geography course. This differs from AK vols. 1–8:

AK 5: 2 times	AK 6: 5 times
AK 7: 4 times	AK 8: 8 times

27. Appearances in the various manuscripts are as follows:

MS Holstein: —	MS Herder 8°: 61	MS Hesse: —
Kaehler: 139	Werner: 136	Messina: 61
Busolt: 72 ("Menschen")	Wolter: 48 (twice)	Barth: 122
Pillau: 221	Dönhoff: —	Dohna: —
Rink: 9: 228, 26; 230, 18; 316, 5		

With the exception of "Herder" and Rink 9: 228, 26, all the occurrences have already been discussed. Since the "MS Kaehler" does not have the term in the relevant place (137); the wording must have been changed by Rink; cf. his singular "Urplatz."

28. Indeed in Hume's essay, "Of National Characters," we find the relevant note. "I am apt to suspect the Negroes to be naturally inferior to the Whites. There scarcely ever was a civilized nation of that complexion, not even any individual, eminent either in action or speculation. No ingenious manufactures amongst them, no arts, no sciences. On the other hand the most rude and barbarous of the Whites, such as the ancient Germans, the present Tartars, have still something eminent about them, in their valour, form of government, or some other particular." In *Essays Moral, Political and Literary* (Oxford: Oxford University Press, 1963), 213.

29. "MS Herder," 61.

30. *Observations on the Feeling of the Beautiful and Sublime*, AK 2: 253.

31. Appearances in the lecture notes are as follows:

Type A_0: Holstein, 130	Type A_1: Herder 8°, 31, 52	Type A_2: Hesse, 83, 105
Type B_0: Kaehler, 374	Type B_1: Messina, 211	Type C: Dönhoff, 44, 162, 188
Type D Dohna, 131		

32. "By taste, I understand / mean the sensual judgement on the perfection or lack of perfection of that which moves our senses. From the deviations in taste by people, one can see that an enormous amount rests on prejudices" ("MS Holstein," 130).

33. "MS Herder 8°," 31, 52f.

34. A corresponding search in the corpus of Kant's printed works, now also digitized, yields 46 occurrences, spread roughly equally over vols. 1–8. The main focus is AK 6, where we find, for instance, "Culture of morality in us. The greatest moral perfection of mankind is to do one's duty and to do it as duty" (*Die Metaphysik der Sitten*, AK 6: 392).

35. "MS Hesse," 89.

36. Read: Africa to America

37. "MS Hesse," 93.

38. "MS Hesse," 94.

39. "MS Hesse," 95.

40. "MS Hesse," 246.

41. "MS Pillau," 360. Cf. "MS Hesse," 94.

42. That is, outside of the wording under discussion here: *Physical Geography* 9: 230, 20–316, 7; "MS Barth," 122; "MS Pillau," 222.

43. "MS Friedländer," 423, 427, 462.

44. Various factors presented elsewhere lead us to identify the "MS Friedländer" as the one that Christian Jakob Kraus was to take to the minister von Zedlitz towards the end of 1778; cf. the letter Kant to Herz, October 20, 1778, AK 10: 243.

45. By way of comparison, the distribution in AK 1–8:

AK 1: —	AK 2: 12	AK 3: 10
AK 4: 10	AK 5: 30, 24 of which in the *Critique of Judgement*	AK 6: 3
AK 7: 31, 29 of which in the *Anthropology*	AK 8: 15	

46. *Anthropology from a Pragmatic Point of View*, AK 7: 220. For comments on the "systematic place" in Kant and on the tradition of 'talent' see Reinhard Brandt, "Kritischer Kommentar zu *Kants Anthropologie in pragmatischer Hinsicht*," *Kant-Forschungen* 10 (1999), 323. Cf. also the drafts (Nachlass) section of AK 15: 232 (R: 534, 534a), 410 (R: 922), 495, 11 (R: 1112).

47. *Anthropologie Friedländer*, AK 25: 556, 3ff.

48. *Anthropologie Friedländer*, AK 25: 626, 11ff.

49. Note that *On the Different Races of Human Beings*, AK 2: 427ff. is based on the second edition, published with numerous changes in 1777; cf. the notes at 519ff. In my opinion, perhaps the most important changes in the second version of the essay consist in the addition of a certain attempt to explain the skin color of the four populations ("races") of the human species in terms of inner, medical and not merely external climatic factors. Evidently Kant adopts such a position already in 1777, which is not clarified in the explanations of the Academy Edition:

> The different colours of the plants are now attributed with good reason to the iron that is in the different saps. Since all animal blood contains iron, nothing prevents us from assigning the different colour of these human races to this same cause. (AK 2: 440, 13ff.)

On the further development in Kant and the possible literary sources, cf. the references of Adickes to *Reflections 1374–1378*, AK 15: 599ff.

50. Cf. especially *Observations on the Feeling of the Beautiful and Sublime*, Section Four, AK 2: 243–56.

51. Cf. *Observations on the Feeling of the Beautiful and Sublime*, AK 2: 252, 4ff., which begins with Asia, then moves to Africa (253, 1ff.) and finally arrives at the "savages" in America (253, 21ff.).

52. The literary form of a separate, especially printed program "Announcement of Lectures" by a full professor in the Königsberg of the eighteenth century is conspicuous. The primary motivation for the publication can be assumed to be an interest in the content: what was it?

53. Cf. *Anthropologie Collins*, AK 25: 8, 27ff.; as well as *Anthropologie Parow*, AK 25: 244, 18ff. In a handwritten remark on *Observations on the Feeling of the Beautiful and Sublime* (AK 20: 44, 12f.) from the mid-1760s, Kant himself wrote that Jean-

Jacques Rousseau is an essential author for the development of his philosophical views, especially in relation to the "human being." The central motif of the *Anthropology*, the "*perfectibilité de l'homme*," goes back to Rousseau (cf. AK 25: li–liii); for this reason, he is the author mentioned most in these lectures. By contrast, his name is not mentioned in a single one of the documents on the *Physical Geography*.

54. Cf. *On the Different Races of Human Beings*, AK 2: 437, 6ff.: "the facial features of the Kalmucks, which in a long series of generations in the same climate, have taken root and become a lasting race, that maintains itself [. . .]."

55. The term occurs twice in the 1777 version; cf. *On the Different Races of Human Beings*, AK 2: 433, 18; 442, 23. In both cases, it is associated with "*Bildung*," formation in an older sense of the word. We should think of an external shape/figure (*Gestalt*) that can be experienced sensuously, that can be more or less completely/perfectly formed that is, recognisable.

56. *On the Different Races of Human Beings*, AK 2: 111. Cf. *The False Subtlety of the Four Syllogistic Figures*, AK 2: 52ff. In the lecture course held at about the same time, there is a statement that fits with this: "The original colour appears to be the white one that arises out of the transparency of the mucous material and small particles of the skin (which proves/shows the translucency of the blood and arteries) since this is a foundation of all and can degenerate through the action of the external air and soil, food into another colour. So Adam appears to have been a white man with fair hair; [. . .]" ("MS Kaehler," 363f.).

57. At *Physical Geography*, AK 9: 316, 4–9.

58. Raphaël Lagier, *Les races humaines selon Kant* (Paris: Presses Universitaires de France, 2004).

6

Translating Kant's *Physical Geography*

Travails and Insights into Eighteenth Century Science (and Philosophy)

Olaf Reinhardt

A brief look at Kant's early publications shows that much of his interest lay in what we would now call "science" in a broad sense, rather than "philosophy"—they are on a theory of winds, whether the Earth is ageing, whether its rotation is slowing, motion and rest and, of course, on earthquakes, after the great Lisbon earthquake.[1] It was an interest he maintained throughout his life, therefore, a knowledge of these papers is essential to an understanding of Kant's work generally. Many of them have not been available in English until the editors of the new Cambridge Edition decided to include them. The longest of the previously untranslated early works was the *Physical Geography*. Since the Cambridge Edition is based on the *Akademie Ausgabe*, David Oldroyd and I were asked to translate the version found there in Volume 9, the version published by Friedrich Theodor Rink. This version first appeared in 1802 based on lecture notes by Kant himself or taken by various students over a lengthy period and annotated by Rink. It was unsatisfactory from the start and received poor reviews.[2] Various attempts to improve on it were made, the most exhaustive being by Erich Adickes, who published two books on it in 1911.[3] Nonetheless, when the Akademie came to publish Kant's lectures in 1923, it used Rink's version, apparently because including Adickes's findings would be "impossible under the current circumstances," partly because it would be too long.[4] Despite the fact that it leaves much to be desired, and that much more material on Kant's lectures has recently appeared in German,[5] the Rink edition at least has the advantage, for the present in any case, in that it is the one quoted in the literature on Kant's lectures.

These early texts give us an insight into a Kant before his great philosophical achievements: they show us a young man anxious to get an appointment at his home university, publishing on a range of topical issues and, as soon as he had attained his permission to teach, offering a course on what was then a fairly new subject, physical geography. They also give us an insight into what was considered "science" at that time, what constituted confirmed knowledge, evidence and proof, how to treat reported facts he deemed unlikely, and how to subvert the censors who thought the world was created in 4004 BCE. In addition to the problems associated with the translation, these are the matters considered in what follows.

When looking at the young Kant, we should bear in mind that, although he rushed into print in the "scientific" essays,[6] it was not only a matter of "publish or perish" even then; rather he also had to "teach or starve." As a *Privatdozent* (1755–70), he depended entirely on fees his students paid to attend his lectures; in other words, he had to be a good and popular lecturer, and he was. He therefore needed a subject to teach that was new, attractive because it was new, and was not taught by anyone else.[7] Physical geography was a good choice and was one of only two subjects that Kant devised himself. The other was anthropology, which, as Werner Stark shows in this volume, essentially grew out of the geography lectures. As we know, the geography was very successful, since Kant taught it with numerous revisions throughout his career. Perhaps one reason for its popularity was that, in a very specific sense, it advanced a new approach to science.

In the mid-eighteenth century, science was still not organized into branches according to subjects because these were not really known, indeed the German word "Wissenschaft" still means any body of knowledge arranged systematically. The natural or physical sciences were not taught as such in universities. There were a number of Academies—the Royal Society in London, the Académie des Sciences in Paris, and the Königliche Akademie der Wissenschaften in Berlin which served to expand knowledge, among other things, by prize competitions on topics of interest, such as the one for which Kant submitted an essay on the rotation of the Earth in 1754. They had meetings, published works and proceedings, and generally tried to promote the sciences. Above all, they were *secular* institutions, unlike many of the universities. Of course, the publishing of books and journals, such as the French *Encyclopédie*, Zedler's monumental work, played a major role in the acquisition and dissemination of knowledge. There were also a number of important "clearing houses" (scholars who acted as bases for the exchange of information): people would mail their findings to such a person, who would then forward copies (without a photocopier!) to others who might be interested. The astronomer Herschel was one of these.

How people actually *did* science can be divided roughly into two categories: those who amassed the raw data, and those who meditated and theorized it.

The former group includes those who published books of reports on travels in various remote parts of the world (i.e., Siberia, the Americas). The latter consisted of "armchair" or "desk" scientists, often university people, clergymen, sometimes landed gentry with a hobby, or even the chief of the Paris police in Diderot's time. Of course, there were a few people working away in laboratories or glass houses, but much of the basics of chemistry and other sciences was still unknown so there was no basis for a system.

A change in all this came about with the publication in 1735 of Linnaeus' *Systema naturae*: not only did it introduce a system of classification, but in order to arrange an object in the system, the object had to be examined closely. This naturally had a huge influence on the advance of science. Despite this, and despite using the Linnean system[8] as a basis for classification in his *Physical Geography*, Kant did not do anything that might be called fieldwork.[9] In any case, Kant was delivering lectures the purpose of which was to enable his students to orient themselves in the world in an educated manner, rather than doing research.

Instead, he read widely. Then, as now, research proceeded by people gathering as much information on a phenomenon as they could by reading about it, perhaps approaching it from a different point of view, interpreting the facts in a new framework, categorizing it in another way, and in the natural sciences, conducting experiments. A problem in Kant's day was that, although the British Royal Society had introduced the idea of peer reviews in 1665, there was no serious evaluation or reviewing process for information that was published (perhaps not unlike the Internet today). It is difficult to know now whether people had a sort of credibility index for publishers or journals in their minds, but in general, the evidence from the *Physical Geography* seems to be that information that had been published was acceptable because it had been published. As a result, at times Kant appears to accept without question statements that seem to us patently absurd. For example, he says that

> the inhabitants of New Holland have their eyes half closed and cannot see into the distance without putting their heads right back against their backs. They have been accustomed to this on account of the many mosquitoes that are always flying into their eyes.[10]

This example is also typical in that it demonstrates how Kant makes a statement, based on reports of others, and then often adds an interpretative explanation that is presumably his own.

Sometimes he is a little cautious: "The people on Formosa [Taiwan] and in the centre of Borneo etc. and whom Rytschkov also encountered among the Turkomen in his "Topography of Orenburg" who have a small hint of a monkey's tail, do not appear to be wholly fictitious."[11]

In the section on the cause of different skin colors in humans, he says:

> Some believe Ham to have been the father of the Moors and to have been punished by God with a black colour, that is now handed down to his descendants. But no reason can be advanced as to why the black colour should be more suited to be the sign of a curse than the white.
>
> Many natural philosophers believe that the colour originates in the epidermis and the black matter with which it is tainted. Still others derive it from the corpore reticulari. Because the colour of human beings goes through all shades of yellow, brown and dark brown, finally becoming black in the torrid zones, it is obvious that climate is the cause. But it is certain that a large number of generations had to pass for it to become handed down and hereditary.[12]

Here the vagueness in the "large number of generations" may refer implicitly to more than the two hundred generations suggested in the biblical version. Furthermore, the expression "natural philosophers" is used to render "Physiker," which to-day means "physicists."

Sometimes, he is scathing: in §75, a section on geographical changes, "proving" that the sea once covered the whole Earth, he explains how it came about that there are shells to be found on mountains high above current sea level and rejects a theory:

> It is ridiculous when La Loubère, in his description of Siam, attributes the presence of these shells to monkeys, who are said to have carried the shells to the tops of high mountains simply to pass the time, as they do at the Cape; or, as another author believes, that Asiatic shells found on European mountains were brought back by the armies who had taken part in the Crusades to the Holy Land.[13]

This leads us to the problems of translation. The guidelines for the Cambridge Edition stipulate that "the most important desideratum of the entire edition is literalness in translation, in the sense of leaving as much of the work of interpretation to the reader as possible. Readability and naturalness of English style are also desiderata, but in an edition the principal purpose of which is scholarly accuracy, they should be subordinated to literalness in cases of unavoidable conflict." Sentence structure and punctuation were to be preserved as much as possible. This means that long sentences, which are much more usual in German than in English, were not broken up. On the other hand, we are dealing here with lecture notes, and Rink's rather than Kant's own text, so the sentences are not as long as in many of the other works. Nor are they are as complex, except for the theoretical sections at the beginning of the *Geography*, particularly as they degenerate into note form towards the end. Word order in German being significantly different from that in English, it was not possible to adhere to the original, particularly in respect of time phrases, to say nothing of verb placement.

A considerable problem was that words have changed their meaning as in the example of "Physiker" just quoted. Many were not as obvious.

"Allein" now means "alone"; it used also to mean a strong version of "however."[14] This can lead to misunderstandings in a context such as "Allein die Natur hat noch andere Kräfte im Vorrath," which can mean either "But nature has other forces in store" or "Nature alone has . . ."; similarly "indessen" can be "in the meantime" or a strong contrastive. Kant uses "blöd" to mean "shy, timid" whereas many people being what they are regard this sort of behavior as stupid, which is what the word now means. "Artig" means well-behaved, and Persians are said to be so; half a page later, the same adjective is applied to a fable and we translated it there as "charming," thus going against the desideratum of always rendering the same German word with the same English one. "Salzsäure" is used in an explanation of the bitterness of sea-water.[15] The dictionary gives "hydrochloric acid," which is accurate but not eighteenth-century English usage, which was "marine acid." "Mergelerde" occurs as a presumed component of alum[16]; this ought to mean "marlaceous" (Al_2O_3), but the contemporary English was "argillaceous earth"—we added an explanatory footnote here. Place names also change or change in their importance. "Mingrelien" is listed with areas such as Georgia and Kashmir[17]; a large atlas reveals that there is a tiny town called Mingrelskaya not far from the Sea of Azov. The Providence Straits are said to separate New Holland from New Guinea; in fact there is a Providential Channel on the northeastern coast of Queensland, some 250 km south of what is now the Torres Strait. Many of the exotic animals listed taxed not only our own resources, but also those of the Sydney zoo and various local and overseas university departments of zoology, for instance, the "Markgraffaultier"[18] had to be left as the "Markgraf Sloth." We also left the "Peruvian sheep-camel," since it is listed under the camels but "raised for its wool and meat," assuming most modern readers would identify it as an alpaca. Sometimes one can be lucky: Astrakhan is described as "eine rechte Pflanzschule schöner Weiber"[19]; our first version was "a breeding ground of beautiful women" until we hit upon "a nursery . . . ," which was one of the meanings in the eighteenth century.

Weights and measures were a problem, as there were so many different ones in the different regions of Germany, and the text even uses a French "toise," an English mile, and other foreign terms. There were similar difficulties with the units and values of currencies. These took quite a lot of research but were generally solved and placed in a separate explanatory note.

At the time Kant gave his lectures, there was pervasive censorship by the government of what was taught in universities. In general, this was applied to religious and political matters, though presumably social issues were also affected. Under the enlightened Frederick II (who ruled from 1740–86), censorship was more relaxed than under his predecessors and successors, and certainly more so than in other European monarchies. Nonetheless, it was still a requirement that lecturers had to nominate to the Minister the textbooks on which their teaching was to be based. Kant was given a special dispensation for the Geography course because there were no textbooks available for this new subject.[20] We know that, despite Frederick's famous statement on religious tolerance,[21] the Ministry was on the watch for anti-religious views—after Frederick's death in 1786, Kant was forbidden to publish texts on religious subjects,[22] which he promised to do and he kept that promise until the death of Frederick William II in 1797. We can also infer a certain nervousness on Kant's part from the fact that he expends considerable effort in the *Universal Natural History* to distinguish his views on the creation of the universe from those of any atheist, all the while avoiding the word "God" as much as possible and using expressions such as "highest being," "greatest wisdom," and so forth.

In the *Geography*, the subject matter makes this kind of disclaimer less necessary, but the concern is still present in the preemptive self-censorship evident in statements about the age of the earth and those on the religions of far-flung countries. The many new lands and peoples being discovered and written about in the mid-eighteenth century tended to lead to a more relative view of one's own culture[23] and religion. Kant evidently took a great interest in this and many religions are recorded briefly in his lectures, almost always neutrally and without scorn, but equally without any mention that Pietism is the one true religion.

The question of the age of the Earth is another. One of the phenomena that raised the issue was the layers of ash and lava on volcanoes, where it was possible to count the eruptions that had occurred in historical records. Thus we find:

> The lava that flows out of Mount Etna contains the same mass as four mountains like Mount Vesuvius. At night it glows like fire and when it cools, it attains the hardness of stone, so that churches can be built of it. But when new lava encounters such a church, the latter melts away. [. . .] Soil does not settle readily on the lava, even though the area beneath mountains where ash is present is very fertile and covered in trees the diameter of which is eighty inches.
>
> But how did the soil come to be on the older lava? The soil gradually generated itself, for this happens even on the smoothest

> stone. The air first carries up dust, and then more and more similar particles accumulate there, until it turns into a real layer of soil, but this must take a very long time. Brydon observed some lava not yet covered in soil and concluded from this that it must still be very young, even though it had been flowing since the Punic War.
>
> When people dig a well in Catania, they come to five or six layers of lava covered with soil, for the formation of which, it is believed, 16,000 years are required.
>
> Moses gives us the age of mankind but not the age of the earth. The earth may have been formed some thousands of years earlier, for we should not allow ourselves to be prevented by Moses' statements from giving consideration to physical evidence. For God, a period like a day is too long for creation; and for the formation of the earth it is too little.[24]

This passage is a good example of the way Kant proceeds from some empirical evidence gathered by others, then tries to account for some associated phenomena, which in turn leads to much bigger questions, and ultimately confronts these with issues about the biblical story of creation. Of course, he is not the first to suggest considering that story as more of a symbolic than literal truth, but the last two sentences in this quotation make no attempt to resolve the issue. Or again:

> Scheuchzer and many other physicists attribute these indications of ancient changes to the Flood; but firstly, this covered the earth for too short a time for it to have been able to cause these changes. A short time, such as Noah's Flood lasted, is not sufficient to have piled up overly large banks of shells, deep layers of soil, or indeed even large rocks. [. . .] The Flood seems to have been merely a universal example of one of these changes, that is, a change in the whole of the dry land into sea and of this back into dry land.
>
> There are undeniable indications that this really happened in some regions of the earth, either before or after, and that many years have elapsed during such changes. The fact that many, indeed all, islands must once have been connected with the dry land, and that the land in between them was changed into sea bed, is evident from the animals that may be found on them. For unless one wants to maintain that God created the land animals separately on every island a long way from the mainland—e.g. on the Azores or the Ladrones [Mariana] Islands—it is impossible to

> understand how they got across to the islands, especially the noxious animals.[25]

While these extracts do not specifically present problems to the translator, they are pointers to the subversive approach adopted by Kant—he presents reported facts in such a way that it is impossible to argue against them in an orthodox way. It is clear that he is asking his audiences to exercise logic and reason on geological matters relating to the age of the Earth in a way that means the Scriptures can be only a symbolic representation. Kant was not being particularly original here; this idea was discussed by his near contemporary James Hutton and by Buffon among others.

There are many problems of authenticity, of precision, of Kant's pedagogical style, of dating the various sections with the whole of the *Geography* as reported by Rink. In particular, the text does nothing to identify the year in which the extracts in the volume were delivered, obviously a crucial problem in a period of such rapid expansion of knowledge of the world—it is hard to believe that Kant would maintain in the 1790s what he had taught in 1756. Nor can he have divided the material and structured the course in the way Rink records it; there is far too little about Europe, for instance. It seems Rink included information on more exotic places disproportionately, presumably to make it more appealing to a general rather than a specialist public. As a result, this text taken in isolation does not reveal very much about Kant as a person or as a philosopher. If it is true that, as Eco maintains, in the first *Critique* "Kant was not interested in *knowledge of* but *knowledge that*"[26] then this would not appear to apply to the *Geography*. Certainly the six introductory paragraphs—which set the bases for acquiring knowledge as such, the difference between knowledge gained by reason and that gained by experience, the way knowledge is classified, the difference between history and geography, and the various branches of geography—are clearly the work of a very ordered and ordering mind, linking it to Kant's critical and methodical thinking. Again, however, as this section is not dated, we do not know whether it precedes the *Critiques* or not; we might imagine, however, that Rink would have taken great care to be as authentic as possible at the beginning of his book, especially as it degenerates into mere notes later. The section following, "preliminary mathematical concepts" refers to matters already dealt with in the *Universal Natural History* of 1755 and so was presumably included in the earliest lectures. This introduces §7, the shape of the Earth, a very logical progression, followed by §8 which begins by discussing the circumference of our planet and moves into a listing of various measures of length. This would appear to be less of Kant and more of Rink. Nevertheless, the overall structure of the *Geography*, being logical and in accordance with Linnaeus's system, is worth reading for an overview of the way the eighteenth century approached a classificatory science. It also has a tendency towards the very unusual, even quirky, such as the entry on:

> The Merman, the Mermaid
> is found in all four continents. The imagination, which is inclined to fables, has constructed a human being living in the sea out of it. But it has little similarity with humans. Its head, which can be mistaken for a human or fish head, with large ears, blunt nose and a wide mouth, is attached to a body covered on the back with a wide, thick fur like that of a flatfish and has hooks like a bat on the sides. Its front feet or fleshy flippers are somewhat like those of a human. This animal has two teats on its breast and a fleshy tail. Because of its fat it is also called the *Sea Pig*.[27]

One wonders what Kant would have made of the platypus. It is not clear whose text this is, but Kant must have listed some of the stranger animals in his lectures, perhaps assuming the common ones were well-known and could thus be skipped over. In any case, the whole text should be read bearing in mind that it is not Kant's text and the other caveats in Werner Stark's essay in this volume.

The fact that Kant took up the subject of geography, essentially a descriptive[28] rather than an interpretative one, seems to be typical of his interest in virtually everything. It is a necessary basis of facts for theorizing. The way he proceeds from a particular example to a general statement demonstrates a philosophical approach to knowledge, for instance in the following, he begins by discussing the thinness of the air at high altitudes, then moves on to the inhabitants:

> The people who live around and on the mountains are said to be very strong and brave and try to assert their freedom in every way possible. But this probably results principally from the fact that in such regions, it is very easy to defend oneself against large armies with only a few people, and also because the mountain tops are uninhabited and uninhabitable; even in the valleys [of mountainous regions], there are fewer riches to be hoped for, so that no one is likely to want to stay in such regions. [. . .]
>
> Thus it appears that the particular character of the inhabitants of mountainous regions does not reside in the quality of the air prevailing there. The noticeable difference between the Highland Scots and the English and the inhabitants of Lowland Scotland results from the fact that the latter receive a very soft upbringing.

> The air in such mountainous regions is said to be the cause of homesickness, especially of the Swiss, since these, when they are in other countries, become melancholy, particularly when they hear their national songs; indeed they pine away if they are not permitted to return to their homeland. But this results partly from the idea people have of the peace of mind which fills people in all countries where the inhabitants live in relative equality; thus especially in Switzerland there is a peace of mind they believe can be found nowhere but on the soil of their homeland. Another reason for this homesickness consists in the greater effort such people have to make to earn their living. [. . .] It is also said that in no country is suicide so common as in Switzerland, although, in general, this tends to affect the rich more; but the Swiss are mostly poor. It has been claimed, however, that the suicides in Switzerland are mainly such people as have previously visited other countries and found a taste for the delights there, and who take their lives because they have to do without these pleasures in their homeland. This change in themselves is also the reason why they all unanimously assert that, on their return, they did not find their homeland the way it was when they left it. Thus they regard a subjective change as an objective one, since they are not capable of perceiving the former.
>
> The homesickness of the Swiss is a yearning or an endeavour [for a goal] they know to be impossible. It is always better to have no hope than an uncertain one; because in that case one ceases to feel any longing and attempts to accept in one's mind a situation in which one has nothing more to hope for. It is precisely for this reason that there is nothing more arduous than to exert one's strength while conscious of the impossibility of attaining the purpose. Homesickness is particularly prevalent in poor regions ill-favoured by nature; because the greater the simplicity of life, the stronger the effects of temperament and desires. Dissatisfaction increases with the latter, especially if one remembers a better way of life or sees how much better it is in other places. Family bonds are stronger the poorer the family is, and the more significant the deprivations nature has placed upon it. On the other hand, the more one is burdened by self-interest, as is the case with luxury, the less solidarity there is among people.[29]

What is interesting here is the way Kant proceeds from an observed scientific fact (thin air) to philosophical issues about the effects of the environment on people, albeit with a somewhat moralizing tone at times.

The *Geography* is the longest of the "early" scientific texts. It covers a large area of information and, together with the next longest, the *Universal Natural History* provides a broad factual basis for understanding the world. It is as though the young philosopher systematically sets out to know as much as possible about the physical world before going on to more fundamental questions about how we know what we know. Without this background, he would not have been able to write his other "scientific" essays and perhaps not the great philosophical works either. Of course there are factual errors and obviously Kant got a lot of things wrong; we should remember, however, that he also got a lot of things right. And it seems to me that one of those is that he was laying the groundwork for a method of inquiry in the *Geography* that would lead him to greater things. So it does have more than mere historical interest—in that it shows how Kant made rational sense of and ordered his world and his own place in it.

Notes

1. See also Jürgen Zehbe, "Einleitung: Die Bedeutung der Naturwissenschaften für die Philosophie Kants," in Immanuel Kant, *Geographische und andere naturwissenschaftliche Schriften* (Hamburg: Meiner, 1985), vii–xxxix.

2. "Der Gewinn für die Wissenschaft scheint aber bis jetzt nicht sehr erheblich, und das Ganze enthält Weniges, was sich einer Auszeichnung verlohnte" (The benefit to science so far does not appear to be extensive, and the whole contains little that is worthy of distinction). *Göttingische Gelehrte Anzeigen* 154 (25 September 1802), 1530.

3. *Untersuchungen zu Kants physischer Geographie* (Studies on Kant's Physical Geography) (Tübingen: Mohr, 1911) and *Kants Ansichten zu Geschichte und Bau der Erde* (Kant's Views on the History and Structure on the Earth) (Tübingen: Mohr, 1911).

4. Paul Gedan, ed., *Kants gesammelte Schriften*, vol. IX (Berlin: de Gruyter, 1923), 509f. Citations from the Akademie edition are hereafter given as AK, followed by volume, page and, where needed, line number.

5. In AK 26, ed. Werner Stark.

6. Cf. the many errors, omissions, and proofreading oversights.

7. A. F. Büsching first introduced lectures on geography at Göttingen in the winter semester 1754/5; Kant began his in the summer of 1756. (Adickes, *Untersuchungen*, 9)

8. In *Physical Geography*, §4, Kant briefly discusses the relationship of his project to that of Linneaus.

9. As Zehbe puts it elegantly, "His direct involvement with nature as with the arts remained well behind his theoretical endeavours." (Zehbe, "Einleitung," xvii).

10. *Physical Geography*, AK 9: 315.

11. *Physical Geography*, AK 9: 315.

12. *Physical Geography*, AK 9: 313f.

13. *Physical Geography*, AK 9: 229.

14. As in *The Magic Flute*: the Queen of the Night sings "Allein, vergebens war ihr Flehn" (But her pleading was in vain) in her first aria "O zittre nicht."

15. *Physical Geography*, AK 9: 204 n. 3.

16. *Physical Geography*, AK 9: 369, 31.

17. *Physical Geography*, AK 9: 311–2.

18. *Physical Geography*, AK 9: 330, 15.

19. *Physical Geography*, AK 9: 403, 17.

20. Nonetheless, Zedlitz, the Minister for Education requested a copy of the text in 1778 (Gedan, AK 9: 512).

21. "Jeder soll nach seiner Façon selig werden" (Let everyone get to heaven in his own way).

22. As was Lessing, though he was not in Prussia. Lessing then wrote a play instead, "Nathan the Wise," which puts Judaism and Islam on an equal footing with Christianity.

23. For instance, "A missionary was surprised that when the Chinese see a rat they rub it between the fingers and smell them with gusto. But by the same token, I ask: why do we now find the smell of musk obnoxious, when everyone thought it smelt so good fifty years ago? How much does the judgement of others do to alter our tastes in accordance with the times!" (*Physical Geography*, AK 9: 320).

24. *Physical Geography*, AK 9: 266f.

25. *Physical Geography*, AK 9: 300–1.

26. Umberto Eco, *Kant and the Platypus* (London: Vintage, 2000), 69.

27. *Physical Geography*, AK 9: 343.

28. "We can equally well call both history and geography descriptions. The difference is that the former is a description in terms of time; the latter in terms of space. [. . .] History is a narrative, but geography is a description" (*Physical Geography*, §4, AK 9: 161).

29. *Physical Geography*, AK 9: 244f.

7

Writing Space

Historical Narrative and Geographical Description in Kant's *Physical Geography*

Max Marcuzzi
Translation: Samuel A. Butler

The question, "what is the subject of geography?" itself presupposes that geography is *one* and that there is *a* discourse or type of discourse proper to this discipline. This question would thus immediately imply another, relative to the discourse itself: *how* does geography uniquely apprehend its object, whatever it might be, such that it constitutes a unitary discipline?

If such a unitary discourse exists it is, it seems, beyond the differences that present themselves as soon as one considers the historical constitution of geographical knowledge. In effect, traditionally, geography is plural, even if at first only because it is constituted, as we know, according to two great traditions. On the one hand, the tradition of a mathematical geography is above all cartography, with Eratosthenes, Ptolemy, and Varenius. On the other hand, it is a so-called comprehensive tradition that derives from Strabo, which attempts to describe the reality of the human world, eventually under the auspices of a project oriented towards ethics and politics. One can distinguish from the outset a physical, mathematical, or political geography, in addition to a moral, theological, or economic geography.

If, in spite of all of this, these diverse forms spring from a single, unitary geography, it must be the case that, beyond their differences, they have an object and an approach fundamentally in common.

Original publication: "L'écriture de l'espace: Récit historique et description géographique dans la *Géographie physique* de Kant," in *Historicité et spatialité, Le problème de l'espace dans la pensée contemporaine*, ed. J. Benoist and F. Merlini (Paris: Vrin, 2001), 117–139.

In the history of geography, it seems that the theoretical determination of this object by Kant has had a nontrivial importance at least because it has participated in the attempt to reunite progressively the two traditions under the perspective of a global educational project, which was the impetus for the teaching and theory of this discipline as proposed by Kant.

According to Kant, geography is useful in many respects:

> Principally, it brings to young pupils[1] and students[2] a concrete knowledge which allows them to supplement their insufficient experience; it is a condition of the possibility of reasoning correctly insofar as it permits them avoid thinking in a void, but rather about facts; it therefore aims their transformation into *knowers* by providing a material for knowledge.
>
> On the other hand, it is valuable as a propaedeutic to knowledge of the world,[3] in particular to men's accession to their role as citizens. Forming the citizen of the world, geography is a condition of the possibility of cosmopolitanism if the latter is not to remain formal and abstract. In a general sense, this knowledge of a *worldly* nature proposes to contribute to making men *prudent*,[4] which is to say capable of acting towards their peers and of caring for their own interests.

For this, the immediate goal envisaged by the instruction of geography is to produce both a systematic and total knowledge of the Earth that might provide the capacity to assign to the correct place all of our particular experiences, for, according to Kant, experience is only profitable if its theoretical place has first been determined in a system. This implies that one anticipates possible future experience by means of a guided mental experience in a scholastic context.

Geographical works of his own epoch had not, it seemed, satisfied these demands, such that we know that Kant himself elaborated the material for his teaching, contrary to the common practice of commenting upon an official manual.[5]

This is why we can propose determining the specificity of geography, as much with respect to its object as to its method, in joint reference to the theory Kant has provided *and* to the manner in which he himself constructed his geographical work. In this respect we propose to sketch in parallel an investigation into the theory *and* into the practice of the discourse of Kant in the *Physical Geography*, more specifically into the relationship between the historical discourse and the geographical discourse as a difference between *narrative* and *description*, or between *explication* and *presentation*. Thus it is the possibility and the success of this differentiation that we want to interrogate, it being understood that it is on this differentiation that the specificity of geography as a unitary discipline depends.

Theory of the Difference Between History and Geography

The specificity of geography with respect to history is presented in its essence in the introduction to the *Physical Geography*. Physical geography is first characterized as *description of the earth* (*Erdbeschreibung*), comprising all phenomena that take place on its surface, including man considered in the physical sense. This *description* (*Beschreibung*) opposes itself to the *narrative* (*Erzählung*) proper to history. The question is thus, first of all, to know what the difference is between *describing* and *recounting*.[6]

But on the other hand, Kant says that

> we can equally well call both history and geography descriptions. The difference is that the former is a description in terms of time; the latter in terms of space.[7]

Thus, there is clearly a hesitation in one sense on the difference between *recounting* and *describing* that seems here to distinguish itself less by intrinsic discursive characteristics than by the principally spatial or temporal form of the object taken up by a discourse which, in both cases, is descriptive. It thus seems that it is the type of object of study that determines the type of discourse deployed, that gives to the discourse—for the moment only in a formal and exterior manner—its character and definition. The difference between describing and recounting would correspond to a difference between two types of description, recounting being a subdivision of description. The question is, however, to know whether a difference proper to the discourse extends that of the objects.

a. Temporality of the objects. Yet, history treats of successions of events, or events considered from the point of view of their succession:

> History concerns the events that have taken place one after another in time. Geography concerns phenomena that occur simultaneously in space.[8]

Simultaneity thus determines the spatial character; but while for Leibniz space is the possible order of coexistents,[9] for Kant instantaneity or simultaneity are temporal in the same sense as succession since all experience happens in time, where there is not, strictly speaking, any reality to simultaneity, in a sense that would correspond to the reality of an atom of time; the spatial as well as simultaneity is apprehended therefore on the basis of a certain order of the successive. Simultaneity, not immediately manifest, cannot directly produce spatial characteristics by economizing an order of succession. Put otherwise, it is because of the manner in which phenomena are apprehended and explored in time, which is to say by the possibility of making the order of perceptions vary, that it is possible to construct the experience of simultaneity as a construction which opens upon the experience of space, or at the very least to the sense of its exploration.

Thus we know that, in the third "Analogy of Experience" in the *Critique of Pure Reason*, the perception of simultaneity is linked to the perception of differences of place in space, such that I only know that things are simultaneous because I can invert the order of their perception, for example "I can direct my perception first to the moon and subsequently to the earth, or, conversely, first to the earth and subsequently to the moon."[10] The certainty of simultaneity is only acquired as a specific type of "reciprocal succession" of phenomena for a subject, and thus it is based on the possibility of traversing space in every direction; space manifests itself thus as this possibility itself.

b. Temporality of the discourses. If one transposes from the discourse the presentation of perception and that of the experience of simultaneity and of succession in the *Critique of Pure Reason*, one sees that the discourse should correspond to experience, all phenomena that can be described reversibly are simultaneous and spatial, and thus arise from geography, and that, on the other hand, all phenomena which cannot be described in this way, and for which the order of description is imposed, are historical.

When, consequently, Kant says that history as narrative (*Historie*) differs from geography in that history is the relation of consecutive events, whereas geography is the relation of phenomena that are produced side by side in space, it seems necessary to understand—since there is, as a corollary to the Copernican revolution, no space in itself—that the order of the description of phenomena cannot directly announce how objects exist, but must state their mode of existence indirectly, beginning with the explication of the manner in which the objects are perceived. In the occurrence, it is the reversibility of the order of perception that makes the difference. Narrative presents an irreversible order, whereas description presents an order that I can invert. Insofar as this is the case, description is a tributary of an order which remains to be specified.

In principle, from there is established a true link between the temporal or spatial form of the object and the type of discourse proper to it. The relative arbitrariness of the spatial route is, for example, manifest with the determination of the prime meridian, for which Kant expresses the wish that one might finally reach agreement and choose between the Greenwich meridian, that of El Hierro, and that of Flores.[11] Since space can be traversed in every direction, geographical description implies that one has first reached agreement on a certain number of conventions to permit understanding. What is more, it is possible to measure the longitude of a place either by measuring its *eastern* longitude or its *western* longitude; the manner in which one begins is indifferent.[12] To establish some order in this domain, Kant reckons it necessary that longitude be determined "either only from the east, or only from the west." In other words, the direction must be fixed even in the full knowledge that it can be reversed, and thus remains arbitrary.

Now, having posed this possible theoretical difference, what is the relationship between history and geography? Considered formally, are their respective discourses radically heterogeneous or can one be articulated by the other?

Kant says that "the history of occurrences (*Geschichte*) at different times—and this is true history (*Historie*)—is nothing other than a consecutive geography."[13] This continuity seems here again to bring the two disciplines together and to homogenize their discourses: history is constructed via the temporal elongation of a description; I grasp the Earth as the totality of things I can describe in an indifferent order (geography), and I prolong this totality towards the past in order to discover the order of its imposed succession (history). Narrative thus consists in following the chronological order of events that succeed each other up to the present of the historian at work. In linking history to geography in this way, narrative to description, Kant sketches out a project of total and absolute restitution of phenomena to the world:

> if one were to describe the events of the whole of nature as it has been through all time, then and only then would one write a real so-called natural history.
>
> If, for example, one were to consider how the various breeds of dogs descended from one line, and what changes have befallen them through all time as a result of differences in country, climate, reproduction, etc., then this would constitute a natural history of dogs. Such a history could be compiled for every single part of nature—for instance, of plants and so forth.[14]

But Kant considers it unrealistic to truly continue geography in order to construct a history of nature: "we can have a description of nature, but not a natural history." For that we lack documents indispensable to the construction of an experience. Without sufficient information, the past can only be constructed in a conjectural mode. This side of the grandiose project of a continuous geography, the articulation between history and geography thus consists more modestly in knowing where historical events take place; and Kant deplores that we still do not know how to complete each *narrative* with the *description* necessary to its comprehension.

Thus it would seem that one has on the one hand a *geography* that exists as a description of nature, and a *history* of nature that is desirable but nonexistent and, without a doubt, impossible to realize because its material is impossible to gather. History and geography oppose one another as the real and the impossible desiderata:

> If one were to journey through nature in such a way that one noted the changes it had undergone through the whole of time, then this would yield a genuine natural history.[15]

But this definition only corresponds to an impossible hypothesis. In effect, Kant says in "On the Use of Teleological Principles in Philosophy" that "natural history can only offer us fragments or shaky hypotheses;" as a science it can "only be presented more in outline than in a work of practicable science (i.e., an activity in which one might find a blank space already marked out for the answers to most questions);"[16] description of nature, or *physiography*,[17] appears on the other hand "in all the full splendor of a great system"[18] which, in order to be complete, should nevertheless include an archaeology of nature,[19] since it is necessary to understand "ancient" geography in order to understand the present state of the earth:

> The description of nature (condition of nature at the present time) is far from sufficient to indicate the ground for the manifold variations. No matter how much one opposes, and rightly so, the boldness of opinions, one must venture a *history* of nature, which is a separate science and which could gradually advance from opinions to insights.[20]

It seems that we might be in a vicious circle before this hesitation of Kant's with respect to the necessity of linking geography and history. First, he posits that geography is susceptible of achieving a systematic form insofar as its object is the entire earth. Then, he affirms that it is only complete when it integrates becoming, and in moving into history, which is impossible: it is only complete (as a system) in moving into that which makes it impossible (as a system).

Histories in Kant's Geography

All the same, if one attends to the text of Kant's *Physical Geography*, it is striking to note that history appears there with regularity, at least in the titles of various chapters: "history of springs and wells," "history of rivers," "history of the atmosphere," "history of the great transformations the earth has undergone and is still undergoing." The question is to know what the sense is of these histories inserted into the general framework of geography, and if with them, the project does not pass from description to narrative.

The "history of springs and wells" seeks to provide an account of their formation by rainwater soaking into the layers of the earth and running out at a low point. The "history of rivers" aims at providing an account of the formation of rivers originating in streams flowing from springs, reuniting their waters. To understand the history of rivers is therefore to understand the "cycle of sea water and the water of the rivers,"[21] stipulating that a part of the cycle passes by rain and snow, and not by subterranean circuits as many authors have incorrectly believed, sometimes inspired by Descartes.[22] The "history of the atmosphere" treats the density of air, the division of the atmosphere into regions and the properties of air—its humidity, its dryness, its mineral salt content, its purity, its salubrity, its color—before treating the causes of winds, which are the difference between warm and cold regions (for example land by comparison to the sea); or the difference in air pressure between two regions (more or less dense air); or the action of subterranean vapors (and in particular sulfurous vapors) that provoke uneven winds; or finally the action of the rotation of the earth.

All these histories are based in geography, insofar as they take up the study of the manner in which zones interact: the interaction of *terra firma* and the sea in the water cycle, the interaction of the hypogeal and the surface for the movement of winds or earthquakes; the interaction of the earth and the atmosphere. However, these chapters are called *historical* in a rigorously Kantian sense because they are consecrated to the study of causes, which is to say processes of formation: the cause of the river is the convergence of streams; springs are the cause of streams; etc. But as the system considered is closed (notably because the earth is round), cause and effect conclude by rejoining one another, in such fashion that their complete system forms a cycle.

All the same, concerning the "history of the great transformations the earth has undergone and is still undergoing," it seems that the situation is different. Earth transforms itself through the fact of earthquakes, the action of rivers, rain, wind and frost, and the sea which "is gradually receding from the coasts of most countries" or, on the contrary, is encroaching upon certain others. What is more, Earth transforms itself under the action of men, who construct dams, drain swamps, fell forests and thus change the climate considerably. What transforms itself here is the earth considered as a whole, in its relation with itself, without reference to another star, the moon, or the sun, as Kant does in order to explain the alternation of the seasons or the tides; there are no more cycles at this level, nor any of the processes which, on the level of narrative, aim at reconstituting simultaneity.

With the study of the "proofs"[23] (*Beweisthümer*) that the sea formerly covered the whole earth (seashells at the summit of high mountains, the shapes of certain valleys that testify to the ancient presence of ocean currents, strata of layers of the ground that testify to the progressive constitution of the earth's crust), it is the duration of the earth considered as a whole, and not the interaction of regions that is considered. And one remarks that beyond hypotheses, Kant posits here a certain number of facts as certitudes (*es ist Gewiß, daß . . .*). It is thus be certain that: the earth was formerly fluid throughout its mass, as this would explain the form that it took in its rotation and the fact that mountains become higher as one approaches the equator; the sea once covered everything, such that the ocean floor constituted the surface of the earth; there have been great subsidences more or less everywhere on earth, due to the fact that the earth is hollow underneath.

These facts make it possible to draw the conclusion that the earth was first a liquid mass and progressively hardened, starting with the surface, taking the form of a sphere flattened at the poles. Over the course of its hardening, heavy masses pushed their way to the center of the earth while lighter masses rose to the surface. Even though this phenomenon in its entirety is totally singular, concerning the whole and manifesting no measure of reversibility, it clearly arises from narrative and history, even if only in a speculative mode.

Finally, concerning man considered in a physical sense, the differences between the races also arise from a history (of which the principle is not restricted to men). In the Introduction of the *Physical Geography*, Kant gives as an example of a history of nature the examination of the manner in which

> the various breeds of dogs descended from one line, and what changes have befallen them through all time as a result of differences in country, climate, reproduction, etc.[24]

One thus understands the relation between different forms of animals or plants by the difference of their respective histories originating from a common line, and this history itself is only rendered comprehensible in its detail by the knowledge of different geographies. Thus, when Kant says, "the lizard and the crocodile are essentially one and the same animal. The crocodile is only an enormous lizard," their difference is related to the difference of locale of the two animals (the latter lives in the Nile, the former on *terra firma* beyond the Nile); but if beyond this difference "they are essentially one and the same animal,"[25] it is probable that a history would reunite them in the past by a common line. And it is thus again that Kant proceeds to explain the difference of human races, for which he makes all of humanity descend from a sole lineage. The comprehension of the differences of the races implies thus the succession of generations and the difference of climates, geography *and* history, so much that to this principle of differentiation is added a principle of irreversibility since Kant says in "Of the different races of human beings" that it can happen that the hereditary modification of a line arrives at a point where it "could no longer provide the original formation of the phylum."[26] This irreversible character in the constitution of differences is still properly historical.

We see therefore that in spite of the affirmation of the difference between historical narrative and geographical description, geographical discourse never totally displaces with Kant a historical discourse which can take several forms.

For one part, the discourse presents itself as historical each time that one gives the cause of a phenomenon and from this fact develops a temporal sequence relating a cause and an effect. However if, for example, one explains the water cycle, historical narrative of the causal type tends to become a description: it suffices in effect for "explaining" this cycle to describe the global movement of the water of the springs on to rivers and the sea, then to clouds, to rain, etc. It thus becomes impossible for an element of the process to take the position of a cause without also being an effect. In such cases, the difference between narrative and description disappears to the extent that the description becomes more exhaustive and arrives at the consideration of the process in its whole by a sort of panoptic presentation that does not impose any order upon the comprehension, since it is indifferent whether the description progresses from the spring towards the sea or inversely. The use of cycles concludes with this result of the neutralization of chronology, and thus the historical dimension, by which it rejoins geographical discourse.

In the case of the constitution of the earth, the formation of the breeds of animals or the races of humans, the temporal moments cannot be escaped, neither the cause nor the origin can be considered as an indifferent moment in a process that one could describe as one wishes. Thus even if geography sometimes reaches towards the confines of history, and if the latter sometimes symmetrically does the same, the difference between the two types of discourses subsists, as does the distinction between the two forms of knowledge to which they correspond respectively.

Of Physics and Chemistry in Geography

Otherwise, description often being complete only when it has become explication, the research into causes that gives rise to this explication leads to exploring another axis besides the temporal and spatial, which is the physico-chemical axis, including physiology and geology. In effect, the phenomena described by geography imply the comprehension of the constitution of the things one investigates. Thus we find with Kant chemical considerations for understanding the relationship between salt and water, or to articulate that heat and cold can be produced by combining certain materials, for example when they ferment; as for physical considerations, they appear beginning with the text of "Preliminary Mathematical Concepts," which opens the physical description of the earth with a reflection on the units and procedures of measurement.

When he studies the earth by elements—water, air, earth, and fire—Kant develops considerations on the nature of them and integrates them into a comprehension of geographical phenomena, as, for example, evaporation to understand the circular movement of water; or the force of attraction, heat, light, electricity, and magnetism to understand phenomena that take place on the surface of the earth, in particular tides and heat. The cause of earthquakes is considered equally relevant to the competences of the physicist or the chemist, insofar as they might be explained by the reaction of pyrite with air, or by the action of the atmosphere, or yet by the movement of subterranean streams. The question, Kant notes, will only be able to be addressed in an experimental manner.

The physicist is called upon again to explain the changing color of chameleons, or the black color of the skin of certain men. The speculation concerning lineages and the modes of transmission of hereditary characteristics in effect provides the principle of differences between the races, but not the physiological reason for such and such a precise phenomenon, which implies that one is looking for another sort of cause. Thus, for the black color

> Many natural philosophers believe that the color originates in the epidermis and the black matter with which it is tinted. Still others derive it from the *Corpore Reticulari*. [. . .]
>
> It appears that the drying out of the vessels that carry the blood and the serous fluid under the skin brings about the absence of beard and the short curly hair on the head, and that the appearance of the black colour is caused by the absorption of the light that passes through the outermost skin into the dried out passages of the *Corpore Reticulari*.[27]

In order to explicate the black color of men, and its difference according to place, it is necessary to have intervene *both* the difference of climates, *and* the theory of lineages, *and* the constitution of the skin and its reaction to heat. The description is only complete on condition of being developed thus from multiple causal explanations. The geographical text is produced at the confluence of all the competences it convokes.

Thus we see clearly that in a general manner Kant has recourse to the competence of the physicist, particularly each time the cause of phenomena noted on the surface of the earth necessitates causal hypotheses. In many respects, the physicist and the historian meet up, each in his own way, in their speculations on the cause. However, as causes here do not integrate themselves in a larger cyclical description, their study never leaves the domain of geography, but rather introduces an exterior support.

Inventory or Description According to Space

That geography thus makes a necessary appeal to diverse competences does not impede it from having a unique character in accordance with which it articulates them, nor from constructing itself around a specific object or method. We have seen that, for Kant, this specific trait is its relationship to space and the type of discourse that this implies. It is thus now a question of articulating what might mean, in a precise manner, the project of a description according to space, which returns to asking where space is in geography, how it intervenes concretely in the discourse which aims at accounting for the "side by side" and simultaneity. In other words, the question is in knowing what a description is.

Yet Kant proves to be extremely discreet as far as what "describing according to space" means and, in reading the text of the *Physical Geography*, one notes that space there is far from being explicitly present equally throughout.

First of all, inasmuch as the *Physical Geography* opens with the presentation of "Preliminary Mathematical Concepts," and inasmuch as these concepts intervene essentially for the calculation of differences of situation and of distances, space is taken into account in geography under the aegis of that which is measured. The general description of the earth according to space aims at describing it, for example, in relating it to the sun and the moon; this enterprise arises properly from cosmography. Space intervenes again to relate places described upon the earth at either natural references (the poles, the polar circles, the tropics, the equator and the zones determined by these references) or conventional ones (longitude). Here, the object "Earth" is constructed as a pure geometrical object upon which we engender points and lines, cut out spheres and circles, put bluntly, that we can construct in pure intuition. Next the relation to space takes the form of descriptions having the character of definitions: descriptions of general forms of water, of the earth and the atmosphere to determine what a continent is, an archipelago, an island, a bay, a gulf, a strait, etc. The differences of quality between the elements are just as often explained in relation to their situation; for example, the degree of salinity of the seas is related to the number of rivers that empty into them, or to the evaporation proper to the climatic zone. Finally, Kant presents the place where all manner of remarkable things can be seen: the presence of phosphorescent microorganisms in the sea, the Sargasso in the middle of the Atlantic Ocean, the driftwood of the seas of the great north, the pumice floating off the Cape of Good Hope. General geography thus provides a localized inventory of remarkable phenomena, and a description of typical forms that can be found in various places on the earth.

The second part of the *Physical Geography* proposes next the particular examination of all that the earth contains, which takes up the realities of the earth in a more specific manner. With men the differences of skin color are related to the climate as well as differences of temperament: they are more beautiful, more industrious, and better endowed with talents in the north than in the south; hardier and more jealous of liberty in the mountains than in the plains. But for all this, Kant is not engaging in a speculation, which would relate forms of culture to climate, and is satisfied in this respect with sketching out an inventory of modifications that men make to their own form, such as differences of dress and of taste. Here space loses its explanatory pertinence in order to limit itself to the localization of that which one cannot help but notice. As for the presentation of animals and plants, there is no general method which, in spite of the announcement made in the Introduction, seems to be at best an amalgam of anatomical or formal classification and spatial localization. Animals are distinguished in the manner of Linnaeus according to the number of their hooves or toes; by the fact of being terrestrial or aquatic, or having a shell; trees are classed by family, by example the palm family, or they are grouped into aromatic woods, or medicinal trees, or trees that bear agreeable fruit, or spice trees. Finally, minerals are classed as metals, nonmetals, salts, rocks, or fossils. It is only under this classification by varied principles that Kant sometimes relates the specimens he describes to the regions where they are found. Geography as a description of things related to places and to the differences between places is thus still present, but in a secondary manner, behind an inventory which, even according to Kant's criteria, is of a rather logical and scholastic sort. In this, the *Physical Geography* only fulfills in an allusive manner the program set out in the Introduction, to always relate beings to places, and only very slightly breaks with anterior classifications.

Finally, the description of countries "according to a geographical order" is taken up by continents. Asia is described beginning with China, then progressing towards the south and west to Arabia, subsequently climbing towards the north and east to Russia, Siberia, and Kamchatka, thus describing a loop that runs from China to China. However, although the Tonkin is explicitly situated to the southwest of China, its situation with respect to the other countries is not furnished by Kant, such that this situation seems presupposed for the circuit to find its geographical orientation. The relationship between countries amongst themselves is impossible to reconstitute from the *Physical Geography*, which does not indicate that it proceeds in a circular direction and which, thus, does not explicitly provide the direction of its circuits, its order. In the same way, to describe Africa, Kant begins with the extreme south, the Cape of Good Hope, and then climbs northeast towards the equator, then northwest, after which he leaves the Canary Islands in this northwest to descend to the countries of Cape Verde, the Gambia River, and the Guinea Coast; next he passes to northeast Africa, Egypt, Abyssinia, and the north coast of Africa. But here again the direction of the circuit is not explicit, and seems rather erratic in all, few indications being given as to the situation of each country, which without doubt is partly explained by the state of explorations, but also shows that Kant was not concerned with going beyond the inventory to description proper. In the best of cases, which is to say when one can easily reconstitute the direction of his circuit, Kant seems to proceed to a spiraling description, as is the case for Asia, leaving from one country in order to return after having passed through a series of intermediate countries: in this case the circle restores to the description the order of a circuit which, in the imagination, can be followed in whatever direction one likes and which can, without loss, be followed in the direction opposite to the linear direction which Kant provides. One can describe Asia in making a loop that descends to the south, the southwest, then climbs back towards the northwest and the northeast, or the reverse.

By these presuppositions, which thus remain largely implicit and are not in the end always followed, this description only truly takes direction if one is able to relate it to a map of the concerned territories. And it is the same way for Europe and America. The text itself thus is not sufficient for producing a description of its object since, insofar as it is a simple inventory, it does not provide the means for organizing in the imagination the manner in which countries are arranged in relation to each other.

Of the Map

This thus raises the problem of the relationship of geography to the map. Kant's text is not accompanied by maps, and we know that his course was done without maps, even though since the Renaissance and the rediscovery of Ptolemy's maps cartography had made considerable progress. The geographical map is not even the object of an analysis. In this respect, one can ask to what extent the text of the description according to space can pretend to give an account of its object by the discourse alone, or if implicitly it doesn't refer back to its obligatory companion of the map, without which the manner in which places are side by side with each other remains largely indeterminate. This returns to an evaluation of the respective performances of the image and the discourse (text and speech), as Lessing engaged it in his *Laocoön* where, as we know, he defended the idea that the arts cannot represent objects that are not homogenous to them in some respect, as "these signs must indisputably bear a suitable relation to the thing signified."[28] From this fact, according to Lessing, poetry can only represent that which itself has a duration (acts, objects, or elements that succeed one another), whereas painting can only represent that which is spatial (bodies).[29] Consequently, the very idea of a description according to space would return, according to this distinction, to the attempt to instate between signs and discourse and the (spatial) reality that they aim to portray a relation which, by the fact of their difference, could not be suitable. The attempt would thus be fated to failure, and pure geographical discourse would be thus revealed to be a forced discourse.

Concerning the utility of maps, Kant first responds in a certain manner in his *Lectures on Pedagogy*, where he presents them as a stimulating element for children, to the extent that they contain and train the imagination in and by precise, nonfictional figures.

However, one can ask oneself if the map is only a transitory element for the formation of the imagination, or if it is presupposed or required constantly for all work on space, either explicitly or implicitly, inasmuch as memorized maps would make orientation in geographical space possible.

In the text "History and Physiography of the Most Remarkable Cases of the Earthquake which Towards the End of the Year 1755 Shook a Great Part of the Earth," Kant says that the western coasts are the most exposed to earthquakes in Italy, Portugal, South America, and Ireland. The cause of this would be that the west and south coasts of almost all countries are more abrupt than the east and north coasts, which would be confirmed by *maps* and information furnished by voyagers (Dampier).[30] Elsewhere, in the presentation of his theory of winds, Kant notes that, for example, in order to understand the winds of the Guinea Coast it is enough to observe the map that Jurin included in Varenius's *General Geography* or that Musschenbroek included in his *Elementa Physica*. So, says Kant, if one has in mind the rule according to which a higher temperature in a region provokes a wind directed towards this region which lasts as long as it remains hotter,[31] looking at the natural, habitual wind from the east on a map one understands/sees (*einsehen*) and explains (*erklären*) in a blink of an eye all the directions of the wind that blows on the sea near the Guinea Coast, as well as tornadoes and all the rest.[32] It thus seems clear that the map is, at least, an instrument that *facilitates* comprehension of the descriptions.

Finally, in the *Anthropology*, the map is cited as a mnemonic device in the context of a "faculty of visualizing the past and the future by means of the power of imagination,"[33] which is to say that it is related to the faculty of memory. Paradoxically, the map is thus brought back to a temporal destination, that of memorization, and not to a spatial function. In this passage of the *Anthropology*, Kant in effect distinguishes several methods of memorization of which one is based in judgment and consists in making tables of systematic classification, such as that of Linnaeus. We recover a forgotten thing by enumerating the elements we remember in order to find those we have forgotten; or we visually represent to ourselves the table of the divisions of a whole, or a map that partitions the totality of a space into provinces situated to the north, in the west, etc.[34] Here the map is thus clearly an instrument that aims at an end other than that of perception and comprehension of space. In the same way that, for the child, it is a useful means to form the imagination in order to domesticate it, here the map is an instrument that aims at recalling the past rather than at representing space in itself. Paradoxically, the image returns to an art of memory.[35] And thus, in spite of a few allusions to maps, it seems that they are not indispensable to the representation of space, that the text is sufficient and succeeds at producing spatial impressions. But up to present, this remains a postulate to confirm and justify.

The question that poses itself now is to know whether, for Kant himself, we find directions permitting justification of the project of a description according to space, which is to say a description capable, by itself, of articulating the spatial character of its object, without recourse to the image—which would justify theoretically Kant's attempt to realize what Lessing considers illegitimate.

In "Concerning the Ultimate Ground of the Differentiation of Directions in Space," Kant noted that our geographical knowledge and our knowledge of the sky would be of no use to us if we could not determine "the entire system of their reciprocal positions, by referring them to the sides of our body."[36] Kant thus does give a precise instruction on the conditions under which a description according to space—"geographical knowledge"—is of use, which is to say is capable of success. It must be referred to our bodies and to a "region." Put otherwise, if we only know "the positions of objects relative to each other," in the manner provided by a purely conceptual analysis of sites of a Leibnizian sort, we will not yet accede to a true knowledge according to space. Besides simultaneity (the order of coexistents), description should thus permit reference of the object to the body "according to the regions." Yet regions are nothing other than determinations of space in function of the relations between parts of space and the sides of our bodies. Consequently, there is no description capable of producing a geographical knowledge except for that which permits, at least in the imagination, situating one's body with respect to the object that it aims to know.

Here again, it is s necessary to bear in mind the fact that there is no geography without corporeal geography, in other words that there is no pure discourse adequate to an object subsisting by itself as a thing in itself, but that, between the discourse and the object, space intervenes such that it manifests itself with the body or bodies as a nonconceptual form which, however, alone makes the object comprehensible by giving sense to the discourse which relates itself to the object by reference to the body. Reference to right and left, above and below is therefore, as much or more than the simple abstract presentation of an order of coexistence, the indispensable intuitive mark for orientation in space, without which, consequently, there is equally no description according to space. In relation to such an exigency, the lack of regular and precise spatial indications in Kant's text would thus seem accidental, in the sense that such a description seems possible, even if in fact these indications hardly received Kant's attention.

This being the case, there remains the question of knowing if it is equally possible to represent the earth as a system of regions oriented only by the resources of discourse, without recourse to a map which would permit, in a worst-case scenario, the organization of space on a human scale in order to understand that which, in reality, unfolds on a planetary scale. The question comes back to asking if, by the discourse, a satisfactory spatial representation can be produced in the imagination which, as the Introduction announces, anticipates future experience, or if spatial representation is always of the order of a certain passivity, and based therefore in an essentially empiricist explication, for which (spatial) experience could not, in fact, be anticipated in any manner, even imaginary.

Kant claims in his Introduction to the *Physical Geography* that it is necessary to expand our experience by geography, beyond our individual finitude such that this acquisition of knowledges might permit the anticipation of our future experience which, Kant says, we will have "in the world."[37] Kant thus evokes the necessity of acquiring a preliminary *concept* (*Vorbegriff*) of all things, of a *overview* (*Abriss*) and of a *plan* (*Plan*). Yet, concerning geography, the knowledges thus anticipated are surely for one part of a conceptual order, but also, as we have seen, of the order of spatialization in the sense of a nonconceptual disposition of the object of geography in space. This object thus cannot be broached in a solely logical manner and one cannot fashion for oneself a simple preliminary *concept*. At one moment or another, the passage from the concept to the sensible is necessary. It is also a sort of *preliminary space*, which would seem to be required, a space which is not yet that of the world, but which would be properly that of the geographer.[38] Consequently, here it is necessary to use a faculty other than pure understanding to apprehend the geographical object: namely, the imagination, as the faculty of spatial intuitions outside of the presence of the object.

The elementary characteristics of orientation in space are, as we know, presented in an exemplary manner in the short text *What Is Orientation in Thinking?* which, in relation to the text, *Concerning the Ultimate Ground*, already cited, displaces space towards the subject, making the subject an a priori form of sensibility. In effect, we know that in the text of *Concerning the Ultimate Ground* space is characterized by the fact that, as in the period that followed the *Dissertation of 1770*, space is not purely conceptual and arises from aesthetics rather than logic; but as distinguished from what would hold in the critical period, space is not linked to the body/subject, which is to say to the sensibility of the latter, but to the body/object taken as a necessary reference and the first, if objective, of all orientation, which is to say of all comprehension of the disposition of objects in regions. While in the short text of 1768 regions are inscribed in absolute space and, as Sylvain Zac notes, "the foundation of symmetrical objects is not found in space, pure intuition, the *a priori* form of sensibility, but in the very reality of extended things,"[39] during the critical period space ceases being amenable to definition by regions based in differences inscribed in things themselves in absolute and original space,[40] and returns radically to the subject. Consequently, the subject seems subsequently to have in itself all the elements necessary for the production of a pure spatial intuition based in the suppositions furnished by concepts. We know that it is thus that are produced mathematical objects, which can be constructed and presented in the in the intuition without having been first intuited empirically. I can thus relate objects to oriented regions that are no longer determined with reference to the body but rather to the sensibility of the subject; and in this sense the experience of (spatial) dispositions can be made in the absence of any empirical referent, as I can represent left and right to myself in a totally abstract manner, in my imagination, without needing thereby to refer to my body (it is, to the contrary, this a priori possibility which makes possible the differentiation of the body according to left and right). The orientation, which ought to make possible a description according to space, is the relation of points of reference proper to the observer with the regions of the world which he apprehends, which signifies first of all the relation of right and left (on the side of the observer) with the four cardinal points (on the side of the explored world).[41] Thus again, we see that *conceptual description according to space* is not sufficient for acceding to the level of geographical study, which is to say for passing from the logical table to geography proper as a genuine description that has "integrated" space. There is only an entry into space if the situated reader, or the listener, or the student—the subject—can (re)produce by his own lights, subjectively, the space explored and described by geography.

Thus, does Kant furnish instructions that permit a forecast of the capacity of the subject to construct a representation of the earth solely from that given in a text, including the instructions required for a description according to space? The question here returns to the evaluation of the performances of the producing imagination, insofar as one only apprehends correctly what one knows how to do, as Kant claims in his *Lectures on Pedagogy*, noting that "one understands a map best when one can draw it oneself," as "the biggest aid to understanding something is to produce it."[42] But, during the event, it is not a map that I am supposed to produce but, in anticipation, an experience of the earth itself, which at least poses a problem of scale as one can imagine that, by its size, the earth and its regions exceed the powers of the imagination. In effect, even if the earth can be measured and has been measured, this measurement corresponds to nothing determined by the imagination, in the sense that, in order to measure the earth in the imagination, it would be necessary to be able to unify in one representation the sum of necessary perceptions or apprehensions in order to be able to construct a whole from them. To construct such a space, it would be necessary for the unities or the impressions first posited to remain sensible while one posits the subsequent ones, such that the "understanding" of the earth or of its parts (continents) would consist in holding together that which is progressively posited by the apprehension. Without this, there is logical knowledge but not aesthetic experience. Yet "all estimation of the magnitude of objects of nature is aesthetic."[43]

At a certain size, the imagination is no longer capable of effecting this sensible synthesis: "for the aesthetic imagination of magnitude there certainly is a greatest [limit]."[44] If I want to grasp a space as such, I must in effect convert it into an intuition by which all parts are retained simultaneously at the end of the construction. This is also what it would be necessary for me to do, although I cannot, for the earth of which, for example, Kant says, the diameter can very well be apprehended in a logical sense (*comprehensio logica*) but of which there is no possible "composition [. . .] in an intuition of the imagination"[45] (*comprehensio aesthetica*); or again, to put it in the language of the Analogies of Experience: I would be incapable of constructing a sequence of perceptions, as the beginning of the series would be lost before its end was produced: thus *a fortiori* it would be impossible to complete the return trajectory which guarantees the simultaneity of the existence of all the elements of the perception; this simultaneity would thus never be acquired under the heading of experience.

On the other hand, the space of the map is largely within the maximum limit of the imagination and is amenable to being grasped and related in an *analogical* mode to the space of which it aims to give an account, such that instead of being disoriented, the (sensible) knowing subject can construct analogically, on a reduced model, his imaginary experience of the greater space, without losing the essence of the relations between places, as the change of scale does not affect these relations with the sensibility of the subject.[46]

The general question of knowing what a geographical description according to space might be presents two aspects, which are (1) a problem of scale (is not the space to be imagined too large?) and (2) a problem of principle (can one, in a general manner, produce spatial representations when beginning with concepts?). Yet we know that for Kant, with math, passing from the conceptual to the sensible form is a performance of which the human mind is capable, contrary to what the empiricists might believe. The schematism, "a hidden art in the depths of the human soul,"[47] realizes this "presentation" of the concept to an intuition. But our imagination is only capable, like our empirical perception, of relating itself to objects whose size does not exceed a certain threshold and which, with respect to the earth, are like reduced models, put otherwise, that have the format of a map, which is the perceptual, imaginary face-to-face encounter adequate for man's experience of the world by anticipation. In fact, in the "Preliminary Mathematical Concepts" that opens the *Physical Geography*, Kant presents a structuration of the earth operated on an imaginary spheroid, which is to say on an abstract, mathematical (geometrical) object serving as a reduced model of the real earth.

Taking up the language of classical rhetoric, Kant calls the presentation of a concept a "hypotyposis;"[48] in the *Critique of the Power of Judgment*, Kant distinguishes two forms of this figure: it is *schematic* when, for a concept grasped by the understanding, the corresponding intuition is given a priori (and such a schematic hypotyposis is produced with "preliminary mathematical concepts"); or else it is *symbolic* when, for a concept that only reason can think, and to which no sensible intuition could be adequate, an intuition is attributed with which the power of judgment can operate in a manner analogical to that in which it would if it were capable of schematizing the concept.[49]

In the case of objects whose size surpasses the capabilities of the imagination, which thus cannot be presented, one must therefore have reference to such symbols. In this sense, description is a symbol. However, the concept to be presented is not in itself necessarily unrepresentable: it is only such because of the fact of the (contingent) limits of our imagination. Earth is in effect not a concept that only reason can think: it is an empirical reality whose corresponding concept is an *empirical concept*;[50] geography thus returns all at once to the example, for its empirical character, and to the symbol, when its object is not directly intuitionable because of its size. Description according to space in a text thus has a character at once exemplary and symbolic which is related to extension in space (in the world) in an analogical way. In this way geography invites us to add to the temporal analogies of experience spatial analogies which would relate the discourse and the world by the intermediary of an intuitive experience in the imagination, which can be an image or a map.

Kant, however, kept the image away from his treatment, and seems paradoxically to have restricted himself in geography to an essentially discursive, which is to say logical, "aesthetic." We can ask about the motive of the choice on the part of the thinker who gave to aesthetics and the imagination a place which up to then had been refused by all of the (rationalist) theorists of objective knowledge.

An indication is perhaps given to us in Kant's texts on poetry and painting. Poetry, in effect, is presented as the art which gives to the concept a presentation that plays with appearance without rendering it misleading.[51] Put otherwise, the concept in the discourse is held in the same way to not mislead or to mislead as little as possible in its production of forms. Painting, which presents a sensible appearance, is characterized by Kant as an art which only gives "the appearance of physical extension" (*gibt nur den Schein der körperlichen Ausdehnung*). Put otherwise, Kant seems, as opposed to Lessing, to hold it as a matter of indifference that painting has an affinity with space and is more homogenous with it than, for example, poetry. When, in §51 of the *Critique of the Power of Judgment*, Kant presents painting as the art of the beautiful description of nature, this description is *Schilderung*, display, and not *Beschreibung*, written, descriptive labor. Subsequently, we understand that maps might be valorized more by the approval they procure as displays than by the knowledge that they might be able to transmit. Whatever Lessing might have said, the canvas, giving only the *appearance* of physical extension, would not prevail over discourse for the description of space. Discourse would justly remain the best means of transmitting knowledge of space, since Kant has no particular reservations about the capacity of language to enlarge the experience of the subject by a *description* of space, of which he never says that it would only give the *appearance* of "physical extension."

Whatever its relation to intuition, geography thus remains, for Kant, a discourse first of all. The price of this option is perhaps that it remains in spite of him more of an inventory than a genuine description. But being otherwise would have doubtless required a constant attention to the proper directions for creating a genuine orientation, a conception of another form of writing, mixing the image with the text, the icon with the legend, a new hieroglyph.

Notes

1. Kant proposes utilizing geographical maps for entertaining children while instructing them, cf. *Lectures on Pedagogy*, trans. Robert B. Louden, in *Anthropology, History and Education*, eds. Günter Zöller and Robert B. Louden (Cambridge: Cambridge University Press, 2007), AK 9: 476. As with all references to the Cambridge Edition, references are to the marginal pagination, which refers to the Akademie edition.

2. Kant, *M. Immanuel Kant's announcement of the program of his lectures for the winter semester 1765–1766*, in *Theoretical Philosophy, 1755–1770*, ed. and trans. David Walford and Ralf Meerbote (Cambridge: Cambridge University Press, 1992), AK 2: 312–3.

3. *Physical Geography*, trans. Olaf Reinhardt (Cambridge: Cambridge University Press, forthcoming), AK 9: 157.

4. On prudence, see for example Kant, *Groundwork of The metaphysics of morals*, in *Practical Philosophy*, eds. Mary J. Gregor and Allen Wood (Cambridge: Cambridge University Press, 1996), AK 4: 416ff.

5. Decree of von Zedlitz of October 16, 1778; see Michelle Cohen-Halimi's introduction to the *Géographie physique* (Paris: Aubier, 1999), 11.

6. [*Trans*: The German *Erzählung* does literally mean "recounting." Marcuzzi usually translates this as *récit*—narrative—but here he uses the word *raconter*. Like the English and French "description" the German *Beschreibung* implies a writing.]

7. *Physical Geography*, AK 9: 160.

8. *Physical Geography*, AK 9: 160.

9. Leibniz, Letter to Nicolas Remond, July 1714, in *Philosophical Papers and Letters*, ed. and trans. Leroy E. Loemker (Chicago: University of Chicago Press, 1956), vol. 2, 1066; *The Leibniz-Clarke Correspondence*, ed. H. G. Alexander (Manchester: Manchester University Press, 1956), 26.

10. *Critique of Pure Reason*, trans. Paul Guyer and Allen W. Wood (Cambridge: Cambridge University Press, 1998), A211/B257.

11. *Physical Geography*, AK 9: 173n.

12. "Thus Philadelphia, for example, would have a longitude of 320 degrees east, although this town is at a distance of only 40 degrees from the first meridian if we were to count the degrees back from the east." *Physical Geography*, AK 9: 173–4.

13. *Physical Geography*, AK 9: 161.

14. *Physical Geography*, AK 9: 162.

15. *Physical Geography*, AK 9: 162.

16. "On the Use of Teleological Principles in Philosophy," in *Race*, ed. Robert Bernasconi (Cambridge: Blackwell, 2001), 39.

17. "On the Use of Teleological Principles in Philosophy," 40, n. 1.

18. "On the Use of Teleological Principles in Philosophy," 39.

19. "Methodology of the Teleological Power of Judgment," in *Critique of the Power of Judgment*, ed. Paul Guyer and Eric Matthews (Cambridge: Cambridge University Press, 2000), AK 5: 428n.

20. "Of the different races of human beings," trans. Holly Wilson and Günter Zöller, in *Anthropology, History and Education*, ed. Robert B. Louden and Günter Zöller (Cambridge: Cambridge University Press, 2007), AK 2: 443.

21. *Physical Geography*, AK 9: 276.

22. On springs, cf. Descartes, *Principia philosophiae*, IV, §64ff.

23. *Physical Geography*, AK 9: 298.

24. *Physical Geography*, AK 9: 162.

25. *Physical Geography*, AK 9: 160.

26. "Of the different races of human beings," AK 2: 430.

27. *Physical Geography*, AK 9: 314. [*Trans*: Marcuzzi uses the term "*physiciens*" at the beginning of this citation to translate the German term "*Physiker*," translated by Reinhardt as "natural philosophers."]

28. Lessing, *Laocoön*, trans. Edward Allen McCormick (Baltimore: Johns Hopkins University Press, 1984), 78.

29. Lessing, *Laocoön.*

30. "History and Physiography of the Most Remarkable Cases of the Earthquake which Towards the End of the Year 1755 Shook a Great Part of the Earth," trans. John Richardson, rev. Stephen Richard Palmquist, in *Four Neglected Essays by Immanuel Kant* (Hong Kong: Philopsychy Press, 1994), AK 1: 459.

31. "Magister Immanuel Kant's New Remarks Offering an Explanation of the Theory of the Winds: by which he invites to his lectures," excerpts translated by Gabriele Rabel in *Kant* (Oxford: Clarendon Press, 1963), 33–4; see AK 1: 492.

32. "Theory of the Winds," AK 1: 493.

33. *Anthropology from a pragmatic point of view*, trans. Robert B. Louden, in *Anthropology, History and Education*, AK 7: 182.

34. *Anthropology from a pragmatic point of view*, AK 7: 184.

35. Cf. Cohen-Halimi, "Introduction," 30.

36. "Concerning the Ultimate Ground of the Differentiation of Directions in Space," in *Theoretical Philosophy, 1755–1770*, AK 2: 380. [*Trans:* The English translation of Kant's text footnotes, but does not translate, the phrase that concludes the sentence in German (*nach den Gegenden*). Marcuzzi, using the French translation by Sylvain Zac in *Quelques opuscules précritiques* (Paris: Vrin, 1970), includes "according to the regions" (*suivant les régions*) in the citation, thus the subsequent reference in the text to "regions."]

37. *Physical Geography*, AK 9: 157.

38. Space being one of the two a priori forms of sensibility, it would not be possible for there to be two spaces. But there can be two manners of relating to one space, a natural and an artificial manner, according to whether one relates to the world or to imaginary substitutes for the world.

39. *Physical Geography*, AK 9: 136.

40. "Differentiation of Directions in Space," AK 2: 383.

41. *What Is Orientation in Thinking?*, in *Kant: Political Writings*, ed. Hans S. Reiss, trans. H. Barry Nisbet (Cambridge: Cambridge University Press, 1991), 238.

42. *Lectures on Pedagogy*, AK 9: 477.

43. *Critique of the Power of Judgment*, AK 5: 251.

44. *Critique of the Power of Judgment*, AK 5: 251.

45. *Critique of the Power of Judgment*, AK 5: 254.

46. "Differentiation of Directions in Space," AK 2: 380.

47. *Critique of Pure Reason*, B181.

48. This figure presents problems in a critical framework from which they did not originally arise as, while from a precritical perspective the sensible could be considered homogenous to the intelligible (the concept being, for Leibniz, a clear idea and the sensible a confused idea), the critical framework (which clearly distinguishes logic and aesthetics) instates a radical difference between them. Consequently, one cannot pass *insensibly* from the concept to the image by obscuration, or from the image to the concept, by illumination. Understanding how a description according to space is possible thus returns to understanding how the hypotyposis can transcend the difference between these two heterogeneous orders.

49. *Critique of the Power of Judgment*, AK 5: 351–2.

50. *Critique of the Power of Judgment*, AK 5: 253–5.

51. *Critique of the Power of Judgment*, AK 5: 321ff.

III

Towards a Cosmopolitan Education

Geography and Anthropology

8

"The Play of Nature"

Human Beings in Kant's *Geography*[1]

Robert B. Louden

A New Course

In the summer semester of 1756—Kant's second semester as a lowly *Privatdozent* at the University of Königsberg—he began lecturing regularly on physical geography, a practice that he continued until he retired from teaching in 1796, and one that grew directly out of his first publications in geography and natural science. Geography was a new academic discipline at this time, and Kant was one of the very first university lecturers to offer independent lectures in this field.[2]

Broadly speaking, Kant's geography lectures match Charles Withers's description of Enlightenment geography as a project that "in one way or another reflected the voyages of exploration and discovery that characterized the period and provided a wealth of information about the natural diversity of the world and its peoples;" an enterprise that was "part of the process of enlightening the world through its inventory and description."[3] But once we look below the surface, we find features of the course that mark it as uniquely Kantian. What were the distinctive features of Kant's geography course, and what were his main aims in presenting these lectures?

First, the geography course was unusual in that it was based not on an official textbook, as was generally required at all Prussian universities at the time, but rather on Kant's own intentionally eclectic collection of materials, to which he added and subtracted over the years. As he notes in his first *Announcement* about the course, published in 1757: "useful information is scattered over numerous and voluminous works, and a textbook suitable for academic use is still lacking. Therefore I decided at the very beginning of my academic teaching to report on this science in special lectures [*in besonderen Vorlesungen*] after the guidance of a summary sketch."[4] And in two later *Announcements* about the course published in 1758 and 1759, he also makes a point of saying that "I have read aloud from my own essays"[5] and "lectured from my own manuscript."[6] Indeed, in October 1778, Minister of Education Karl Abraham von Zedlitz specifically exempted Kant's physical geography course from the traditional required-textbook regulation:

> The worst compendium is certainly better than none, and professors may improve upon the author as much as they can, but lecturing from one's own notes [*das Lesen über Dictata*] must absolutely be stopped. However, from this Professor Kant and his lectures on physical geography are exempted, because it is known that no entirely suitable textbook is yet available.[7]

At any rate, what several recent commentators have bemoaned as the "intellectual and political embarrassment;" the "unbelievable hodge-podge of heterogeneous remarks, of knowledges without system, of disconnected curiosities"[8] in Kant's physical geography lectures seems to have been part of his intent all along. One of his key aims was simply to summarize current developments in the emerging field of geography. As he remarks in the 1757 *Announcement*, he wanted to explore topics with students "not with that completeness and philosophical exactitude in areas that is the business of physics and natural history, but with the reasonable curiosity of a traveler who seeks out everywhere what is noteworthy, peculiar, and beautiful, and then compares his collected notes and reflects on his plan."[9]

This last quotation points to a second distinctive feature: Kant's geography course was designed to be popular and entertaining—not dry and rigidly academic. For instance, in the 1765–66 *Announcement* for the course, he relates that he wants to make geography "into an entertaining and easy compendium [*Inbegriff*]"; a "great diversity of entertaining, instructive, and easily understood knowledge," which will be well suited to a "sociable century."[10] Similarly, at the end of the Introduction to Rink's edition of the *Physical Geography* lectures (which, according to Adickes, stems from a 1775 lecture),[11] Kant remarks that his lectures "serve for our own enjoyment [*Vergnügen*] and provide rich material for social conversation."[12]

At the same time, the goals of popular entertainment and enlightenment through science are by no means mutually exclusive. Kant also envisioned his geography course as a kind of popular science for laypeople, and this popular science goal constitutes a third distinctive feature. For instance, he opens the 1757 *Announcement* by announcing that "the rational taste of our enlightened time" has brought us to a point where we are now "no longer in danger of losing ourselves in a world of fables [*Welt von Fabeln*] instead of attaining a correct science of natural curiosities [*natürliche Merkwürdigkeiten*]."[13] And in the 1765 *Announcement* he stresses that the first part of the course, which explores "the natural relationship that holds between all the countries and seas of the world, and the ground for their connection, constitutes the real foundation [*das eigentliche Fundament*] of all history, without which history is scarcely distinguishable from fairy tales [*Märchenerzählungen*]."[14]

Fourth, Kant's physical geography lectures were also intended to be useful and pragmatic. Indeed, though readers normally associate the term "pragmatic" with Kant's 1798 work, *Anthropology from a Pragmatic Point of View*, if his own declining health had not prevented him from editing the geography lectures on his own,[15] he could just as well have entitled them *Physical Geography from a **Pragmatic** Point of View*. For instance, in the 1758 *Announcement*, he describes the course as a "useful [*nützlich*] and agreeable science,"[16] while in 1765 he states that the course should prepare students "and serve them for the exercise of practical reason;" adding later that he has recently introduced changes in the lectures "which are of even greater utility [*noch gemeinnütziger*]."[17] At the beginning of the 1775 *Announcement*, he calls the geography course "more of a useful entertainment [*nützliche Unterhaltung*] than a laborious business,"[18] and at the end, in one of his most famous descriptions, he calls geography a form of

> useful academic instruction [*nützliche akademische Unterricht*] . . . which I may call the preliminary exercise in the *knowledge of the world*. This knowledge of the world serves to procure the *pragmatic* element for all otherwise acquired sciences and skills, by means of which they become useful not merely for the *school* but rather for *life* and through which the accomplished apprentice is introduced to the stage of his destiny, namely, the *world*.[19]

Similarly, in the Introduction to Rink's edition of the lectures, Kant stresses that he intends to show students "how to make our knowledge *practical*[20] [*das **Praktische** zu geben*]. And this is *knowledge of the world*."[21] And he concludes the Introduction by stressing that "the usefulness [*der Nutzen*] of this study is very extensive."[22]

Related to this emphasis on usefulness and pragmatic applications is a fifth feature that I call the *orientation-aim* of the lectures: Kant wanted to give students an empirically informed orientation toward the world at large; a sense of what to expect after they leave the narrow confines of home and school.[23] For instance, in the Introduction to Rink's text he criticizes the cosmopolitan tourist who thinks he has acquired *Weltkenntnis* simply because he has been fortunate enough to travel widely: "there is more to knowledge of the world than just seeing the world. He who wants to draw utility [*Nutzen ziehen*] from his travels must have a plan [*ein Plan*] in advance of his travel."[24] Granted, as he acknowledges later, "through travel one can broaden one's knowledge of the external world, but this is of little use if one has not already received a certain preliminary exercise through instruction [*durch Unterricht eine gewisse Vorübung*]."[25] Kant's geography course—along with the later anthropology course, which to a certain extent grew out of the geography lectures, before splitting off as an independent course in 1772—was supposed to provide students with this preliminary exercise. Without this preparatory orientation, the results of the cosmopolitan's travels will form merely an unsystematic aggregate;[26] a hodgepodge that "can yield nothing more than fragmentary groping around and no science."[27] This need to orient oneself by means of an accurate conception of the world and its inhabitants is stressed particularly in the 1765 *Announcement*, when Kant describes "the *second* part" of the geography course,

> which considers the *human being*, throughout the world, from the point of view of the variety of his natural properties and the differences in that feature of man which is moral in him [*was an ihm moralisch ist*]; a very important and also highly stimulating consideration, without which universal judgments about the human being would scarcely be possible, and where the comparison of human beings with each other and with the moral state of earlier times furnishes us with a comprehensive map of the human species [*eine große Karte des menschlichen Geschlechts*].[28]

Finally, related to the orientation-aim is what Kant himself calls the *cosmological* aim of his geography lectures. Both parts of *Weltkenntnis* (nature and the human being) are to be treated "cosmologically;" that is, from a broad, holistic perspective rather than a partial or local one. The cosmological goal is to acquire an overall sense of nature as a systematic and integrated whole, so that we may better find our way in the world and in our interactions with other people. In the 1775 *Announcement*, for instance, Kant stresses that both nature and the human being "must be considered *cosmologically*, namely, not with respect to the noteworthy details that their objects contain (physics and empirical psychology) but with respect to what we can note of the relation as a whole in which they stand and in which everyone takes his place."[29] The main stress of the adjective "cosmological" is on the need to understand the whole and how the parts relate to the whole. Without this broader perspective—which Kant believes must precede travel and social interaction rather than issue out of it—students are lost. Without a preparatory cosmological outline, any information about the world and its products and inhabitants gleaned from travel will constitute merely an unsystematic aggregate; a "fragmentary groping around" that can never yield science or true understanding. As Kant states in the Introduction to Rink's text, "we must know the objects of our experience *as a whole* so that our knowledge does not form an *aggregate* but rather a *system*; for in a system the *whole* is there before the parts, on the other hand in an aggregate the *parts* are there earlier."[30] At the same time, this cosmological whole is arrived at empirically—it contains not the "strict universality [*strenge Allgemeinheit*]" that Kant identifies as one of the "sure signs [*sichere Kennzeichnen*] of an *a priori* cognition" in the first *Critique*, but only "empirical universality."[31]

Human Beings in Kant's Geography and Anthropology: A Distinction without a Difference?

In 1772, Kant began to separate anthropology from geography, offering an annual course in each subject after this point. But drawing a clear line between these two new Kantian disciplines is easier said than done. One might suppose that geography simply concerns "earth-description" (*geo*, earth + *graphein*, to write; *Erdbeschreibung*)[32] and that anthropology is the study of human beings (*anthropos*, human being + *logos*, word, reason; *Anthropologie*) and leave it at that. And Kant himself does occasionally distinguish the two disciplines in this manner. For instance, in the the *Pillau* anthropology lectures (1777–78), he states: "In physical geography we consider nature, but in anthropology the human being, or human nature in all of its situations. These two sciences constitute knowledge of the world."[33]

However, because human beings are also part of nature and because they also live on the surface of the earth, this way of distinguishing the two disciplines won't work. "Geography," as defined by both Kant and contemporary geographers, also considers human beings.[34] For instance, in the 1757 *Announcement* he states that "physical geography examines only the natural condition of the earth, and what is located on it: seas, continents, mountains, rivers, the atmosphere, the human being [*der Mensch*], animals, plants, and minerals."[35] And later in this same *Announcement* he notes that he intends to take up

> in the geographical way of teaching all the countries of the earth, in order to display the inclinations [*die Neigungen*] of human beings, as they flow from the region in which they live; the variety of their prejudices and way of thinking [*Vorurteile und Denkungsart*], in so far as all of this can serve to make the human being more closely acquainted with himself; a brief idea of their arts, commerce, and science, a description of the products of the countries, . . . in a word, everything which belongs to physical geography.[36]

In Rink's edition of the *Physical Geography*, human beings are also featured prominently. In the Introduction, Kant states that his course will deal with five different types of geography: *mathematical, moral, political, economic* [*merkantilische*] and *theological*.[37] All but the first of these five geographies (which deals with "the shape, size, and motion of the earth, and its relationship to the solar system in which it is located"[38]) clearly concern human beings. And at the beginning of the Second Part of these lectures, there occurs a ten-page section entitled "Concerning the Human Being."[39] Human beings also make frequent appearances in other parts of Rink's text. For instance, they are described as one of the principal causes of changes in the shape (*Gestalt*) of the earth,[40] the importance of knowledge concerning coastlines, depths of oceans, causes of winds and storms, etc. to sailors is stressed,[41] various trees and plants are discussed in terms of their usefulness to human beings,[42] and his concluding discussion of the countries of the earth is largely a comparative survey of human cultures.[43]

In short, in all descriptions and records of Kant's physical geography lectures that are currently available to us, human beings play a prominent role. It is simply not the case that Kant's physical geography course deals exclusively with the earth itself, nor is it the case that his anthropology course deals exclusively with human beings. (For instance, in the *Anthropology* he occasionally discusses non-human animals.[44]) Insofar as Kant's physical geography and anthropology lectures are both concerned with human beings, there is considerable overlap between the two disciplines.

Perhaps we can then distinguish them by emphasizing the *particular manner* in which they treat human beings? Emil Arnoldt, for instance, claimed that Kant's

> physical geography chiefly takes into consideration uncivilized human beings; the anthropology, civilized human beings, and . . . the difference between the consideration of the human being in the physical geography and in the anthropology in general can thus be formulated as follows: physical geography considers human beings primarily from the outside, anthropology from the inside.[45]

But we have seen already that this is false. Kant planned to investigate various aspects of "civilized" human life (e.g., economics, politics, ethics, theology) in his geography lectures, and his anthropology lectures by no means consider only "civilized" human beings. For instance, in the *Anthropology* Kant discusses the patience and courage of the Indians of America, and asks later whether their forehead, "overgrown with hair on both sides, is a sign of an innate feeble-mindedness."[46] *Contra* Arnoldt, "the line of demarcation between the physical-geographical and anthropological consideration of the human being" is by no means "easy to draw."[47]

Similarly, it is often asserted that Kant's anthropology considers human beings *pragmatically*, whereas his physical geography treats them non-pragmatically—as

May puts it, "in a physical and 'external' sense" that is concerned only "with unreflective behaviour;" "as a part and product of nature, viewed from the perspective of the natural environment."[48] Or, as another commentator remarks, in Kant's geography lectures the human being is considered not "as a free being, but rather an inhabitant of the earth like plants, animal, and minerals—it considers man as one type of 'thing' on earth."[49] But this attempt to demarcate the two disciplines, while closer to the truth than Arnoldt's, is also problematic for at least two reasons.

First, as noted earlier, Kant's physical geography lectures are also pragmatic. Geography (like anthropology) "serves to procure the *pragmatic* element for all otherwise acquired sciences and skills" and is "useful not merely for *school* but rather for *life*;"[50] geography (like anthropology) is a "useful and pleasant science;"[51] geography (like anthropology) is intended to "prepare and serve students for the exercise of a practical reason."[52]

Second, when Kant defines "pragmatic anthropology" in *Anthropology from a Pragmatic Point of View* he is not distinguishing geography from anthropology, but rather the *physiological* anthropology of Ernst Platner (1744–1818) and other "philosophical physicians" from his own anthropology. Physiological anthropology, Kant asserts, "concerns the investigation of what *nature* makes of the human being; pragmatic, the investigation of what *he* as a free-acting being makes of himself, or can and should make of himself."[53] However, all versions of Kant's anthropology lectures contain extensive discussions of human characteristics that stem from their biological composition rather than their free choices. For instance, much of the material on individual temperament and (physical) character, as well as the subsequent material on the character of the sexes, the peoples, and the species, is not about free human action but rather about what nature makes of human beings. What Kant chiefly objects to in the physiological approach to anthropology is its "eternally futile inquiries as to the manner in which bodily organs are connected with thought"[54]—that is, its ungrounded speculations concerning the causal processes of human thought. What he proposes in its place is an empirical account of human nature that emphasizes pragmatic applications. But insofar as major portions of the anthropology lectures view human beings as products of nature, we also cannot distinguish Kantian geography from anthropology by asserting that the former treats human beings as products of nature while the latter does not. Both sets of lectures at times treat human beings as products of nature. Similarly, when Kant in his geography lectures sets out to examine "the differences in that feature of the human being which is *moral* in him [*was an ihm moralisch ist*],"[55] it would seem that he must at least occasionally consider free human choices.

Several commentators have also tried to distinguish Kant's geography from anthropology by means of his doctrine of outer and inner sense. May, for instance, writes:

> this distinction [viz., between outer and inner sense] is of crucial importance for . . . [Kant's] separation of anthropology from geography, since the world as object of outer sense is nature, and hence the concern of geography, whereas the world as the object of inner sense is man conceived as soul or self, and is the concern of anthropology.[56]

Similarly, Emmanuel Eze contends that "while anthropology studies humans or human reality as they are available to the *internal* sense, geography studies the same phenomena as they are presented or available to the *external* sense."[57] And there is at least one Kantian text that supports this reading. In Rink's edition of *Physical Geography*, Kant states: "The world as *object of outer sense* is **nature**; as *object of inner sense*, however, it is **soul** or *the human being*. . . . *Anthropology* teaches us *knowledge of the human being*; *knowledge of nature* we owe to *physical geography* or *description of the earth* [*Erdbeschreibung*]."[58]

However, this attempt to demarcate geography from anthropology also fails for at least two reasons. First, what we said earlier about Kant's treatment of temperament, the sexes, peoples, and the human species at large in the anthropology lectures is also relevant here. His discussion of these topics primarily involves data gathered from outer sense (viz., the way that external objects distinct from ourselves—in this case, other human beings—are made available to us in intuition). Much of the material in the anthropology lectures does not involve inner sense (viz., the way our own mental states are made available to us in intuition) at all.

Second, Kant's famous argument in the first *Critique* that inner sense and outer sense are necessarily bound up with each other spells additional trouble for any attempt to distinguish geography from anthropology by means of inner versus outer sense. "Inner experience itself," he notes,

> depends on something permanent, which is not in me, and consequently must be outside me, and I must consider myself in relation to it; thus for an experience in general to be possible, the reality of outer sense is necessarily bound up with that of inner sense: i.e., I am just as certainly conscious that there are things outside me to which my sensibility relates, as I am conscious that I myself exist determined in time.[59]

Contra May and Eze, it is simply not the case that Kant's anthropology is concerned exclusively with human beings' "self-consciousness"[60] or their "psychological, moral, internal aspect"[61] and that his geography is not.

Nor can we successfully distinguish geography from anthropology by asserting that the former deals exclusively with "unconsciously held mores, with man's 'second nature' as it conditions his behaviour unconsciously, through the peculiar exigencies of place and custom."[62] For Kant also discusses unconsciously held mores in his anthropology lectures. The French nation, for instance, "is characterized among all others by its taste for conversation,"[63] while the Spaniard "displays in his public and private behavior a certain *solemnity*,"[64] and the German "submits most easily and permanently to the government under which he lives."[65] Indeed, Kant's chapter on "The Character of the Peoples" is at bottom nothing but a mélange of human customs and mores. Granted, his aim here is to put this allegedly empirical data to pragmatic use: to show his listeners "what each can expect from the other and how each could use the other to his own advantage."[66] But, again, his geography is also pragmatic. It too aims to show us how to apply our knowledge of the world and its inhabitants to practical purposes.

In sum, there is no bright, clear line to be drawn between Kant's physical geography and anthropology lectures, particularly when we use human beings as the intended line of demarcation. Obviously, the anthropology lectures are much more concerned with human beings than are the geography lectures, but the study of human beings is not the exclusive domain of either discipline. The most we can say is something like the following: in the geography lectures, human beings are treated primarily (though by no means exclusively) as "*things* in the world" that are "products belonging to the play of nature;"[67] whereas in the anthropology lectures, the human being is considered primarily (though by no means exclusively) as "a *citizen of the world*"[68] who is "a free-acting being"[69]—a creature who is not completely determined by his natural environment and regional habitat, and who thus, "as a rational being endowed with freedom,"[70] can, at least to some extent, make his own character and determine his own way of life.

Other Geographies: Human Culture and Natural Environment

In the Introduction in Rink's edition, Kant states that physical geography is "the ground [*Grund*] of all other possible geographies."[71] Specific geographies mentioned under the category of "other" include *mathematical geography*, "in which the shape, size, and motion of the earth as well as its relation to the solar system are treated;"[72] *moral*, "in which the different customs and characters of human beings according to different regions [*Sitten und Charakteren der Menschen nach der verschiedenen Gegenden*] are discussed;"[73] *political*, in which "laws that are connected with [*beziehen . . . auf*] the nature of the soil and of the inhabitants" are treated;[74] *economic* [*merkantilische*] *geography* or geography of trade [*Handlungsgeographie*], in which is indicated "why and from what source one land has in excess that which another lacks;"[75] and *theological*, where theological principles that undergo changes "according to differences of soil [*nach der Verschiedenheit des Bodens*]"[76] are considered.

Although mathematical geography does not necessarily require the presence of human beings (the earth has a certain shape, size, and motion regardless of whether human beings are present), the remaining four other geographies all focus exclusively on certain aspects of human culture in relation to the natural environment. And in each case Kant makes the surprising claim that the cultural practice itself—or at least the specific part of it that concerns him in these lectures—is causally determined by the natural environment. (Political laws "connect with the nature of the soil," theological principles change "according to differences of soil," moral customs and characters "vary according to region," etc.) What does Kant say about these other four geographies after the Introduction? How, if at all, does he support his ambitious determinist claims with respect to them?

Unfortunately, in the Third Part of the Lectures, where Kant discusses "the principal natural curiosities of all lands according to geographical order,"[77] he offers no detailed presentation of any of these four geographies. For the most part, his brief discussions of moral, political, economic, and theological geography are simply blended into the general country entries. Thus in the entry on Ceylon we are told that "the women throw their children away or give them away, if they fancy they were born at an unlucky hour;"[78] and in the entry on Madagascar, we are informed that the inhabitants "have no divinity [*Gottheit*] other than a cricket they feed in a basket, into which they put evil things. . . . This they call their '*Oly*.' "[79] Similarly, in the entry on Italy, Kant notes that the inhabitants "are jealous, vengeful, and secretive; but otherwise inventive, clever, and political;"[80] while in the section on South America, he states that the Tapeje tribe in Brazil "have no concept of God, no word that designates him, go naked, devour captured enemies, although not with such cruel torture as the Canadians, drill a hole through their lips and put a kind of green jasper into the opening. . . ."[81]

The exceptions to this minimalist treatment occur in the somewhat longer discussions of China, Arabia, the Russian Territory, and Africa. For instance, the entry on China includes separate sections on "Customs and Character of the Nation," "Eating and Drinking," "Compliments," "Agriculture, Fruits, and Manufacturing," "Concerning the Sciences, Language, and Laws," "Religion," and "Marriages." In the first section, we are told that the Chinese "is of an uncommonly serene demeanor,"[82] but that he is also "cowardly, very industrious, very obsequious, and devoted to compliments to an excessive degree."[83] In the section on "Religion," we are told that religion "is treated in a fairly cold manner [*ziemlich kaltsinnig*]. Many believe in no God; others, who accept a religion, do not concern themselves much with it."[84]

Similarly, under the entry on Arabia there is a separate section on "Religion." Here Kant focuses primarily on Mohammed's life, stating that while Mohammed "admitted that he could not perform any miracles [*keine Wunder tun könne*] [,] . . . he is reputed to have split the moon into two parts."[85] And the entry on the Russian Territory includes separate sections on "The Character of the Nation in Siberia" and on "Religion." In the former Kant proclaims that "the laziness [*die Faulheit*] in these lands is astonishing,"[86] and in the latter he states that "if one excludes the Russians of these regions and the Mohammedans, the other peoples have nothing to do with any divinity other than the devil; for, although they maintain that there is a supreme God, he lives in heaven and is far too far away. However, devils rule the earth."[87] Finally, in his opening discussion of Africa, Kant includes a short section entitled "Products of the Land," where he states that "cinnabar and some gold is found there,"[88] and that "the wine is splendid."[89]

What should we make of Kant's abrupt but frequent forays into moral, political, economic, and theological geography? Answer: not much. His discussions of these other geographies are extremely undeveloped and unsystematic. In his 1765 Announcement for the geography course, Kant states that he intends to condense "that part of the subject which is concerned with the physical features of the earth [*physische Merkwürdigkeiten der Erde*], to gain the time necessary for extending my course of lectures to include the other parts of the subject which are of even greater utility. This discipline will therefore be a *physical, moral*, and *political* geography."[90] However, in none of the various geography lecture transcriptions currently being edited by Werner Stark for inclusion in the German Academy edition does Kant offer us an extended moral and political geography. Physical geography always looms larger than moral and political geography.[91] Likely causes for this failure to enlarge the cultural geography dimension of the lectures include both Kant's increased preoccupation, after the "critical turn" in 1770, with "pure" philosophy (which resulted in a diminished interest in empirical studies); and, after 1772, his new anthropology course, which effectively displaced cultural geography.

Furthermore, nowhere in these discussions—or anywhere else—does Kant make the slightest effort to support the strong environmental determinism to which he commits himself in the Introduction to Rink's edition of the *Geography*. Regardless of whether the Chinese are or are not cowardly, industrious, and obsequious; regardless of whether the inhabitants of the Russian Territory are or are not lazy and do or do not have anything to do with any divinity other than the devil; regardless of whether the inhabitants of Madagascar do or do not have any divinity other than a cricket; and regardless of whether the Tapeje of Brazil do or do not have a concept of God; Kant provides no arguments that such beliefs and dispositions are caused by the natural environment.

At the same time, it is instructive to compare these comments about cultural practices with similar one founds in the *Anthropology*. In some cases the remarks are virtually interchangeable. For instance, in the *Geography*, as we saw earlier, Italians are described as "jealous, vengeful, and secretive;"[92] while in the

Anthropology Kant bemoans their "*bad* side:" "knifings, bandits, assassins taking refuge in hallowed sanctuaries, neglect of duty by the police, and so forth."[93] And in the *Geography* Kant informs us that the inhabitants of Greece "have greatly declined from their previous good character,"[94] while in the *Anthropology* he laments "the fickle and groveling character of the modern Greek."[95] This interchangeability constitutes additional support for our earlier claim that when it comes to human beings, there is no bright, clear line of demarcation between Kantian geography and anthropology. Kant sometimes makes virtually identical comments about the same human beings in both sets of lectures.

Nevertheless, there are some significant differences. The discussion of peoples in the *Anthropology* is very western-Eurocentric. The only featured people are the French, English, Spanish, Italians, and Germans. After the separate sections on each of the "big five," Kant concludes his chapter on "The Character of the Peoples" with a few brief, dismissive remarks regarding the Russians, Poles, Turks, Greeks, and Armenians.[96] In Rink's edition of the *Geography*, on the other hand, Kant claims to provide entries on "*all* lands according to geographical order,"[97] and his entries on some non-European lands (e.g., China) are much more detailed than the entries on European countries. Indeed, in the *Geography*—despite the announced focus on "all lands"—Kant neglects to even include entries on England and Germany.

Also, in *Anthropology from a Pragmatic Point of View* Kant abandons the environmental determinism that he asserted in the Introduction to Rink's edition of the *Geography*. At the beginning of the chapter on "The Character of the Peoples," he warns readers that "climate and soil [*Klima und Boden*] . . . cannot furnish the key here."[98] But in its place he inserts an even darker biological determinism. In the *Anthropology* Kant professes to be describing the "innate, natural character [*angeborener, natürliche Charakter*]"[99] of peoples—yet another sign that his *Anthropology*, contrary to received opinion, does not simply consider human beings as "free-acting beings."[100]

Race: Distortions in the *Große Karte*

However, the part of Kant's geography lectures where human beings are featured most prominently comes in a section called "Vom Menschen"—Concerning the Human Being. All extant versions of the geography lectures, including not only Rink's edition but also the various student and auditor transcriptions that are currently being edited by Werner Stark, contain a variant of this important section.

Race is the main topic in "*Vom Menschen*," and it has understandably cast a pall over all of Kant's work in geography. Two of the most notorious and frequently-criticized passages from Kant's *Geography* appear in this short section. First, we are informed that "humanity is in its highest degree of perfection in the white race. The yellow Indians already have a somewhat lesser talent. The Negroes are much lower, and lowest of all is a part of the American races."[101] Second, we are told that

> the inhabitant of the temperate zone . . . is more beautiful in body, harder working, more witty [*scherzhafter*], more moderate in his passions, more intelligent [*verständiger*] than any other kind [*Gattung*] of people in the world. Consequently, these people have at all times taught the others and vanquished them by the use of weapons.[102]

This particular part of the lectures, according to Adickes, dates from 1758–59, and at this early stage of his career Kant appears to have held that climate was the major causal factor behind race. For instance, in Subsection §3, "Opinions as to the Cause of this Color" (viz., black), he states, "it is obvious that the *hot* climate is the cause. But it is certain that a large number of generations had to pass for it to become handed down and hereditary."[103] Speculation about the role of climate in determining not only human skin color but also mental and moral capacities was rampant during the Enlightenment, and geography's unfortunate association with race also stems from this concern with climate. The most influential author here was "the celebrated Montesquieu, a great champion for the climate,"[104] who, in Book XIV of *The Spirit of the Laws*, set out to show that "the temper of the mind and the passions of the heart" vary significantly according to climate, and that "the laws ought to be in relation both to the variety of those passions and to the variety of those tempers."[105] But Montesquieu was not the only Enlightenment advocate of climatology. Buffon also held that "climate is the principal cause of the varieties of mankind" and "the chief cause of the different colors of men."[106] In his *Anthropology*, Kant pays tribute to Buffon as "the great author of the system of nature,"[107] and in his first major statement on race, he points to "*Buffon's* rule, that animals which produce fertile young with one another (whatever difference in shape there may be) still belong to one and the same physical species [*physische Gattung*]" as the correct "definition of a natural species [*Naturgattung*]."[108]

Kant's most developed accounts about race occur in three separate essays: 1) *Of the Different Races of Human Beings* (1775), 2) *Determination of the Concept of a Human Race* (1785), and 3) *On the Use of Teleological Principles in Philosophy* (1788).[109] The full title of the first essay was *Of the Different Races of Human Beings, in Announcement of the Physical Geography Lectures in Summer Semester 1775*, and it functioned as a kind of advertisement for his geography lectures of that year. The second and third papers were first published as freestanding essays in journals, and were not officially connected to the geography course. Overall, the contour of Kant's thinking about race is a gradual move away from climatological causes and toward biological ones. However, climate continues to play a secondary role even in his later statements about race, functioning as a circumstantial cause that triggers a more fundamental biologically predetermined development. At any rate, the geography lectures, while they contain some of Kant's earliest and most notorious remarks about race, do not constitute his most systematically developed statements on the topic.

Nevertheless, the geography lectures are never entirely detachable from Kant's work on race—discussions of race continue to appear in these lectures, year after year. The "*Vom Menschen*" section, for instance, appears both in the earliest surviving version of the geography lectures, the so-called *Holstein* manuscript of 1758–59 (although it differs in certain fundamental respects from the version in Rink's text),[110] as well as in the last version, the *Dohna* manuscript of 1792.

When he is careful, Kant restricts the concept "race" to the single physical heritable trait of skin color. For instance, in the 1785 essay, he divides the human species into four races "with respect to their skin color,"[111] and then argues that "no other characteristic property is *necessarily hereditary*"[112] in these four races except the skin colors of white, yellow, black, and copper-red. Unfortunately, he is not always careful, and does not consistently adhere to this modest "skin-color-only" conception of race. In the third essay of 1788, for instance, he claims that the native American is "too weak for hard labor, too indifferent for industry and incapable of any culture [*unfähig zu aller Kultur*] . . . and ranks still far below even the Negro, who stands on the lowest of the other steps that we have named as differences of the races."[113]

Kant himself was well aware that there were two very different faces to his race theory—a narrower physical one and a broader moral and intellectual one. In a revealing letter to Johann Jacob Engel, publisher of *Der Philosoph für die Welt* (The Worldly Philosopher), in which a revised and expanded version of the first race essay was published in 1777, Kant suggests that his broader moral characterization of race will be of more interest to the average reader: "the attached principles of a moral characterization [*moralische Charakteristik*] of the different races of the human species will serve to satisfy the taste of those who do not pay particular attention to the physical [*physische*] characterization."[114]

But the moral characterization of race seems also to have satisfied the taste of the Sage of Königsberg, for it is not clear that he ever abandoned it. And it has proved to be a major interpretive problem ever since, for both Kantians and un-Kantians alike. How are we to reconcile the Kant who claims that native "Americans and negroes cannot govern themselves. Thus they serve only as slaves"[115] with the Kant who proclaims that "the human being and in general every rational being *exists* as an end in itself, *not merely as a means* to be used by this or that will at its discretion?"[116] At present, three competing interpretive strategies are evident. One group accuses Kant of "inconsistent universalism"[117]—in some places he says that *all* human beings must be treated as ends in themselves, while in others he asserts that nonwhite races should serve as slaves for whites. But the universalist strand, according to this first group, is more central to Kant's core commitments than the racist strand. A second group accuses Kant of "consistent inegalitarianism," and holds that the racist strand is in fact dominant. Kant, on Charles Mills' reading, "makes whiteness a prerequisite for full personhood and . . . limits nonwhites to "subperson" status."[118] So when Kant says that all persons are ends in themselves, he in fact means only whites.

A third group acknowledges that Kant's position on personhood is inconsistent in his earlier writings, but holds that in his late works he abandons racial hierarchy and finally espouses both a consistent universalism and a consistent egalitarianism. Sankar Muthu, for instance, writes: "In Kant's later years . . . the hierarchical and biological concept of race disappears in his published writings. . . . In his last published discussion of race, . . . Kant makes no arguments about the preeminence of whites or Europeans over other human races."[119] More recently, Pauline Kleingeld, while taking issue with certain aspects of Muthu's ameliorative account, has also argued that "during the 1790s Kant restricts the role of the concept of race, and drops his hierarchical account of the races in favor of a more genuinely egalitarian and cosmopolitan view."[120]

The geography lectures, which virtually span Kant's entire writing career, provide us with a way to test the claims of this third group. Did Kant change his mind about race over the years, and does he in fact make no mention of racial hierarchy in his works from the 1790s?

In the version of "*Vom Menschen*" that occurs in the last of the geography lectures currently available, the *Dohna* lectures of 1792, Kant repeats a version of Hume's infamous footnote about Negroes,[121] which he had also cited with approval in his 1764 work, *Observations on the Feeling of the Beautiful and Sublime*.[122] The *Dohna* version reads:

> *Hume* says, that of the many thousands of Negroes who have gradually been freed, there is no example of one who has distinguished himself with a special skill. Something essential in the character of the Negro is a kind of vanity, arrogance—this is why no freed Negro cultivates the land, he prefers to live in a monkey-house or to become a servant.[123]

Clearly, in the *Dohna* lectures of 1792 Kant still "employs a hierarchical account of the races." Isn't the assertion that "Kant changed and improved his position [on race] during the 1790s"[124] therefore false?

Kleingeld acknowledges that in *Dohna* Kant "still endorses Hume's claim that blacks are naturally inferior," and as a result she qualifies her position concerning Kant's second thoughts about race by saying that he changed his mind "most likely *after* 1792" and that he "makes no mention of racial hierarchy anywhere in his *published* writings of the 1790s."[125] But I think the *Dohna* passage should lead us to be more circumspect in making any pronouncements about Kant's second thoughts on race. While it is certainly true, as Kleingeld and others have pointed out, that Kant explicitly condemns colonialism in publications of the 1790s such as *Toward Perpetual Peace*[126] and *The Metaphysics of Morals*,[127] nowhere in either his published or unpublished works of the 1790s does he issue any sort of explicit *mea culpa* and acknowledge to readers that he has changed his mind about race. Accordingly, I think it is more probable that Kant, like Thomas Jefferson and other leading Enlightenment intellectuals, while firmly opposed to colonialism, also continued to hold that "the blacks . . . are inferior to the whites in the endowments both of body and mind."[128]

There is much to admire in Kant's geography lectures—particularly his emphasis on "how to make our knowledge *practical*"[129] and his goal of introducing students to the stage of their "destiny, namely the *world*."[130] But his heavy reliance on racial stereotypes when attempting to provide students with "a comprehensive map of the human of the human species"[131] results only in a severely distorted map of humanity. Far better is the moral map of Kant's *Anthropology* and history essays, which describes the vocation of the human species as the "progressive organization of the citizens of the earth into and toward the species as a system that is cosmopolitically united."[132] If and when our species is cosmopolitically united, local contingencies of climate and soil will play a smaller role in people's lives.

Notes

1. I would like to thank Eric Watkins for sharing a draft of Olaf Reinhardt and David Oldroyd's translation of Rink's edition of the *Physical Geography* (forthcoming in Kant, *Natural Science*, ed. Eric Watkins [Cambridge University Press]); Werner Stark, for offering me access to electronic versions of the geography lectures that he is currently editing for publication in the German Academy Edition; and Joseph S. Wood, for giving me a crash course on geography. Finally, thanks also to Eduardo Mendieta and Stuart Elden, for providing me with the opportunity to do some work on Kant's geography lectures.

2. Paul Gedan writes: "Kant in Königisberg and Gatterer in Göttingen were the first university lecturers who offered independent lectures in geography, thereby introducing what was until then an undervalued science in universities into the circle of academic disciplines" Immanuel Kant, *Physische Geographie*, 2nd ed., ed. Paul Gedan (Leipzig: Dürr'schen Buchhandlung, 1905), vi. See also Erich Adickes, *Kant als Naturforscher*, 2 vols. (Berlin: de Gruyter, 1925), II: 388; Margarita Bowen, *Empiricism and Geographical Thought: From Francis Bacon to Alexander von Humboldt* (Cambridge: Cambridge University Pres, 1981); Richard Hartshorne, *The Nature of Geography* (Lancaster: Association of American Geographers, 1939; reprinted, 1961, with corrections by the author), 38; and Joseph A. May, *Kant's Concept of Geography and its Relation to Recent Geographical Thought* (Toronto: University of Toronto Press, 1970), 4, 51–2.

3. Charles W. J. Withers, "Geography," in *Encyclopedia of the Enlightenment*, ed. Alan Charles Kors (Oxford: Oxford University Press, 2003), 2: 115. See also David N. Livingstone and Charles W. J. Withers, eds., *Geography and the Enlightenment*, (Chicago: University of Chicago Press, 1999).

4. *Entwurf und Ankündigung eines Collegii der physischen Geographie nebst dem Anhange einer kurzen Betrachtung über die Frage: Ob die Westwinde in unsern Gegenden darum feucht seien, weil sie über ein großes Meer streichen*, Akademie edition of *Kants gesammelte Schriften* (hereafter AK, followed by volume and page number), 2: 4.

5. *Neuer Lehrbegriff der Bewegung und Ruhe und der damit verknüpften Folgerungen in den ersten Gründen der Naturwissenschaft, wodurch zugleich seine Vorlesungen in diesem halben Jahre angekündigt werden*, AK 2: 25.

6. *Versuch einiger Betrachtungen über den Optimismus von M. Immanuel Kant, wodurch er zugleich seine Vorlesungen auf das bevorstehende halbe Jahr ankündigt*, AK 2: 35.

7. As cited by Werner Stark in "Immanuel Kants physische Geographie—eine Herausforderung?," http://staff-www.uni-marburg.de/~stark/ws_lese4.htm, 2. See also Karl Vorländer, *Immanuel Kant: Der Mann und das Werk*, 3rd ed. (Wiesbaden: Fourier Verlag, 2003), 2: 56–7. Zedlitz wrote to Kant in February 1778, noting that he had enjoyed reading a rudimentary transcription of the geography lectures, and requesting a better copy. See Stark, "Immanuel Kants physische Geographie," 2–3; May, *Kant's Concept of Geography*, 73.

8. David Harvey, "Cosmopolitanism and the Banality of Geographical Evils," *Public Culture* 12:2 (2000): 529–564, 532, and Roger-Paul Droit, "Kant et les fournis du Congo," *Le Monde*, February 5, 1999, cited by Harvey, 532.

9. *Entwurf und Ankündigung eines Collegii der physischen Geographie*, AK 2: 3.

10. *Nachricht von der Einrichtung seiner Vorlesungen in dem Winterhalbenjahre von 1765-1766*, AK 2: 312–3.

11. Erich Adickes, *Ein neu aufgefundenes Kollegheft nach Kants Vorlesung über physische Geographie* (Tübingen: J. C. B. Mohr, 1913), 10. Adickes argued that the beginning part of Rink's text (*Physische Geographie*, AK 9: 156–273) stemmed from Kant's 1775 lecture notes, while the later part (AK 9: 273–436) stemmed from 1758/59 lectures.

12. *Physische Geographie*, AK 9: 165.

13. *Entwurf und Ankündigung eines Collegii der physischen Geographie*, AK 2: 3.

14. *Vorlesungen in dem Winterhalbenjahre von 1765–1766*, AK 2: 312.

15. In the Preface to *Anthropology from a Pragmatic Point of View*, Kant writes: "As for physical geography, it is scarcely possible at my age to produce a manuscript from my text, which is hardly legible to anyone but myself," *Anthropology from a Pragmatic Point of View*, ed. and trans. Robert B. Louden (Cambridge: Cambridge University Press, 2006), AK 7: 122 n. I have recently prepared a new translation of this text.

16. *Neuer Lehrbegriff der Bewegung und Ruhe*, AK 2: 25.

17. *Vorlesungen in dem Winterhalbenjahre von 1765–1766*, AK 2: 312.

18. *Von den verschiedenen Racen der Menschen*, AK 2: 429.

19. *Von den verschiedenen Racen der Menschen*, AK 2: 443.

20. Adickes, in *Ein neu aufgefundenes Kollegheft nach Kants Vorlesung über physischer Geographie*, 43, suggests "pragmatische" rather than "praktische" here. Cf. Gedan's note at AK 9: 515.

21. *Physische Geographie*, AK 9: 158.

22. *Physische Geographie*, AK 9: 165.

23. For related discussion, see Paul Richards, "Kant's Geography and Mental Maps," *Transactions of the Institute of British Geographers* 61:1 (March 1974): 1–16. Richards rightly draws attention to Kant's conviction that geographical space constitutes "a locational 'framework of knowledge' " (8), and he also draws some interesting parallels between Kant's conviction and more recent work on cognitive or mental maps. However, my own view is that he over-theorizes Kant's position. It is primarily a *pragmatic* and *practical* orientation that Kantian physical geography offers students: a sense of what to expect once they leave their local communities—not an abstract " 'schema for co-ordinating with each other absolutely all things externally sensed' " (7).

24. *Physische Geographie*, AK 9: 157; cf. *Anthropology from a Pragmatic Point of View*, AK 7: 120.

25. *Physische Geographie*, AK 9: 158.

26. Cf. *Physische Geographie*, AK 9: 158.

27. *Anthropology from a Pragmatic Point of View*, AK 7: 120.

28. *Vorlesungen in dem Winterhalbenjahre von 1765-1766*, AK 2: 312–3. For related discussion, see my essay "Anthropology from a Kantian Point of View: Toward a Cosmopolitan Conception of Human Nature," *Studies in the History and Philosophy of Science* Part A 39 (4), 2008: 515–522.

29. *Von den verschiedenen Racen der Menschen*, AK 2: 443; cf. *Physische Geographie* AK 9: 157.

30. *Physische Geographie*, AK 9: 158.

31. *Critique of Pure Reason*, trans. Paul Guyer and Allen W. Wood (Cambridge: Cambridge University Press, 1998), B4.

32. Cf. *Physische Geographie*, AK 9: 157.

33. *Anthropologie Pillau*, AK 25: 773; cf. *Von den verschiedenen Racen der Menschen*, AK 2: 443, *Physische Geographie*, AK 9: 156.

34. E.g., the geography entry in J. J. Johnston, Derek Gregory and David M. Smith, ed., *The Dictionary of Human Geography*, 3rd ed. (Oxford: Blackwell, 1994), begins: "Geography can be formally defined as the study of the Earth's surface as the space within which the human population lives . . ., or simply as the study of the Earth as the home of people." And the first definition under "geography" in Erin McKean, ed., *The New Oxford American Dictionary*, 2nd ed. (Oxford: Oxford University Press, 2005) reads: "the study of the physical features of the earth and its atmosphere, and of human activity as it affects and is affected by these, including the distribution of populations and resources, land use, and industries."

35. *Entwurf und Ankündigung eines Collegii der physischen Geographie*, AK 2: 3.

36. *Entwurf und Ankündigung eines Collegii der physischen Geographie*, AK 2: 9; cf. *Vorlesungen in dem Winterhalbenjahre von 1765-1766*, AK 2: 312–3.

37. *Physische Geographie*, AK 9: 164–5.

38. *Physische Geographie*, AK 9: 164.

39. *Physische Geographie*, AK 9: 311–20.

40. *Physische Geographie*, AK 9: 296–8.

41. *Physische Geographie*, AK 9: 306–8.

42. *Physische Geographie*, AK 9: 356–65.

43. *Physische Geographie*, AK 9: 377–436.

44. *Anthropology from a Pragmatic Point of View*, AK 7: 135, 266, 269n., 327.

45. Emil Arnoldt, *Kritische Excurse im Gebiete der Kant-Forschung* (Königsberg: Verlag von Ferd. Beyer's Buchhandlung, 1894), 343.

46. *Anthropology from a Pragmatic Point of View*, AK 7: 257, 299.

47. *Kritische Excurse im Gebiete der Kant-Forschung*, 343.

48. May, *Kant's Concept of Geography*, 115, 70, 65.

49. Alix Cohen, "Kant's Critique of the Human Sciences" (PhD diss., Cambridge University, 2005), 123.

50. *Von den verschiedenen Racen der Menschen*, AK 2: 443.

51. *Neuer Lehrbegriff der Bewegung und Ruhe*, AK 2: 25.

52. *Vorlesungen in dem Winterhalbenjahre von 1765–1766*, AK 2: 312.

53. *Anthropology from a Pragmatic Point of View*, AK 7: 119.

54. Kant to Herz, toward the end of 1773, AK 10: 145; cf. *Anthropology from a Pragmatic Point of View*, AK 7: 119, 176.

55. *Vorlesungen in dem Winterhalbenjahre von 1765–1766*, AK 2: 312, my emphasis.

56. May, *Kant's Concept of Geography*, 108; cf. 72. See also Arnoldt, *Kritische Excurse im Gebiete der Kant-Forschung*, 343.

57. Emmanuel Chukwudi Eze, "The Color of Reason: The Idea of 'Race' in Kant's Anthropology," in *Postcolonial African Philosophy: A Critical Reader*, ed. Emmanuel Chukwudi Eze (Cambridge: Blackwell, 1997), 103–140, at 106.

58. *Physische Geographie*, AK 9: 156–7.

59. *Critique of Pure Reason*, Bxl–xli n.

60. May, *Kant's Concept of Geography*, 113.

61. Eze, "The Color of Reason," 105.

62. May, *Kant's Concept of Geography*, 115.

63. *Anthropology from a Pragmatic Point of View*, AK 7: 313.

64. *Anthropology from a Pragmatic Point of View*, AK 7: 316.

65. *Anthropology from a Pragmatic Point of View*, AK 7: 317.

66. *Anthropology from a Pragmatic Point of View*, AK 7: 312.

67. *Anthropology from a Pragmatic Point of View*, AK 7: 120.

68. *Anthropology from a Pragmatic Point of View*, AK 7: 120.

69. *Anthropology from a Pragmatic Point of View*, AK 7: 119.

70. *Anthropology from a Pragmatic Point of View*, AK 7: 285.

71. *Physische Geographie*, AK 9: 165; cf. *Vorlesungen in dem Winterhalbenjahre von 1765–1766*, AK 2: 312.

72. *Physische Geographie*, AK 9: 164.

73. *Physische Geographie*, AK 9: 164.

74. *Physische Geographie*, AK 9: 164.

75. *Physische Geographie*, AK 9: 165.

76. *Physische Geographie*, AK 9: 165.

77. *Physische Geographie*, AK 9: 377.

78. *Physische Geographie*, AK 9: 394–5.

79. *Physische Geographie*, AK 9: 411.

80. *Physische Geographie*, AK 9: 423.

81. *Physische Geographie*, AK 9: 430.

82. *Physische Geographie*, AK 9: 378.

83. *Physische Geographie*, AK 9: 378. Kant's extended discussion of China is in part a reflection of strong Enlightenment interest in China initiated by Jesuit missionaries and intensified by Leibniz and Wolff. For discussion and references, see the section on "German Philosophers and China" in my essay, " 'What does Heaven Say?': Christian Wolff and Western Interpretations of Confucian Ethics," in *Confucius and the Analects: New Essays*, ed. Bryan W. Van Norden (New York: Oxford University Press, 2002), 73–93.

84. *Physische Geographie*, AK 9: 381.

85. *Physische Geographie*, AK 9: 399.

86. *Physische Geographie*, AK 9: 402.

87. *Physische Geographie*, AK 9: 402.

88. *Physische Geographie*, AK 9: 409.

89. *Physische Geographie*, AK 9: 410.

90. *Vorlesungen in dem Winterhalbenjahre von 1765-1766*, AK 2: 312.

91. Cf. Hartshorne, *The Nature of Geography*, 39.

92. *Physische Geographie*, AK 9: 423.

93. *Anthropology from a Pragmatic Point of View*, AK 7: 317.

94. *Physische Geographie*, AK 9: 422.

95. *Anthropology from a Pragmatic Point of View*, AK 7: 320.

96. For discussion, see "Peoples," in Robert B. Louden, *Kant's Impure Ethics: From Rational Beings to Human Beings* (New York: Oxford University Press, 2000), 87–93.

97. *Physische Geographie*, AK 9: 377, my emphasis.

98. *Anthropology from a Pragmatic Point of View*, AK 7: 313.

99. *Anthropology from a Pragmatic Point of View*, AK 7: 319.

100. See *Anthropology from a Pragmatic Point of View*, AK 7: 119.

101. *Physische Geographie*, AK 9: 316.

102. *Physische Geographie*, AK 9: 317.

103. *Physische Geographie*, AK 9: 314.

104. Henry Homes, Lord Kames, *Sketches of the History of Man*, 3rd ed. (1788), ed. James A. Harris (Indianapolis: Liberty Fund, 2007), I: 39. After summarizing Montesquieu's position, Kames states: "It is my firm opinion that neither temper nor talents have much dependence on climate" (I: 40).

105. Montesquieu, *The Spirit of the Laws* (1748), trans. Thomas Nugent (New York: Hafner, 1949), 221.

106. Georges-Louis Leclerc, Comte de Buffon, *A Natural History, General and Particular* (1748–1804), in *Race and the Enlightenment: A Reader*, ed. Emmanuel Chukwudi Eze (Oxford: Blackwell, 1997), 22, 27.

107. *Anthropology from a Pragmatic Point of View*, AK 7: 221.

108. *Von den verschiedenen Racen der Menschen*, AK 2: 429.

109. Translations of each of these essays are included in Kant, *Anthropology, History, and Education*, ed. Günter Zöller and Robert B. Louden (Cambridge: Cambridge University Press, 2007).

110. E.g., the first statement on race cited earlier (AK 9: 316) is missing entirely, but a version of the second statement (AK 9: 317) is present. See 126–7 in the *Holstein* manuscript.

111. *Determination of the Concept of a Human Race*, AK 8: 93.

112. *Determination of the Concept of a Human Race*, AK 8: 94.

113. *On the Use of Teleological Principles in Philosophy* AK 8: 176.

114. Kant to Engel, July 4, 1779, AK 10: 239. Kant is referring here to a draft of what eventually became his second race essay, first published in the *Berlinische Monatsschrift* 11 (1785): 390–417.

115. *Reflexion* 1520 ("Character der Race), AK 15: 878.

116. *Groundwork of The metaphysics of morals*, in *Practical Philosophy*, eds. Mary J. Gregor and Allen Wood (Cambridge: Cambridge University Press, 1996), AK 4: 428.

117. In describing these three interpretive strategies I have borrowed from Pauline Kleingeld, "Kant's Second Thoughts on Race," *The Philosophical Quarterly* 57 (2007): 573–92. Advocates of this first strategy, Kleingeld notes, include Robert B. Louden, Thomas McCarthy, and Thomas E. Hill Jr. and Bernard Boxill.

118. Charles W. Mills, "Kant's *Untermenschen*," in *Race and Racism in Modern Philosophy*, ed. Andrew Valls (Ithaca: Cornell University Press, 2005), 169–93, at 170. (Kleingeld also lists Emmanuel Eze and Robert Bernasconi as advocates of this second interpretive strategy.)

119. Sankar Muthu, *Enlightenment Against Empire* (Princeton: Princeton University Press, 2003), 184. See also Susan Meld Shell, *The Embodiment of Reason: Kant on Spirit, Generation, and Community* (Chicago: University of Chicago Press, 1996), 387 n. 23.

120. Kleingeld, "Kant's Second Thoughts on Race," 573. Cf. 575, 586, 592. *Contra* Muthu and Shell, Kleingeld argues—as I have also done in the present essay—that in *On the Use of Teleological Principles* Kant continues to make assertions about "the preeminence of whites or Europeans over other human races."

121. Hume, "Of National Characters;" in *Essays Moral, Political, and Literary*. ed. Eugene F. Miller (Indianapolis: Liberty Fund, 1985), 208 n. 10.

122. *Observations on the Feeling of the Beautiful and Sublime*, AK 2: 253.

123. *Die Philosophischen Hauptvorlesungen Immanuel Kants. Nach den neu aufgefundenen Kollegheften des Graften Heinrich zu Dohna-Wundlacken*, ed. Arnold Kowalewki (Munich: Rösl and Cle, 1924), 105.

124. Kleingeld, "Kant's Second Thoughts on Race," 575.

125. Kleingeld, "Kant's Second Thoughts on Race," 577 n. 10, 575, 586 (my emphases in the last two quotations).

126. Of 1795—see AK 8: 358.

127. Of 1797—see AK 6: 266, 353.

128. Thomas Jefferson, *Notes on the State of Virginia*, Query XIV; in *Writings* (New York: Library of America, 1984), 270.

129. *Physische Geographie*, AK 9: 158.

130. *Von den verschiedenen Racen der Menschen*, AK 2: 443.

131. *Vorlesungen in dem Winterhalbenjahre von 1765–1766*, AK 2: 313.

132. *Anthropology from a Pragmatic Point of View* AK 7: 333.

9

The Pragmatic Use of Kant's *Physical Geography* Lectures

Holly L. Wilson

Although there is not much written about Kant as a educator, there really ought to be more attention given to this area, since Kant very thoughtfully developed new courses that were not being taught in the German universities at the time, and he developed these courses to meet specifically pedagogical purposes. Kant's physical geography lectures are a prime example of his creative pedagogy. Kant developed these lectures not to advance science so much as to bring about his students' capacity to think for themselves and hence instantiate the great impulse of the Enlightenment era in which Kant wrote, taught, and thought. But his aim was even greater than this. He wanted to teach students the useful skill of prudence, and for that they needed to learn how to think for themselves and develop a sense of initiative. The physical geography lectures as they developed over time and came to be associated with the anthropology lectures were intended to prepare students for acquiring this very useful skill. The pragmatic use of physical geography lies in this preparation for developing prudence.

Background to the Physical Geography Lectures

One might easily make the assumption that Kant intended his scientific works to be contributions to the scientific search for truth of his day, but a careful reading of such works reveals a different purpose. G. Gerland argues that the purpose of *Allgemeine Naturgeschichte und Theorie des Himmels* was not to explain the whole universe on mechanical principles in order to prove that it can be done without the hypothesis of God, but rather to prove that the whole universe is governed by a lawfulness whose source must be God.[1] When Gerland looks at Kant's essays on earthquakes he comes to a similar conclusion: Kant was not writing these scientific essays to contribute to science, but rather to establish the mechanistic view of nature for the sake of proving God's existence. Such an interpretation is supported by the obvious intent of Kant's *Der einzig mögliche Beweisgrund zu einer Demonstration des Daseins Gottes* written in 1763. Although I agree with Gerland that the purpose of Kant's "scientific" essays is not to further the science of his day, I want to depart from his interpretation of their actual purpose, because I think their actual purpose is relevant to the interpretation of the purpose of Kant's physical geography lectures.

In contrast to Gerland, I think Kant is eager in the *Naturgeschichte* to establish that the universe can be explained purely mechanically so that he can exclude God as an explanation for particular events. It is not that he wants to prove that the universe can be explained without the hypothesis of God, but rather he wants to make the point that the appropriate place for attributing God's action is not to individual events, but rather to the very lawfulness of the whole. This is a significant difference. Kant is not trying to prove that God exists, since he is assuming God does exist. Rather, he is making a point about where God's causal activity resides, namely in the lawfulness of the universe not in the particular events that one experiences within the world. In other words, God's will is revealed not in particular events like earthquakes, but rather in the lawfulness of the whole.

Such a perspective makes sense of why Kant would write about the causes of an earthquake. In a society that is superstitious, the cause of an earthquake would be attributed to God, and one may draw conclusions about God's will based on that. Perhaps God is angry with the people and wants to punish them (as, for instance, in the case of Sodom and Gomorrah). In the superstitious worldview, God then wills that the people stop whatever sin they are engaged in. To combat this superstitious view of God, Kant is eager to show that there are purely mechanical causes for such natural events and hence they cannot be attributed to God. The very fact that earthquakes occur is relevant for our understanding of how God has lawfully set up the earth. Kant explains,

> Human beings are in the dark when it comes to guessing the intentions God had in mind for the government of the world. But we are in no uncertainty when it comes to the application, of how we should use this providence appropriately for our purposes. The human being is not born to build eternal houses on this stage of vanity. Because his whole life has a far more noble end, as evidenced by all the devastations which the unsettled state of the world allows us to see, of which the greatest and most important seem to be there to remind us; that the goods of the earth cannot provide to satisfy the drive for happiness![2]

Earthquakes do reveal something of God's intent for human beings, but not in the particular event of this earthquake, but rather in the lawful and causal relation of any earthquake to the lawful intents of human beings. Namely, all earthquakes destroy the comfort of human beings who universally desire comfort and happiness. Kant wants us to draw the conclusion that God's intent in making earthquakes a part of the earth and universe was to make human beings aware of the fact that their lives have a greater destiny than comfort. This is not attributing their cause to God in a particular event, but rather it is viewing earthquakes as purposive for human growth and development.

What we can draw from this that will be relevant to the understanding of the physical geography is that Kant finds it important to establish that the earth and all the natural events on the earth can be explained mechanically so that his students will no longer use God as an explanation for particular events. Kant's purpose is then not really to establish a proof for the existence of God, but rather to train the judgment of his students to look for natural causes rather than to resort to the simple but misleading explanation that God caused the event. Kant does "science" not to remove God from the picture but rather to put God's causality in the proper framework, namely God's work is evident in the laws of nature rather than in a suspension of those laws.

When we compare Kant's physical geography to other geographies in this same period we find that Kant is not just giving a description of the earth and its features, but he is also taking pains to explain the causes of natural events. The word "because" occurs in almost every paragraph of the work. Frequently Kant explains the cause of phenomenon: the moon causes an eclipse of the sun on the earth;[3] the numbers of harbors in Norway cause the blossoming of trade;[4] ginseng in too large doses causes fever;[5] the Demokalo spider causes insanity;[6] the population decline in Spain is caused among other things by persecution of the Jews and Mohammendans;[7] in Iceland the melting snows cause torrents;[8] decayed bodies in Russia are caused by natural causes;[9] different climates cause skin color differences;[10] goiter is caused by the water in Tyrol and Salzburg;[11] mental laziness is "probably brought about by brandy, tobacco, opium, and other strong things";[12] too great or too little perspiration cause thick viscous blood;[13] the causes of the variations in temperament are similar to causes of the variations in animals;[14] the saltiness of sea water is caused by "the gradual washing down of salt from plants and growing things which carry a small amount of common salt";[15] etc.

Neither Johann Jacob Schatzen's nor Johann Hübner's geographies are concerned with explaining the causes of natural events.[16] That Kant pays such careful attention to natural causes can perhaps be explained by his reliance on Bernhard Varenius's *geographie generalis*, but not entirely. Kant certainly used many other sources, which do not emphasize natural causation in order to put his own physical geography together, so it is evident that he chose to emphasize natural causation. This then could be explained by the inference that he is attempting to contribute to the science of this day. But I think not, given what Kant explicitly says about his intent in giving the physical geography lectures and given what seems to be his general motivation for writing his "scientific" works. Kant's emphasis on natural causation is done for pedagogical reasons—namely in order to bring about a greater capacity for critical thought and for self-initiated action among his students—what Kant calls "thinking for oneself" and "prudent action," respectively. The pragmatic use of the physical geography resides in these two purposes. To see this more evidently, I will review the three announcements Kant gave for his physical geography lectures, in which his intents for the course are made explicit.

Entwurf und Ankündigung eines Collegii der physischen Geographie (1757)

In this first announcement of his lectures, Kant makes it clear that the lectures are useful for his students because they will appeal to the interests of a traveler. Paul Menzer reports that the Herder papers have Kant saying that the standpoint of the observations in the physical geography are "worthy of the reflections of a traveler" and bring about a "noble pleasure."[17] Indeed, in the *Entwurf and Ankündigung eines Collegii der physischen Geographie*, Kant says he wants to limit the content of the physical geography lectures not only in their completeness but also in their exactness in order to appeal to the natural curiosity of a traveler.[18] People travel for pleasure. But if you cannot travel or do not want to travel, one may also, as Kant does, use books to travel in one's mind. Kant enjoyed reading travel descriptions very much and it is easy to see why he thought his students would enjoy this course as well.

Yet Kant's opening words in the *Entwurf* make clear that the physical geography lectures are meant to appeal not only to those who like to travel, but to everyone. With the words "the sensible taste of our enlightened times," Kant is pointing to the universal appeal of the content of his lectures within an environment of enlightenment. He thinks that the taste of this time, "has presumably become so common that one can presuppose that there will be few who can be found, for whom would be indifferent to know the peculiarities of nature, which occur outside of their horizon in other regions of the terrestrial globe."[19] With this statement, Kant is placing these lectures in the context of the Enlightenment. The Enlightenment for Kant was the replacing of the superstitious worldview with trust in one's own powers of reason. Kant defines the essence of enlightenment in his essay "*What is Enlightenment?*": "Enlightenment is man's emergence from his self-imposed immaturity. Immaturity is the inability to use one's understanding without guidance from another."[20] So when Kant refers to these "enlightened times" he is talking about the worldview in which one is challenged to think for oneself. In teaching the physical geography, Kant is intending to teach students to think for themselves and think critically by being exposed to the scientific understanding of natural causality that occurs in experience. Physical geography had the advantage over metaphysics in that it was a science based on experience and could train the judgment of students.

Kant informs his students in this announcement that the physical geography can protect against superstition and fairy tales and this is one of its great advantages. He explains that

> It is not to be taken as a small advantage, that the gullible admiration of the caretakers of unending fantasies has made way for a careful examination, through which we are in a position to acquire sure knowledge from credible witnesses, without the danger of erring in a world of fables, and instead achieve a correct science of natural occurrences.[21]

Physical geography, if it contains a careful account of natural causality, can teach students to exercise their judgment and critically distinguish between a true account of the world and fairy tales, not only those brought forth by religion, but also by those brought forth through ungrounded metaphysics. And Kant encourages his students to decide for themselves whether it is allowable for a person to call himself learned and still be ignorant of physical geography.[22]

Nachricht von der Einrichtung seiner Vorlesungen in dem Winterhalbenjahre von 1765–1766

Almost 10 years later Kant published another announcement of his lectures, and the *Nachricht* from 1765 adds to our understanding of what Kant intended for the physical geography lectures and what use he saw for his students. The pedagogical context of the physical geography lectures is made explicit in section 4 pertaining to the physical geography lectures: Kant tells his potential students that, even at the beginning of his teaching career, he perceived a great neglect in the education of young people studying at universities to be the fact that they were learning to reason without sufficient historical knowledge. The problem with this neglect, Kant explained in the introductory section of the *Nachricht,* is that students develop loquacity too early on in their education, which is actually blind. Universities, he bemoans, are teaching students beyond their maturity level because professors are teaching students concepts without teaching them the relationship of those concepts to principles and consequences.

They are teaching the art of reasoning without beginning with a sound understanding. Kant holds that a student needs to become an informed human being before becoming an intelligent person and certainly before becoming a learned person. The universities are doing just the opposite, and students are imitating learnedness without sound understanding. He complains that science clings to them, like a shirt, but does not originate in them. Their disposition is, as a result, unfruitful, and they give only the illusion of possessing wisdom.

The trouble lies, of course, with the way subjects are being taught dogmatically. Philosophy is taught as a finished system, not as something that one discovers through inquiry (*zetetically*). Because students are simply memorizing what they are learning before they have the understanding to be able to grasp it, they are not gaining insight into how to philosophize. Kant maintains that students should learn *how* to think, not just learn thoughts. Kant suggests that professors should not carry students by having them memorize concepts, but rather lead them to develop their own thinking so that in the future they can skillfully find their way in life. This makes even more sense to Kant when he notices that many students do not aspire to the heights of education and learnedness and hence should be able to take something from academia that can make them more skilled and prudent in life. Kant is convinced that physical geography is useful for such students because it gives unity to their knowledge and gives them something learned to speak about in a very sociable society. If successful, such students will not have to say "What I know, is not proper, and what is proper, I don't know."[23] Kant thinks that geographical knowledge will prepare students for practical reason and also for further learning in the future. Kant wants his students be able to use their knowledge and he wants them to become life-long learners.

The question now is how the content of the physical geography lectures brings about this purpose of teaching students to think for themselves and how such content is useful for students. If physical geography is just a description of different parts of the earth and different people and animals on the earth, how can it bring about the capacity for students to think for themselves? Certainly students could also just memorize the various descriptions and be no further along in establishing a sound understanding that will serve them well in life.

Kant defines the sound understanding in the *Anthropology* as comprising not only the increase of knowledge through experience, but also the enlarging of the experience itself.[24] The physical geography can easily meet both of these criteria. Students are gaining knowledge through the reliable experience and reports of travelers who have gone to the far reaches of the earth and have come back to share their knowledge. They are gaining knowledge through an understanding of the causal network that underlies all natural events and things. They know that wherever they were to travel the same natural laws would be effective. They are learning that all the earth is under a common system of natural laws. And they would be able to anticipate and predict certain events because of the lawfulness of nature. They are broadening their horizons by learning about different people in different environments. They are learning that all human beings, despite their differences, are the same species since they can procreate with one another. They are seeing for the first time, what it means to be a human being as a species among other species on the earth. Human beings as a species share things in common that go beyond national and religious boundaries. They are broadening their horizons beyond their provincial world of experience in that they are seeing that different peoples do different things and have different values. This will allow them to question whether what they do is the truly right thing to do or the most productive thing to do. They are also learning that different people have different desires and want different things. And other human beings have to contend with different environmental challenges (whether those come from water, mountains, desert, or ice). Certainly such information could stimulate thought about one's own way of life.

Von den verschiedenen Racen der Menschen

However, Kant became aware that something more was needed to enhance this learning and Kant tells his students in the third announcement of his physical geography lectures that he plans to expand the course into a second half which will deal primarily with human beings. Kant knows that students need to know not only about human beings in faraway lands, but also about their next-door neighbor. After all, most of them will be interacting with other people who are near them to a greater extent than to people far away. So in 1772, Kant announces his lectures on anthropology along with the physical geography lectures. In the *Von den verschiedenen Racen der Menschen*, Kant articulates his understanding of what the two fields have in common.[25] Both of them, he tells his students, are preliminary studies in the knowledge of the world (*Weltkenntniß*). Such disciplines are meant to take all the accomplishments of the other sciences and skills and give them a pragmatic character so that they are not just useful for the university, but also for life. These two disciplines will help students find their place and destiny on the stage of life. Kant emphasizes that they should not differentiate the two based on their respective objects (physics and empirical theory of the soul), but rather see their commonality in that they are both cosmological disciplines, which will allow students to find their place in the cosmos. What Kant is pointing to here is the pedagogical intent of the two lectures. They are not intended to be two different sciences about two different objects (nature and the human being), but rather both are to be pragmatically oriented toward developing a sound understanding in students, which will help them orient themselves in life. That Kant kept teaching the two courses year after year, one after the other, indicates that he thinks they are necessary for each other.

The Anthropology from a Pragmatic Point of View

The *Anthropology from a Pragmatic Point of View*, which was based on Kant's lectures on anthropology, adds something to the understanding of what it means for the two disciplines to be useful and of pragmatic worth. First of all, Kant defines clearly what makes this discipline of anthropology pragmatic. Second, Kant presents his theory of human nature, which is comprised of four natural predispositions. The *Anthropology* is meant to develop ones technical skills, social skills (prudence), and wisdom. In light of these goals for the lectures, one is brought back to the question of how physical geography is necessary to prepare one for the goals of the discipline of anthropology. The importance of the physical geography lectures is highlighted when one understands that what makes anthropology useful is that it develops prudence. We come back to the point that the physical geography lectures contain not only descriptions of phenomenon on the earth, but also a methodological concern for the natural causes that make things occur. There is something about the natural causal nexus that is important to the development of the skill of prudence.

Kant tells his students right from the start that the *Anthropology* will be from a pragmatic point of view rather than a physiological point of view. The difference between these two disciplines is evident in the example Kant uses of memory. A physiological anthropology will teach what memory is; a pragmatic anthropology will teach how to increase one's memory. Thus, what makes anthropology pragmatic is the use that it brings with it. It is useful to have more memory. It may not be very useful to know what memory is if one cannot remember the person's name who promised to help complete a project. Knowing how to use a mnemonic device will increase one's capacity for memory. Insofar as the anthropology teaches students how to increase their skills, it contributes to the useful development of the technical predisposition. Anthropology can also tell one what one ought not to do.[26] Primarily, pragmatic anthropology teaches "what can be made of man" or "what man as a free agent makes, or can and should make, of himself."[27]

But the content of the anthropology is also about developing prudence. Kant defines prudence [*Klugheit*] as the skill for using other human beings as a means to one's own ends.[28] This is a very positive skill and in no way is diminished because prudence does not constitute moral action. Further, using another human being to one's end is not only permissible when it is with their consent, it is admirable that one is able to gain the consent of the other. Kant certainly does not mean by "use another person to one's own end" that we should use other people without their consent. Kant distinguishes between the prudent person and the cunning person. It is the cunning person who uses others without their informed consent. The prudent person knows that it is important to gain the informed consent of another for their own lasting happiness. Kant defines cunning or craftiness [*Arglist*] as "the skill in cheating others."[29] Cunning people are also those who "surrender all use of their reason and submit, passively and obediently, to the precepts of a holy man" because they are doing so to shirk having to repent and because they are looking for someone else to blame.[30] Elsewhere, Kant defines cunning as the "head for intrigue."[31] Kant explains why cunning [*Arglist*] is different than prudence, which it resembles superficially. Cunning people need trusting people, but these people can be deceived only once, and hence cunning people will defeat their own purposes in the long run.

On the other hand, prudence [*Klugheit*] is not primarily the skill in being cautious, which it is defined as today. In fact, Kant tells us that the person who is inordinately concerned with losing his life (too cautious) will never be happy.[32] Prudence is the capacity to integrate all of one's ends toward happiness, and toward lasting happiness. Our happiness is dependent upon other people cooperating in our ends. Hence, prudence is the skill in getting others to cooperate in our ends.[33] Such a skill is extremely useful. Kant counts politeness as a subtype of prudence. Offensiveness is imprudent.[34] People who are well-mannered are prepared for pleasing others and making themselves well-loved and admired.[35] One gains the cooperation of others in this way. On the contrary, the proud arrogant man turns people away.[36] Those who have a mania for domination are imprudent because it is a precarious way to use others as a means to one's own ends since it arouses their opposition.[37] It is clear in such examples that Kant believes that prudence is not using others as an unwilling means to one's own ends. Kant states that the desire for power to exercise influence over other men comes very close to prudence, since it is a type of using others for one's own purposes, but Kant denies that it is prudence since it is really like possessing another person as a "mere tool of our will."[38] Clearly, prudence is not using others as tools and without their consent.

Kant really does mean that prudence is a skill in using others. He writes in the *Lectures on Ethics*, "Man can certainly enjoy the other as an instrument for his service; he can utilize the other's hands or feet to serve him, though by the latter's free choice."[39] We might want to define prudence, then, in more contemporary language, as the ability to gain the willing cooperation of others in our ends, or as the capacity to motivate others to cooperate with us. Using such language avoids the confusion that arises from Kant's definition of prudence as using others as a means to our own ends.

Now the question arises as to how prudence can be taught and specifically what it is about the pragmatic anthropology and the physical geography that promote the acquisition of the skill of prudence? Kant's students were more likely than not being used by others for their own ends, and were in a state of immaturity where they let others lead them, rather than making decisions for themselves. In *What is Enlightenment?*, Kant identifies those who make decisions for his students: "I have a book to serve as my understanding, a pastor to serve as my conscience, a physician to determine my diet for me, and so on, I need not exert myself at all . . . The guardians who have so benevolently taken over the supervision of men have carefully seen to it that the far greatest part of them . . . regard taking the step to maturity as very dangerous . . ."[40] To teach the skill of prudence, Kant would have to break the hold of passivity in his students. He would have to teach them to be good critical thinkers so that they would question what they are learning rather than memorize it. He needs to teach them to think for themselves. In addition, he would have to awaken their initiative so that they would become aware of their capacity to engage others in their ends. Kant thinks that thinking freely does indeed increase one's ability to act freely.[41]

Physical geography, as a discipline that exposed students to the world and different conditions, would cause students to compare their own experience with that of others. That may well have the effect of teaching students to question what has been handed down to them. Then, as a discipline that taught students to seek natural causes for natural phenomenon, it would encourage students to look beyond the religious worldview in which all causality is attributed to God. Not only would looking for specific causes for natural phenomenon stimulate thought and thinking for themselves, it would also encourage them to be, themselves, causes. It would encourage them in their belief in their own ability to initiate a new set of conditions. Thinking, on the contrary, that God is the cause of natural events impedes and undermines a person's initiative because we can never have God under our control. We can however have natural causes under our control to a much greater extent. Learning to think of nature as a set of natural causes encourages a student in thinking that he is capable of controlling and using nature to his own ends. This furthers a student's sense of initiative. Learning to think through natural causality in a systematic way as one does in the physical geography, one learns how to think for oneself. There is a systematic methodology that Kant is teaching his students. Once the method is learned, the student can begin to take the method of looking for natural causes and apply it to new circumstances and events. In this way, a student is learning to think independently and is being prepared for exerting his own initiative.

Thinking for oneself and thinking one can bring about a new set of conditions would both be necessary before one could begin to develop the skill or prudence. One cannot conceive of using others as means if one does not even think of oneself as capable of initiating action. One cannot think of using others if one does not understand them enough to be able to exercise control or influence. This would mean that physical geography would be the essential preliminary discipline to the development of a human being's pragmatic predisposition. The anthropology would fine-tune prudence and teach students to think about the different kinds of capacities human beings have, the different kinds of human beings, and the different ways in which human beings could be used.

Coming back to the initial metaphor Kant used to speak of the perspective of physical geography, we find that the traveler is not a bad metaphor for the person who is prudent. The traveler is one who is not so enmeshed in situations that he cannot extricate himself enough to think of ways in which he could use others. The traveler is more of an observer than actor, and it seems precisely that prudence requires that one become an observer of sorts. The traveler notices without being noticed. One who knows human beings and how to use them, observes them without letting on that he is observing.[42] A prudent person does not ever reveal that she is using others as means to her own ends, since this would only arouse resistance and refusal to cooperate. But not to be an observer of humankind is only to open oneself to being used by others without one's consent and hence made into a fool. And finally, the traveler must think for herself. She has no one to guide her, so she must make decisions for herself and initiate action. Hence, in using the traveler as the guiding image of the physical geography lectures, Kant is preparing his students for prudence, a very useful skill.

We saw that Kant's purpose in lecturing on physical geography was not to further the science of his day, but rather to further the ability of his students to learn a skill that would be useful in their lives. That skill was primarily thinking for themselves. Thinking for oneself is the essential condition for prudent action. Prudence is a very useful skill. It is necessary for achieving happiness in life. Happiness is a universal desire of all human beings. Hence, the physical geography was not just a pleasant discipline meant to entertain students, but a serious pedagogical discipline meant to lead students to the final ends of their lives by helping them develop the skills they will need to be happy.

Notes

1. G. Gerland, "Immanuel Kant, seine geographischen und anthropologischen Arbeiten," *Kantstudien* 10 (1905): 1–43; 417–547, here 446.

2. *Geschichte und Naturbeschreibung der merkwürdigsten Vorfälle des Erdbebens, welches an dem Ende des 1755sten Jahres einen großen Theil der Erde erschüttert hat,* Akademie Edition of *Kants gesammelte Schriften* (hereafter AK, followed by volume and page number), 1: 460.

3. *Physische Geographie,* AK 9: 182.

4. *Physische Geographie,* AK 9: 192.

5. *Physische Geographie,* AK 9: 363.
6. *Physische Geographie,* AK 9: 395.
7. *Physische Geographie,* AK 9: 425.
8. *Physische Geographie,* AK 9: 427.
9. *Physische Geographie,* AK 9: 427.
10. *Physische Geographie,* AK 9: 313–4.
11. *Physische Geographie,* AK 9: 315.
12. *Physische Geographie,* AK 9: 317.
13. *Physische Geographie,* AK 9: 317.
14. *Physische Geographie,* AK 9: 317.
15. *Physische Geographie,* AK 9: 201.
16. Johann Jacob Schatzen, *Kern der Geographie, das ist, kurtze und deutliche Beschriebung unsere Erd-kugel,* (Leipzig, 1762); Johann Hubner, *Vollständige Geographie,* 2 vols., 6th ed. (Hamburg: Conrad König, 1748).
17. Paul Menzer, *Kants Lehre von der Entwicklung in Natur und Geschichte* (Berlin: Georg Reimer, 1911), 79.
18. *Entwurf and Ankündigung eines Collegii der physischen Geographie*, AK 2: 3.
19. *Entwurf and Ankündigung eines Collegii der physischen Geographie*, AK 2: 3.
20. "An Answer to the Question: What is Enlightenment?," in Immanuel Kant: *Perpetual Peace and Other Essays,* trans. Ted Humphrey (Indianapolis: Hackett Publishing Company, 1983), AK 8: 35.
21. *Entwurf and Ankündigung eines Collegii der physischen Geographie*, AK 2: 3.
22. *Entwurf and Ankündigung eines Collegii der physischen Geographie*, AK 2: 4.
23. *Nachricht von der Einrichtung seiner Vorlesungen in dem Winterhalbenjahre von 1765–1766*, AK 2: 313.
24. *Anthropologie in pragmatischer Hinsicht*, AK 7: 140.
25. *Von den verschiedenen Racen der Menschen*, AK 2: 443.
26. *Anthropologie in pragmatischer Hinsicht*, AK 7: 214.
27. *Anthropologie in pragmatischer Hinsicht*, AK 7: 119.
28. *Anthropologie in pragmatischer Hinsicht*, AK 7: 201.
29. *Anthropologie in pragmatischer Hinsicht*, AK 7: 205.
30. *Anthropologie in pragmatischer Hinsicht*, AK 7: 200.
31. *Anthropologie in pragmatischer Hinsicht*, AK 7: 198.
32. *Anthropologie in pragmatischer Hinsicht*, AK 7: 239.
33. *Anthropologie in pragmatischer Hinsicht*, AK 7: 201.
34. *Anthropologie in pragmatischer Hinsicht*, AK 7: 211.
35. *Anthropologie in pragmatischer Hinsicht*, AK 7: 244.
36. *Anthropologie in pragmatischer Hinsicht*, AK 7: 272
37. *Anthropologie in pragmatischer Hinsicht*, AK 7: 273.
38. *Anthropologie in pragmatischer Hinsicht*, AK 7: 272.
39. *Lectures on Ethics*, eds. Peter Heath and J. B. Schneewind, trans. Peter Heath (Cambridge: Cambridge University Press, 1997), AK 27: 384.
40. "An Answer to the Question: What is Enlightenment?," AK 8: 35.
41. "An Answer to the Question: What is Enlightenment?," AK 8: 41.
42. *Anthropologie in pragmatischer Hinsicht*, AK 7: 121.

10

The Place of the Organism in Kantian Philosophy

Geography, Teleology, and the Limits of Philosophy

David Morris

> In [physical geography], we are concerned with the setting of nature, the earth itself, and those regions where things are actually encountered. But what matters [in contrast] in the system of nature is not birthplace [*Geburtsort*], but similarity of form.[1]

> I understand by an analytic of concepts . . . [not an analysis of the content of concepts, but the] *analysis of the faculty of understanding* itself, in order to research the possibility of *a priori* concepts by seeking them only in the understanding as their birthplace [*Geburtsort*] . . .[2]

It is easy to think that Kant's critical project aims to hermetically seal philosophy within the limits of reason, as this alone would give philosophy a rigorous foundation. But Kant's project continuously explores problems that involve reason with something beyond its limits. Indeed, the central problem of the transcendental dialectic in the *Critique of Pure Reason* (*KRV*) is that reason tends to overstep its own limits in pursuing ideas of God, and so on. In this case, while reason's overstepping concerns things beyond reason's limit, it is prompted by ideas generated seemingly from *within* reason's limits, via imagination. In contrast, the *Critique of the Power of Judgment* (*KU*)[3] studies teleological judgments prompted by the particularities of organisms. Unlike judgments of causality, teleological judgments do not determine their objects, rather they proceed and are warranted by reflection on the particular system of causal relations through which the understanding cognizes organisms. To this extent, in teleological judgment the understanding and thence reason "lean on" particularities arising *beyond* reason's limit. In Kant and the subsequent tradition of German idealism, this "leaning" leads to questions about the harmony of reason and nature. The issue of reason "leaning on" outside particularities is only amplified when we see how it resonates with Kant's point in the *Physical Geography* (*PG*) that an organism's birthplace (*Geburtsort*) is important to it, especially if this point is understood in light of the discussion of the teleology and development of organisms in *KU* (and related writings[4]). Birthplace is important because, according to Kant, particularities of organisms are shaped by the places in which they develop. Comprehension[5] of organisms therefore requires attention to birthplaces—to particularities beyond even the organism's limit. Geography and biology thus involve reason in particularities well beyond reason's limit that stem from place as a domain of particularities (vs. space as universal manifold that is one and the same everywhere).

As we shall see, this opening of the organism to its birthplace is no mere empirical accident, but inheres in the very logic of organisms. To comprehend an organism as having a purpose is to comprehend something that closes itself as a system, but only through its relation to outside particularities, to places. I suggest that this logic of the organism—which keeps cropping up in Kant's earlier empirical and later critical writings—in fact echoes the logic of reason's overstepping its limits in the *KRV*'s dialectic.

This seemingly strange echo between organisms, geography, and dialectic is rooted in the issue of imagination. As May notes, geography has a significant place in Kant's philosophical system partly because it is important to education and is thus propaedeutic for philosophical knowledge.[6] May highlights the link between geography and philosophy by referring to some passages from Kant's book on *Education*. There Kant writes that "Children very generally have a very lively imagination, which does not need to be expanded or made more intense by the reading of fairy tales. It needs rather to be curbed and brought under rule, but at the same time should not be left quite unoccupied."[7] Here we must recall that the imagination is central to *KRV* because its spontaneity allows the understanding to go beyond the given to produce schemata that can subsume an intuition of manifold particulars under universal concepts. But if imagination is to be productive of knowledge, its spontaneous overstepping of the limits of the given must be limited by rules imposed upon the imagination.

What is significant for us is that immediately after the above passage in *Education*, Kant lauds geography, specifically maps, as ideal for inviting the free but ruled play of the imagination. Why is this? Kant writes that "There is something in maps which attracts everybody, even the smallest children."[8] Indeed, we can see that maps attract us by inviting us to imagine ourselves in other places. This is what children love to do with maps. But the map provokes imagination in relation to a place that it maps. So reading, drawing, or reproducing a map entails imagining another place while limiting one's imagination according to the mapped place and according to rules of scale, topography, etc., through which such a place is represented. (Even if that place is imaginary, like Narnia, it still imposes rules upon one's maps of it.) In mapping, the imagination is thus both receptive to place and productively spontaneous in creating rules as condition for representing place. In other words, in our receptivity to place, we find an important empirical (experiential) resource of particularities that are of potential import to philosophy itself, since these intrude on reason in ways that prompt and cultivate rule-bound imagination. Yet these particularities precisely emerge from beyond reason's limit.

My suggestion is that that the imagination's role in the system of reason is kin to the role of place in supplying particularities crucial to the empirical education of imagination and to the birth of the organism. Transcendental dialectic is crucial to delimiting the system of reason, but in this dialectic reason learns to withhold from overstepping its limits only insofar as imagination generates ideas that invite such overstepping: in transcendental dialectic, reason is, as it were, receptive to ideas that it wants to map, yet it must not take them as pointing to anything real. But to have this role, imagination must generate particularities that exceed reason, otherwise it could not invite overstepping. Imagination is thus like a place inside reason that generates excessive particularities crucial to the genesis of the system of reason, in the way that outside place is crucial to the generation of the organism as organic system.

I draw out these points by first sketching the role of limits and place in Kant's critical philosophy and in his discussion of organisms in *PG*. Sections two and three deepen the connection between place, limit, and organism, through a detailed reconstruction of Kant's critique of teleological judgment that shows how, for Kant, places beyond the organism's limits can be inherently important to the organism as self-limiting and developmental. The conclusion shows how this gives new insight into Kant's dialectic and his remarks about the "birthplace" and "epigenesis" of reason. Recent scholarly debate on the latter intriguingly suggests that Kant draws models of the system of reason from biological models of development.[9] My point is that for Kant this would suggest a sort of internal opening or excess within reason, via imagination.

At the outset, I should emphasize that my reading of Kant—which explores reason as a system, and links the issue of system with issues of limits, insides, and outsides, and finds an echo of these issues in Kant's geography and biology—is informed by looking back at Kant from the tradition of later German idealism, especially Schelling and Hegel. Whether or not these themes are connected for Kant himself as they are for this later tradition is a question requiring further investigation. What I want to point out, though, and what I find extraordinary, is the way that Kant's philosophy keeps turning to concrete issues such as geography, place, and organisms, and the way that these concrete issues echo in his critical philosophy. In attending to the imagination as the site where something concrete and beyond the limits of philosophy erupts, I am also inspired by Heidegger's discussion of the role of imagination in Kant[10]—but I am approaching the imagination from a much more concrete direction, from within Kant's discussion of geography and organisms, as intersecting with his critical philosophy.

Place and Limit in Kant's Philosophy

Kant's critical project aims to secure philosophy and science by inscribing them within the limits of reason. The project faces two challenges. The first is internal to philosophy, namely reason's inherent tendency to overstep the understanding's and thence reason's limits. This challenge, studied in *KRV*'s dialectic, is transcendental and partly stems from the imagination's role in reason.

The second challenge, studied in the critique of teleological judgment in *KU*, is empirical. Empirically, the world exhibits more purposiveness than reason can itself constitute. This excess purposiveness is manifest in general rules and patterns of nature (particular biological organizations and rules of physics) that appear harmonious with reason. This harmony invites reason to explain these rules and patterns. But any explanation from within reason's limits is in fact illegitimate, because reason can only determine abstract, not particular, laws, and the organism exceeds the forms of mechanical explanation legitimated by universal laws. The critique and dialectic of teleological judgment thus hinges on the difference between universal laws deployed in determinative judgments, which can legitimate mechanical explanation of inorganic phenomena; and general rules and principles of reflective judgments, which afford only comprehension (not explanation) of organic phenomena.[11]

Organisms—like maps in Kant's *Education*—thus invite reason to overstep its limit: their very form begs rational comprehension, even while strict explanation of them exceeds reason's limits. But in doing so organisms—also like maps—precisely limit reason's overstepping. It is the very limits of the organism—its articulate boundaries and divisions between organs—that at once invite and rule comprehension. Much as maps invite children into imaginative accounts of the interrelations of places, the organism's visible parts invite children to give imaginative accounts of the interrelation of organs (especially the child's own organs). So organisms, like maps, are inherently helpful in the education of a ruled imagination.

Organisms, like maps, thus limit imaginative overstepping—but the limit is coming from something external to reason. This is our point above that the organism opens reason to particularities beyond its limit, a point that deepens below as we see that the organism inherently opens to a place beyond its limit.

To begin placing this point within Kant's project, we need to remind ourselves how reason's reflective regulation of itself is key to reason countering its own tendency to overstep limits, and how this overstepping and reflective regulation work on two different registers, the transcendental and the empirical. *KRV*'s transcendental-dialectical overstepping of limits is countered by the critical realization that some ideas are merely regulative in their employment. The empirical overstepping of limits in natural science (studied in *KU*) is similarly countered by the critical realization that some principles of judgment are again strictly regulative rather than determinative in their employment: they specify subjective rules that aid comprehension, rather than laws that determine phenomena. More precisely, such principles work on a subjective (rather than a determinative and thence objective) level, by way of a reflection on what is empirically given, in which we realize that the particularities of the given can be comprehended by us only as ordered in a certain way, without yet claiming that that ordering is constitutive of the given thing itself.[12]

The key point for us is that in the case of transcendental-dialectical overstepping, what regulates reason are ideas specified and generated *within* reason's limit. In contrast, in comprehending organisms, what regulates reason is a teleological principle grasped in reflecting on our mapping of the functional anatomy of the organism. Such a principle is thus grasped in response to particulars *outside* reason's limit. Kant's geographical works ramify this point by noting that places (*Örter*) beyond the organism enter into these particulars.

This connection to place arises because the organism's development can depend on outside places. Kant's attention to this organism-place connection in *PG* and related works stems from a critical, and methodological point, namely that comprehension of organisms requires something more than natural description (*Naturbeschreibung*): it requires natural history (*Natursgeschichte*).[13] To comprehend an organism is to grasp particulars distinctive of it and its function. This first of all requires grasping and identifying the organism in terms of the particulars distinctive of its species. So comprehension first requires classification of organisms, and part of geography's importance for knowledge (and thence philosophy) is as propaedeutic to such classification. Geography is propaedeutic because natural description (as opposed to natural history) is driven by merely formal criteria of systematicity that reason imposes from within its own limits. For natural description, any distinction of an organism will do so long as it lets reason generate a systematic formal classification. But the formal distinctions convenient for reason's classification system may turn out to be accidental to what really makes an organism distinctive, thus subverting comprehension. For example, absence of red blood has long been taken as an essential distinction of invertebrates, but scientists have recently discovered a species of deep-sea clam that *has* red blood.[14] While red blood seems a convenient marker of the vertebrate/invertebrate distinction, it is in fact accidental to that distinction: it only marks a descriptive similarity between organisms. Such marks only allow reason to draw what Kant calls scholastic divisions. Scholastic divisions are helpful for systems of memorizing the different kinds of organisms,[15] but, because they are picked out by reason, they can be accidental and miss the natural distinctions that really make organisms distinct. So Kant turns to the organism's lineage: natural distinctions will be crucial to the actual life, reproduction, and descent of the organism from previous generations; they allow reason to draw what Kant calls natural divisions—divisions that are there in nature itself.[16]

But natural divisions require that reason open itself to the natural history of organisms, and thence to organisms, as what Kant calls natural purposes (*Naturzwecken*). What marks a distinction as natural (vs. accidental) is its role in a *purpose* intrinsic to the organism. It turns out that the deep-sea clam's blood is red because it has evolved a variant of haemoglobin that can derive energy from reduced hydrogen sulphide spewed from the undersea volcanic vents in the place in which the clam lives. The natural distinction here is not the red color, but the presence of haemoglobin (which accidentally happens to be red). Where other invertebrates get by without it, haemoglobin is vital to this clam's life—because of its dwelling place near deep-sea volcanoes.

For Kant, attention to natural divisions and natural purpose inherently opens the organism—and thereby reason—to differences of place. If we gather a variety of organisms and describe them merely in terms of characteristics they present here and now; without regard to their place of origin, we are liable to divide them scholastically, according to what Kant calls a natural system (see *PG* quote below). But if we pay attention to the places organisms come from, two things may happen.

First, we may find that a kind of organism, such as our deep-sea clam, is unique to a place. This uniqueness of place, together with the descriptive similarity of organisms that marks them as being of a kind, warrants the judgment that the kind of organism forms a unique lineage. In effect, a difference that is physiologically extrinsic to the organism—*that it is from this place and/or unique to it*—becomes the basis for drawing a natural division, since it marks the lineage of the organism, by way of a connection between natural geography and history that inheres in the fact that organisms (as singular beings) are particular to places. This sort of connection between place and species identity is often marked in common or even scientific names, for example, Canadian hemlock (*Tsuga canadensis*).

Second, we may be able to connect the organism's living place or birthplace—a difference seemingly *extrinsic* to it—with something intrinsic to it. For example (given our present knowledge), presence in a terrestrial organism of haemoglobin capable of deriving energy from reduced hydrogen sulphide licenses the inference that the organism lives near undersea volcanic vents. And since this intrinsic difference in haemoglobin marks an extrinsic difference in the place in which it lives, it connects with the organism's natural history and thence its lineage. So, intrinsic differences that link with extrinsic differences of place might be especially helpful in drawing natural (vs. scholastic) divisions.

The above overlaps between organisms, natural division, places, natural history, and natural purposes are behind an important passage of the introduction to the *Physical Geography*, which is worth citing at length:

> Division of knowledge according to concepts is logical; according to time and space it is physical. By means of the former [logical division], we obtain a natural system (*systema naturae*), as for example that of Linnaeus. With the latter [physical division], we obtain a geographical description of nature.
>
> If, for example, I say that the species "cattle" is one of the kinds of four-footed animals . . . then this is a division I make in my head: it is a logical division. The *systema naturae* is, as it were, a kind of register of the whole, wherein I situate all things, each in the class to which it belongs, even if on earth they are to be found in widely separated areas.
>
> In accordance with the physical division, however, things are considered in terms of the positions [*Stellen*] they occupy on earth. The system indicates their position [*Stelle*] in the classification. The geographical description of nature, on the other hand, seeks after [*nachweisen*] the positions [*Stellen*] on earth where those things are actually to be found [*jene Dinge auf der Erde wirklich zu finden sind*]. Thus, for example, the lizard and the crocodile are basically the same animal. The crocodile is only an enormous lizard. But the places [*Örter*] where the two are found are different. The crocodile lives on the Nile; the lizard on land, including in our area. In sum, we are concerned with the setting [*Schauplatz*] of nature, the earth [*Erde*] itself, and those regions [*Gegenden*] where things are actually encountered. But what matters [in contrast] in the system of nature is not birthplace [*Geburtsorte*], but similarity of form.[17]

Morphologically, with respect to merely descriptive similarities, reason could very well classify the crocodile as a lizard, just bigger. But once we note that it lives on the Nile, and not here, we begin identifying it as having its own lineage, and begin attending to markers of it as belonging to this lineage and not others. The places in which organisms are found—as extrinsic yet intimate to organisms—become a matrix for a division of nature that avoids the problems of a merely natural system. But to do this "geographical description" reason needs to seek after and study (*nachweisen*) the actual places in which things are found (not just abstractly indicate their position, as in physical division). (That is, physical division merely needs to note the latitudes and longitude of things, locate them in a universal space, whereas geographical description needs to study the concrete particularities of the places in which they are found.) Geography, as the matrix for comprehension of organisms, is thus important to systems of knowledge and indeed philosophy—a central point of Kant's introduction to *PG*, which aims to justify philosophical interest in geography.

Behind this passage from Kant's geography is a deeper argument for a link between the organism and its birthplace that also supports *PG*'s general appeal to an intimate connection between the intrinsic and extrinsic characteristics of organisms. The organism-birthplace link is rooted in the teleology and development of the organism. To grasp this link—and its ties to issues of systems overstepping their limits—let us first reconstruct the thrust of Kant's critique of teleological judgment in the second part of *KU* and then tie this to his geographical works.

Kantian Teleology and the Organism as Overstepping Its Limits

Kant's critique of teleological judgment begins with an initial tension that leads to a central problem governing the rest of his argument. The initial tension is endogenous to the very project of critical philosophy and is tied to the issue of overstepping limits previously discussed. The tension emerges from within the limits of reason, or rather from reason's attempt to comprehend particularities beyond its limit—nature—on the basis of conceptual resources warranted by reason working within its limit. As we will see, the logic of this tension forces Kant to conceive purposes as all and only intrinsic to organisms as *self-limiting* systems. But this very logic also forces Kant to acknowledge that the process of organic self-limiting spills over its limit. *PG*'s link between intrinsic and extrinsic characteristics is thus inherent in the very logic of organisms as spilling over their limit.

The initial tension in Kant's critique of teleological judgment is this. On the one hand, transcendental principles give us "good reason to assume . . . a subjective purposiveness of nature in its particular laws,[18] for comprehensibility [*Faßlichkeit*] for the human power of judgment. . . ."[19] That is, reason has warrant to *subjectively think* that there are purposive forms in nature that appear *as if* adapted to our subjectivity and judgments, such that nature appears *as if* an object ordered by technique or craft—and as thereby comprehensible to us.

On the other hand, it is *only as if* nature is purposive. Whilst the *KRV* shows that nature is subject to universal laws (since causal ordering is a transcendental condition of any experience), this specification is so abstract as to leave room for nature to be structured in a "thousand ways." So there is no a priori warrant for determining that nature must have a purposive form and thereby be comprehensible to us. Yet we can in fact comprehend nature, and comprehending nature means experiencing it as having a purposive order.[20] As Kant puts it, teleological judging thus "belongs to the reflecting, not to the determining power of judgment," yet it serves as a principle "for bringing [nature's] appearances under rules where the laws of causality about mere mechanism do not suffice."[21] That is, we judge that something is purposive not because our act of judgment itself determines that thing as purposive (as is the case in causal judgments), but because, in reflectively encountering the particularities of that thing (and in reflecting on our encounter with it) we find that comprehending the thing entails bringing it under a rule that grasps a particular kind of order in it.

We must now ask: what sort of rule/order does something manifest such that it warrants being judged purposive and what sort of object actually warrants such a judgment? The rule and warrant cannot be purely a priori, precisely because the power of teleological judgment is reflective, not determinative of its object. So the sort of order in question cannot devolve from reason itself but must be manifest in the particularities of a natural thing beyond the limit of reason. Kant stipulates that such a thing—a natural purpose—manifests a purpose that is material (not merely formal, as in the circle's harmonious form) and objective (not merely subjective, as when *we* find a purposiveness without objective purpose in something beautiful).[22] But it must manifest purpose without introducing a new sort of causality into nature. So purpose must be manifest in terms of mechanical cause and effect relations ("mechanism"), for this causality alone can be legitimately judged in experienced, natural objects. (It is, for example, illegitimate to claim that our teleological judgments are warranted by an experience of subjective intentions/purposes lodged *within* natural objects, as might be claimed, e.g., by a vitalist doctrine.)

This articulation of the problem behind teleological judgment lets us appreciate how Kant's core insight regarding teleology is carefully calibrated to respond to this problem. Here is his insight:

> Experience leads our power of judgement to the concept of an objective and material purposiveness . . . only if there is a relation of the cause to the effect . . . such that we can understand the relation as lawful only insofar as we find ourselves capable of subsuming the idea of the effect under the causality of its cause, as the underlying condition of the possibility of the cause.[23]

To unpack this rather dense sentence, what warrants teleological judgment is a peculiar material object, one that can be comprehended by us as lawfully[24] governed only if we conceive the effects that the object causes . . . as conditioning the very causes that produce these effects. Put more simply, teleological judgment is warranted only in cases where conceiving the object as lawful entails conceiving causes that already depend on their effects (contrary to the order of mechanical causality). Why would this kind of dependence of causes on effects warrant teleological judgment? To comprehend a cause that depends on its effect is to comprehend an object as itself ordered to producing an effect—and this means grasping a purpose in an object. Cases where we are forced to conceptualize causes as depending on their effects (where, reflectively, we find ourselves deploying such concepts and rules in comprehending things) thus warrant teleological judgment.

We must be precise here, though. We cannot judge that there is, in the object itself, a cause that depends on its effect. (This would illegitimately introduce into nature a new sort of causality, since, in cases of mechanical causality, effects depend on causes, not vice versa.) It is just that (in the case we are considering) if *we* want to comprehend the object as governed by a law, we find (on reflection) that we must conceptualize a cause that depends on its effect. That's the only way we can make sense of it. Nonetheless, and this is the key point, it is the very chain of causality, *legitimately experienced in the object itself,* that enjoins us to suppose—*if we are to comprehend nature*—a cause that depends on its effect.

Here we are in the middle of a very fine balancing act at the limit of reason. As reflective, teleological judgment is warranted by the transcendental demand that we comprehend things—but by this demand as responsive to mechanical causes in an object. On the one hand, the rule that we apply has a form that reflects us and the limits of reason (a kind of simplicity of rules and orders that we can follow and that allows comprehension), but nonetheless, it is the object beyond the limits of reason that itself demands and thus manifests this rule. (Compare, here, the case of the map ruling the child's imagination.)

The question remains: What sort of object genuinely warrants teleological judgment and demands that we apply it? Kant's reply is governed by the constraint that the object itself must be the source of this demand (since nothing within reason can warrant it a priori). Relative purposes (as Kant calls them) do not do this. The sea's withdrawal from the land causes sandy soil to be deposited, which effect is good for (and thus relative to) spruces that thrive in sandy soil. But this warrants no judgment that the cause (the sea's withdrawal) depends on the effect (nourishing spruce). After all, all sorts of other effects might have been and are caused by the sea's withdrawal. And in any case it is only when we comprehend that the sandy soil serves the spruce's purpose that we could even comprehend deposition of sandy soil as tied to any effect beyond just moving sand around. Judging the sea's withdrawal as purposeful, therefore, already entails judging a purpose in the spruce; it is only relative to this that the sea's withdrawal can be judged purposeful. The relative purpose that we can judge in the sea thus begs the question of what warrants a judgment of purpose in the first place.[25] While it may be possible to say that the sea, relative to the spruce, stands as a purpose, it is not the case that we can say that the sea is possible *only* as a purpose; if we consider the sea in itself, it is not serving a purpose. The sea in itself does not warrant a teleological judgment.[26]

The logical problem with the sea and other examples of relative purposes is that in these cases causes and effects are *external* to one another. The causes and effects can exist or happen independent of one another, they are beyond one another's limits. So there is no warrant for saying that the cause is *necessarily* dependent on the effect and therefore no warrant for saying that the cause is in itself a purpose. As Kant puts it, relative purposes involve *extrinsic* purposive relations, and these warrant only judgments of hypothetical purpose (if the spruce is purposive, then sandy soil is purposive for it), rather than judgments of categorical purpose.[27]

Organisms solve this logical problem of the externality of cause and effect. This is because an organism is both "*cause and effect of itself*,"[28] on a species and individual level. Individuals that constitute the species are the cause of the individuals who constitute the species (it is spruces that cause spruces to come to be). The species is thus the cause and effect of itself. And this cause is *necessarily* dependent on the effect, because there would be no species (as cause) if the species did not keep producing its mortal members (as effect). The point is sharper on the individual level: The individual spruce causes the sandy soil to change into spruce material. We must understand this cause (the spruce's nutritive process as actualized in its organs) as based on and governed by its effect (producing the spruce organs—which alone actualize the nutritive process as cause). Here we do not run into the problem that the cause may be producing this effect only accidentally or relative to some further term . . . because the effect is in turn the cause of its cause. The spruce material produced by the spruce makes the organs that feed the spruce that make spruce material . . . ad infinitum.[29]

The organism, in other words, manifests a closed loop of causes and effects. It is the closure of this loop—and consequent shift from extrinsic relations to relations intrinsic to the terms of a closed loop—that warrants judgment that in the object itself there is a cause that *must* be comprehended as depending on its effect. *What warrants the judgment of purpose is a causal closure intrinsic to the organism.*

This closure becomes even clearer in Kant's central formulation of natural purposes. As Kant famously writes, in order for a thing to be a natural purpose, it is necessary first "that its parts (as far as their existence and their form are concerned) are possible only through their relation to the whole" and second "that its parts be combined into a whole by being reciprocally cause and effect of their form."[30] In such a system, parts and wholes reciprocally depend on one another for their structure, maintenance, and existence. We can find the sea apart from the spruce tree. But we cannot find functioning spruce leaves outside the spruce and its circulatory system, which depends for its function on the leaves and in turn feeds the spruce and the circulatory system that depends on them. Such a system is both "an *organized* and a *self-organizing* being,"[31] a self-producing being, and such "*[a]n organized product of nature is that in which everything is an end and reciprocally a means.*"[32]

In sum, Kant's effort is to analyze teleological judgment as a reflective judgment operative in and crucial to our comprehension of nature. But for Kant this sort of judgment must be warranted by nature itself in its objective mechanical, causal ordering. Kant finds his warrant in self-limiting objects beyond reason's limit that, as self-limiting, provoke and warrant a kind of comprehension that, in reflecting back on itself, finds that such comprehension entails conception of causes that (internally) depend on their effects. The crucial point for this chapter is that teleological judgment is warranted all and only in cases of self-limiting (closed-loop) systems. Kantian teleology, in its very conception, pertains to *systems that define their own limits.*

Kant's principle of teleology is marvelously and powerfully anticipative of an important recent conception of the organism, namely Francisco J. Varela's concept of autopoiesis. Varela too conceptualizes the organism as a system in which the parts and the whole reciprocally depend on and produce one another, both with respect to their form and their existence.[33] But Varela seems to go further than Kant when he argues that such a system importantly depends on building and maintaining a *boundary* that *partitions* the organism from its environment. However, the above reconstruction shows that the very logic of Kant's argument is such that a reciprocal relation of parts and wholes may be judged as intrinsically (and not just relatively) purposive only insofar as it describes a self-limiting and thence bounded system that defines itself in face of an environment. (This bounding, we could note, is the first line drawn by the "organism as map"—and its maintenance entails the mapping of further lines, namely those between the organs, which organs maintain the organism/environment boundary.)

Yet Varela and his followers are very much attuned to the way that the logic of organismic self-bounding puts in question any easy boundary between the inside and the outside of the organism. This is because the organism's building and maintenance of a boundary precisely entails that the organism be open to drawing in outside material, from which alone the organism forges the inside-outside boundary. So, the autopoietic organism is not only a self-limiting system (as Kant would agree): the very logic of its self-limiting is such that this limitation is achieved only by the organism overstepping its limits.[34]

But Kant himself is aware that the teleology of organisms is such that they overstep their limits: (1) In his own example of the growth of organisms, the closed loop of the organism takes in outside material (this is why food producing causes are important as relative purposes.[35] (2) In his own example of the species being the cause of itself,[36] the closed loop of causes extends across individuals. Similarly in *KU* §82, Kant admits that the only legitimate case of judging that extrinsic relations between discrete things are purposive, is in the case of relations between organisms of different sexes—male and female creatures are intrinsically related to one another across their limits. (3) Kant, in an observation that echoes Bergson,[37] notes that because we can graft scions from one kind of tree onto another and arrive at a living hybrid, "one can regard every twig or leaf of one tree as merely grafted or inoculated into [the one tree], hence as a tree existing in itself" that is parasitic on the one tree. We can equally think of the tree as one limited thing, or a plurality of limited things linked across their limits. In general, Kant realizes that the closed loop of the self-delimiting organism in fact depends for its closure on *not* being closed, on crossing over between the organism and its environment, between individuals of different generations and sexes, and even between parts that are, in their function, only ambiguously closed.

In other words, judging a purpose in an organism logically entails comprehension of the organism as at once delimiting itself within a broader place, yet being open to that place and things contained in it. This is our first step to the argument behind *PG*'s linkage between the organism and its birthplace. The second step involves Kant's epigenetic view of development, to which we now briefly turn.

Place and the Epigenesis of the Organism

KU's line of argument naturally leads Kant to a concern with development, because accounting for the development of organisms in terms of mechanism could reconcile teleology and natural causality. But Kant's claims about development in *KU* §81 plunge us into a heated and complex scholarly debate. In a nutshell, the debate concerns Kant's position on a spectrum of biological positions contemporary to Kant. At one end of the spectrum, development is conceived as the "evolution" of an already preformed germ (*Keime*). (Evolution here means "unfolding"—it does not yet have its "Darwinian" connotations). At the other end, development is conceived as "epigenesis," the production of a genuinely new organism via a purely mechanical process that operates above and beyond any already given germ or genesis (hence "epigenesis"). The prevailing and most convincing claim is that Kant lies somewhere between these poles: on the one hand, Kant rejects strict preformationism as incoherent (because, e.g., it requires an appeal to the supernatural and cannot explain hybrids); on the other hand, he rejects purely mechanical epigenesis, because as he famously puts it, we do not yet have a Newton who could "make comprehensible even the generation of a blade of grass" according to mechanical processes.[38] Arguably,

Kant advocates a "generic preformationism": a germ, irreducible to mechanism, must already be given if the organism is ever to come to be, but this germ does not specify the individual to the last detail, it is generic to the species, and includes generic predispositions (*Anlagen*); as mere predispositions, *Anlagen* are such that mechanical factors can play on them to plastically shape the organism.[39]

This claim about the organism is crucial to the organism-birthplace link in *PG*. And it is also crucial to issues concerning comprehension of organisms discussed in Kant's pieces on the teleological principle (*TP*) and the different races of men (*RM*), both of which (like *PG*) are concerned with comprehending the difference of human "races." The key point for us is that in *TP* and *RM* Kant aims to defend "monogenesis," the claim that all humans have descended from one stock, contra "polygenesis," which claims that human races constitute fundamentally different species. Once again we enter an area of intense scholarly debate, here focused on Kant's concept of race.[40] But we can say that Kant defends monogenesis both for reasons of explanatory economy and coherence, and because the claim that humans are split by fundamental biological differences would undermine Kant's cosmopolitanism and claims about the centrality of rational will to human being.

What is important to us is that "generic preformationism" allows Kant to claim that humans are one species *and* to account for their differences—but only by appealing to places as shaping different races. In this context Kant also remarks on place and climate as effecting, e.g., the formation of feathers in birds, which fits with his position (about, e.g., crocodiles) in *PG*. But here we can deepen the point by referring back to our analysis of *KU*: *KU* conceptualizes organisms as *active* self-organizers. So, logically speaking, it would have to be something beyond the organism's limit that causes its predispositions to vary in development. A wholly active developmental predisposition (*Anlage*) would in fact be a preformationist germ (*Keime*). And what factor within a wholly self-activating and closed organism would let it vary in its development? The active organism must therefore be passive to something outside it—to place. And we have already seen that the very logic of the organism, as warranting teleological judgments, entails that organisms as actively self-limiting must be passively open to place.

Conclusion: Place and the Epigenesis of Reason

We have seen that when reason tries to thoroughly comprehend the organism, it must comprehend it in relation to the organism's place—a central claim of *PG*. The *KU* exposes the underlying logic of this point: organisms are self-limiting and therefore bounded against the openness of place; but organisms are also linked with outside place as providing relative purposes that are internally crucial to the organism as intrinsic purpose. *KU* §81, *TP*, and *RM* argue that this link with place is crucial to the organism's development, helping secure *PG*'s claim. The upshot is that when reason comprehends an organism, its reflective application of the teleological principle spills beyond the limits of the organism as self-limiting system—reason must attend to place. This gives a different route to our initial point about geography's importance in Kant's philosophical system: geography fills in what is missing within the limits of the organism itself.

Earlier I noted that the map, in its relation to place, invites the limited overstepping of limits that is crucial to the logic of Kantian imagination. And I noted that the organism invites a similar play of the imagination. We can now see that the organism does so in its relation to place, via the organism's *own* overstepping of its limits. The organism is not just like a map, but like a map that points beyond itself to the place that it maps. The map invites *our* imaginative receptivity to the place beyond it to which it points. But the crocodile or clam with haemoglobin *itself* demands that we ask: what place is it from? *It* demands that we seek after and study (*nachweisen*) its place. Organisms are themselves systems that overstep their limits into a place beyond.

Issues of rational systems overstepping their limit inhere in Kant's critical project. Curiously, reason finds organisms beyond reason's limit embodying this rational issue of overstepping limits. In the organism, reason finds a self-limiting system like itself, and in the organism's relation to place, reason finds a system like itself that organizes itself in face of something beyond it, through a passive/active relation to place that echoes imagination as receptive/spontaneous.

Here we can turn to Kant's famous remark that the categories arise from a "system of the *epigenesis* of pure reason."[41] Scholars take this remark to be drawing an analogy between the genesis of the categories, and the genesis of the organism, such that it becomes crucial whether epigenesis, in Kant's conception, is wholly productive a posteriori or whether it retains an a priori preformationist moment. And the answer has been sought in Kant's relation to biology of his time (see the previous discussion), such that Kant's view of the epigenesis of reason might turn on empirical claims and discoveries about biological epigenesis.[42] I suggest that the reason-epigenesis link in fact turns on a logic deeper than the merely empirical, on the logic of systems inherently open to something beyond their limit. Reason finds this logic exemplified both in organisms, and in reason, insofar as the categories of the understanding do not seal reason within its limits, but dialectically open it to something beyond it—in a way that involves precisely the play of imagination that opens at the limits of maps and organisms. The key point here is that in our analysis, the overstepping of the organism's limit is *not* empirically subsequent to the organism, but is inherent in its very logic—the organism forms itself out of a place beyond it.

If this were the case with reason, if the dialectical overstepping of limits that understanding's finitude invites were not empirically subsequent to reason, but transcendentally inherent in its very logic, then we could make new sense of Kant's discussion of the epigenesis of reason and his related remark (in our epigraph) about the understanding as birthplace of the categories. Kant's discussion would not reduce to a mere analogy between reason and biological development, a reduction that troubles efforts to read Kant as articulating a *transcendental* and not merely empirical argument.[43] Rather it would point to a systematic incompletion that opens at the limit of both reason and organisms, an incompletion that affiliates reason and organisms and that is at play in both reason's dialectic and reason's comprehension of organisms. And this would mean that the understanding as birthplace of reason is not fully preformed and given, but open to a place beyond reason.

Here we would have to suggest that reason's opening to a place beyond it involves imagination as active/passive, receptive/spontaneous. This is fitting, given our initial interweaving of maps, organisms, and imagination, and given, e.g., Heidegger's provocative claims about the centrality of imagination to Kant's philosophy.[44] But this would also fit, e.g., with Casey's emphasis on Kant's philosophy as open to place,[45] e.g., in Kant's analysis of left and right hands as incongruent counterparts that are comprehensible yet not reducible to abstractions explicable with reason's limit. Most of all, this would invite an intriguing connection between Kant, and Schelling's and Hegel's development of Kant.[46] For what is suggested here is that the birthplace and epigenesis of reason, the self-organizing of the categories, arises through a logic of desire, of systems, by their very own logic, stepping beyond themselves, in such a way that the movement of the system of reason finds its echo in the movement of nature and organisms, without yet reducing reason to an organic model.

Notes

1. *Physical Geography*, AK 9: 160. Citations are given using the pagination in the *Akademie Ausgabe* of *Kants gesammelte Schriften* (AK, followed by volume and page number). The translations used here are drawn from Ronald L. Bolin, trans., *Immanuel Kant's Physical Geography* (MA Thesis, Indiana University, 1968) and Olaf Reinhardt, trans., *Physical Geography* in *Natural Science* (Cambridge: Cambridge University Press, forthcoming), with modifications.

2. *Critique of Pure Reason*, A66/B90–1. Translations of the *Critique of Pure Reason* are from Paul Guyer and Allen W. Wood, trans., *Critique of Pure Reason* (Cambridge: Cambridge University Press, 1998), with modifications.

3. Translations of *Critique of the Power of Judgment* are from Paul Guyer and Eric Matthews, trans., *Critique of the Power of Judgment* (Cambridge: Cambridge University Press, 2000).

4. The most important of these are: "Über den Gebrauch teleogischer Principien in der Philosophie," AK 8: 157–84, translated as "On the Use of Teleological Principles

in Philosophy," in *Race*, ed. Robert Bernasconi (Malden: Blackwell, 2001); "Von den verschiedenen Racen der Menschen" of 1775, AK 2: 427–44, translated as "On the Different Human Races: An Announcement of Lectures in Physical Geography in the Summer Semester of 1775," in *Philosophy and Race*, ed. Robert Bernasconi and T. Denean Sharpley-Whiting (Albany: State University of New York Press, forthcoming).

5. On comprehension vs. explanation, see note 10.

6. Joseph A. May, *Kant's Concept of Geography and its Relation to Recent Geographical Thought* (Toronto: University of Toronto Press, 1970), 133.

7. *Education*, trans. Annette Churton (Ann Arbor: University of Michigan Press, 1960), 78.

8. *Education*, 78.

9. See note 39.

10. See note 45.

11. Following *Critique of Judgment*, §61, I distinguish reflective judgment as (merely) comprehending (*fassen*) things from determinative judgment as explaining (*erklären*). I also distinguish the universal (*Allgemeinheit*), as determined a priori, and the general (*general*), as determined empirically (see *Critique of Judgment*, §§7–8, AK 5: 213).

12. On reflective judgment see *Critique of Judgment*, introduction, §V and John H. Zammito, *The Genesis of Kant's Critique of Judgment* (Chicago: University of Chicago Press, 1992).

13. See *On the Use of Teleological Principles in Philosophy*, AK 8: 161–3 for an articulation of the *Naturbeschreibung/Natursgeschichte* distinction; also see the *Physical Geography* excerpt below. See Mark Fisher, "Kant's Explanatory Natural History: Generation and Classification of Organisms in Kant's Natural Philosophy," in *Understanding Purpose: Kant and the Philosophy of Biology*, ed. Philippe Huneman (Rochester: University of Rochester Press, 2007) and Phillip R. Sloan, "Kant on the History of Nature: The Ambiguous Heritage of the Critical Philosophy for Natural History," *Studies in History and Philosophy of Biological and Biomedical Sciences* 37 (2006): 627–48 for helpful discussions of the historical context and importance of this distinction.

14. Tony Koslow, *The Silent Deep: The Discovery, Ecology and Conservation of the Deep Sea* (Chicago: University of Chicago Press, 2007), 95.

15. *On the Different Human Races*, AK 2: 429.

16. See *On the Different Human Races*, AK 2: 429. See Sloan, "Kant on the History of Nature" for a discussion.

17. *Physical Geography*, Introduction, §4, AK 9: 159–60.

18. Note that "law" in this passage and the following discussion in fact concerns general rules of nature, not (confusingly) universal laws determined by reason. On laws vs. rules see *Critique of Judgment*, AK 5: 360, 378.

19. *Critique of Judgment*, AK 5: 359.

20. *Critique of Judgment*, AK 5: 359–60.

21. *Critique of Judgment*, AK 5: 360.

22. *Critique of Judgment*, §62.

23. *Critique of Judgment*, §63, AK 5: 366–7.

24. See note 18.

25. *Critique of Judgment*, §63.

26. See the opening sentence of *Critique of Judgment*, §64.

27. *Critique of Judgment*, §67, AK 5: 378.

28. *Critique of Judgment*, AK 5: 370.

29. *Critique of Judgment*, §64.

30. *Critique of Judgment*, AK 5: 373.

31. *Critique of Judgment*, AK 5: 374.

32. *Critique of Judgment*, AK 5: 376.

33. On autopoiesis and its relation to Kant, see Francisco J. Varela, Humberto R. Maturana, and Ricardo Uribe, "Autopoiesis: The Organization of Living Systems, Its Characterization and a Model," *BioSystems* 5 (1974): 187–96; Francisco J. Varela, "Organism: A Meshwork of Selfless Selves," in *Organism and the Origins of Self*, ed. Alfred I. Tauber (Dordrecht: Kluwer, 1991); Andreas Weber and Francisco J. Varela "Life After Kant: Natural Purposes and the Autopoietic Foundations of Biological Individuality," *Phenomenology and the Cognitive Sciences* 1 (2002): 97–125; and Evan Thompson, "Life and Mind: From Autopoiesis to Neurophenomenology. A Tribute to Francisco Varela," *Phenomenology and the Cognitive Sciences* 3 (2004): 381–98.

34. This is especially apparent in Dorion Sagan and Lynn Margulis, "Epilogue: The Uncut Self," in *Organism and the Origins of Self* and this overstepping of boundaries is also emphasized by endosymbiotic theories of the organism, which emphasize how organisms evolve by incorporating other organisms into themselves; see Lynn Margulis, *Symbiosis in Cell Evolution: Life and its Environment on the Early Earth* (San Francisco: W.H. Freeman and Company, 1981); Lynn Margulis and Dorion Sagan, *Acquiring Genomes: A Theory of the Origins of Species* (New York: Basic Books, 2002).

35. *Critique of Judgment*, AK 5: 367–9.

36. *Critique of Judgment*, AK 5: 371.

37. Henri Bergson, *Creative Evolution*, trans. Arthur Mitchell (Mineola: Dover, 1998; original edition, 1907), 16.

38. *Critique of Judgment*, AK 5: 400.

39. For an entry into the debate see Phillip R. Sloan, "Preforming the Categories: Eighteenth-Century Generation Theory and the Biological Roots of Kant's A Priori," *Journal of the History of Philosophy* 40 (2002): 229–53; John H. Zammito, " 'This Inscrutable *Principle* of an Original *Organization*': Epigenesis and 'Looseness of Fit' in Kant's Philosophy of Science," *Studies in History and Philosophy of Science* 34 (2003): 73-109; Sloan, "Kant on the History of Nature." For helpful background, see Robert John Richards, *The Meaning of Evolution: The Morphological Construction and Ideological Reconstruction of Darwin's Theory* (Chicago: University of Chicago Press, 1992) and Justin E. H. Smith, *The Problem of Animal Generation in Early Modern Philosophy*. (New York: Cambridge University Press, 2006).

40. For an entry point, see Robert Bernasconi, "Who Invented the Concept of Race? Kant's Role in the Enlightenment Construction of Race," in *Race*.

41. *Critique of Pure Reason*, B167.

42. See Günter Zöller, "Kant on the Generation of Metaphysical Knowledge," in *Kant: Analysen-Probleme-Kritik*, ed. Hariolf Oberer and Gerhard Seel (Würzburg: Königshausen & Neumann, 1988) and note 39 for an entry point.

43. Here is the issue. Sloan, "Preforming the Categories" gives a careful study of Kant's texts in relation to the biological controversies of his time. He shows how the biological distinction between *Keim* ("germ") and *Anlage* (roughly, "disposition"), which

is crucial to debates about preformation vs. epigenesis in embryology, can shed light on texts where Kant, e.g., discusses the origin of the categories in terms of an "epigenesis of reason." Indeed, Sloan very helpfully shows how Kant's concepts are nuanced and influenced by the biological models from which Kant draws his terms.

What is worrisome is Sloan's suggestion—which is quite a different matter—that the terms deployed by Kant retain an essentially biological meaning, such that Kant's account of reason is bedded in biology. E.g., Sloan at one point notes that the categories "would still be biological properties"; he approaches Kant as applying an "epigenetic thesis" from biology to the study of cognition (245); and he concludes that "If my thesis is correct, Kant did indeed ground the categorical a priori on a contemporary version of biological preformation" (251–2). This might suggest that biological concepts are applied in Kant's critique of reason more or less wholesale, and become a driving force in it.

No doubt biological debates in Kant's time offer Kant conceptual tools for articulating his arguments and analysis, and inflections in Kant's shifting positions echo inflections in the resources offered by biological debates. But this does not entail that Kant's claims reduce to, are grounded in, or are driven by the biology of his time. One of the central things that philosophers do is take up existing terms and concepts and appropriate them for new purposes, giving them new meaning. When Aristotle uses the Greek word for lumber, *hyle*, to discuss what we now call matter, the word takes on a new and independent meaning. To be sure, the Kantian terms that Sloan discusses retain threads of connection with their biological originals, partly because Kant is interested in the biological issues in their own right. But there is a very big difference between saying that changing biological results drive or ground Kant's shifting philosophical claims, and saying that Kant's argument (and shifts in it) are enabled by or find an outlet or motive in shifting biological results. In the former case we would have to ask why Kant turns to biology (and not, say, chemistry) to draw grounding concepts for his study of reason; the answer is not obvious, other than that Kant is questioning the origins of something with a complex structure, as does embryology—but does that analogy in itself warrant the application of biological concepts? The suggestion here is that Kant's appropriation of biology is motivated by an analysis of reason, independent of biology, that reveals a shared logic (of overstepping limits) in the system of reason and in living systems. In this case biology would be a resource for appropriating concepts relevant to such systems—but the appropriation of biological terms would be mediated by a conceptual framework there in advance of biological concepts (even if, paradoxically, the biological concepts are helpful or even crucial to articulating the prior conceptual framework). Critical philosophy and transcendental argument, not biology, would remain in the driver's seat. But critical philosophy would have to overstep itself into biology to appropriate the concepts needed to articulate itself.

When Kant says that there is an epigenesis of reason, he is not reducing this process of reason to something biological or to a biological model, he is not drawing a mere analogy; he is critically noting that reason, like the organism, is a self-limiting system that nonetheless becomes articulate through its relation to something beyond its original limit. Sloan's exposition of the biological background helps us understand that it is this critical point, and not some other one, that Kant is making when he deploys the language of epigenesis. But we have to insist that Kant's point is critical, not biological; and the suggestion here is that Kant's point is not based on a mere analogy, but if anything, on a systematic logic that is at work in both the system of reason and living systems.

44. See Martin Heidegger, *Kant and the Problem of Metaphysics*, 5th ed., trans. Richard Taft (Bloomington: Indiana University Press, 1997). I am indebted to Emilia Angelova for my understanding of this point.

45. Edward S. Casey, *The Fate of Place: A Philosophical History*, (Berkeley: University of California Press, 1997).

46. Frederick C. Beiser, "Kant and *Naturphilosophie*," in *The Kantian Legacy in Nineteenth-Century Science*, ed. Michael Friedman and Alfred Nordmann (Boston: MIT Press, 2006) offers support for this connection in light of *Critique of Judgment*'s critique of teleology.

IV

Kant's Geography of Reason

Reason and Its Spatiality

11

Kant's Geography of Reason[1]

Jeff Malpas and Karsten Thiel

Kant's interest in geography marks him out as unusual among philosophers. Few have engaged with geographical concerns in any significant way, yet not only did Kant lecture extensively on the subject, but the lectures he gave every year between 1756 (only two years after his inauguration) and 1796, and which were published in 1801 under the title *Physische Geographie*, represent the very first lectures on the subject within a modern university setting.[2] From the perspective of the history of geography, Kant thus occupies a founding role in the development of geography as an academic discipline. From a purely philosophical perspective, however, Kant's geographical engagement is also significant, since it directly connects with and informs key elements in his own philosophical project. In this respect, Kant can be construed as one of the pioneers, perhaps the very first, in the project of a "philosophical topography"—a project that aims to explore the manner in which space, and also place, figure in human knowledge and experience as both the object of such knowledge and experience, and as part of its very structure.[3] This is especially so in that most famous of his works, the *Critique of Pure Reason*, first published in 1781, which inaugurates a trilogy of critical investigations dealing with pure, practical, and judgmental reason—with metaphysics, morals, and aesthetics. While the first *Critique* takes metaphysics, and not geography, as its focus, it does so in ways that nevertheless constantly invoke the geographical and direct us towards the topographical, and that establishes the critical project itself as "geographical" in character.

As the prefaces to both the first and second editions of Kant's great work make clear, the concern of the *Critique of Pure Reason* is with the reform and possible revival of metaphysics. The two prefaces disagree, however, over exactly how this is to be achieved. According to the preface to the first edition, the task of the *Critique* is to relocate metaphysics in its proper place—to achieve an appropriate "placing" of metaphysics. Mislocated, improperly placed, metaphysics becomes itself a place of strife—"the battlefield of endless controversies."[4] The battle that is waged here is one between, on the one side, those who aim to "rebuild [metaphysics], though never according to a plan unanimously accepted among themselves," and, on the other, those "nomads who abhor all permanent cultivation of the soil."[5] As Kant explains later, in "The Transcendental Doctrine of Method," the philosophical nomadism at issue here provides a "resting-place [*Ruheplatz*]," but "not a dwelling-place for permanent residence [*Wohnplatz*],"[6] and this Kant clearly views as an inadequacy, while the attempt to engage in philosophical building can achieve nothing without an agreed plan. The result is that metaphysics is left unsettled, unplaced, and uncertain.

The preface to the second edition seems more modest in the ambitions it sets out—and, at first glance, those ambitions no longer seem to be expressed in geographical terms. Rather than looking to establish a proper "dwelling-place" for metaphysics—somewhere it can settle—the preface to the second edition seems content simply to bring some stability to metaphysics, to set it into a "persistent state [*beharrlicher Zustand*]."[7] Yet once one looks further into how this is to be achieved, one sees that Kant's conception of the task at hand, even when construed in this apparently more modest fashion, continues to be explicated in terms of geographical concepts. The Kantian stabilization of metaphysics is to be undertaken by establishing the proper boundaries within which it can maintain itself—boundaries that inhibit it from knowledge of the absolute.[8] Kant's characterization of his project in the preface to the second edition of the *Critique of Pure Reason* is given, then, in terms of a certain task of "mapping"—a mapping that will define the realm of knowledge by marking it off from the realm of uncertainty. It is thus that, in the section of the *Critique* titled "The ground of the distinction of all objects in general into phenomena and noumena," a section that remains, although with some changes in both editions, Kant develops his famous image of "the land of truth" as an island beset by dangerous and shifting seas, surrounded by mists and fog-banks.[9]

Yet although it is commonplace to note the geographical orientation of Kant's work in the first *Critique*, as well as elsewhere in his writing, seldom is any sustained attention given to exactly how that geographical orientation actually plays out in Kant's inquiry, or even to what extent and in what ways it is indeed integrated into that inquiry.[10] Our claim is that the manner in which the geographical intersects with the philosophical Kant's work is crucial for understanding the Kantian critical project as well as suggesting connections between the geographical and the philosophical that go beyond the Kantian framework alone. Consequently, in this essay, rather than explore the lectures on geography as such—the material presented in the *Physische Geographie*—we intend to make use of the ambiguity in the term "Kant's Geography" to explore the idea of a Kantian *philosophical* geography, particularly as that is developed in the *Critique of Pure Reason*. What we aim to show is not only the way in which geography may intersect with philosophy, but also how two different conceptions of the geographic are in play in Kant's thinking, and the manner in which the critical project that is undertaken in the *Critique,* and elsewhere, is itself dependent upon the prioritization of one of these.

Kant refers directly to geography when describing his task in the "Transcendental Doctrine of Method"—the much smaller second part of the *Critique of Pure Reason* (the first part being the "Transcendental Doctrine of Elements"). In the section entitled "On the impossibility of a skeptical satisfaction of pure reason that is divided against itself," Kant refers to David Hume as one of the "geographers of human reason,"[11] but he describes Hume's geographical achievement as incomplete since the geography of reason that he sets out "merely limits [*einschränken*] our understanding without drawing boundaries [*begrenzen*] for it."[12] Referring back to the characterization of philosophical nomadism that appears in the first preface, and that is also invoked at the very beginning of the "Transcendental Doctrine of Method,"[13] Kant describes Humean skepticism as providing only a "resting-place"—a kind of temporary surveyors' camp—rather than "a dwelling place for permanent residence."[14]

There can be little doubt that Kant himself aims to carry the geographical task beyond Hume, and that Kant therefore included himself among the ranks of the "geographers of reason"—although, unlike Hume, Kant's approach aimed at the establishment of a more permanent place of habitation. Kant's reference to geography here is thus significant in that it shows that Kant does not see himself as the only geographer in Western philosophy, and so does not use the idea of a "geography of reason" as a way to separate himself from his predecessors. Nonetheless, Kant also defines his own contribution to the geographical project of philosophy in a way that does make that contribution quite distinctive—a distinctiveness that depends on recognizing the difference between the marking of a mere limit and the drawing of a boundary.

The distinction between limit and boundary is itself one that Kant explicates in further detail in the so-called "Conclusion" to the *Prolegomena to any Future Metaphysics*, published in 1783, under the heading "Determining the Boundary of Pure Reason" (so-called because the section seems more properly to summarize ideas from the *Critique*, especially the "Transcendental Dialectic," rather merely from the *Prolegomena*—a point that, as we shall see below, turns out to be quite important). In one of his most explicit discussions of the difference between boundaries and limits, Kant explains that

> Boundaries [*Grenzen*] (in extended things) always presuppose a space [*Raum*] that is found outside a certain fixed location [*Platz*], and that encloses that location; limits [*Schranken*] require nothing of the kind, but are mere negations that affect a magnitude insofar as it does not possess absolute completeness . . . In all boundaries there is something positive (e.g., a surface is the boundary of corporeal

> space, yet is nonetheless itself a space; a line is a space, which is the boundary of a surface; a point is the boundary of a line, yet is nonetheless a locus in space), whereas limits contain mere negations.[15]

The manner in which Kant characterizes the idea of boundary invokes the geographical, even if the concept of limit as employed here does not, but Kant's characterization is also explicitly topographical inasmuch as the idea of boundary is directly connected with the idea of a location or place—"Boundaries . . . always presuppose a space . . . outside a certain fixed location, and that encloses that location." The characterization may even be said to echo Aristotle's characterization of place (*topos*) in *Physics* V as "the limit [*peras*] of the surrounding body, at which it is in contact with that which is surrounded,"[16] where limit ought itself be understood, as Edward Casey argues, in terms of a notion of boundary.[17] Thus, in his discussion of Aristotle in his philosophical history of place, *The Fate of Place*, Casey writes:

> This discussion leads us to distinguish between *boundary* and *limit* . . . limit, like shape, belongs primarily to what is limited and only secondarily to what does the limiting (e.g., a container) . . . To be a boundary, by contrast, is to be exterior to something or, more exactly, to be around it, enclosing it, acting as its surrounder. As such a boundary belongs to the container rather than the contained—and thus properly to place conceived as the inner surface of the containing vehicle.[18]

Casey takes the idea of boundary to be closely connected with the concept of place, and one might argue that the difference between Kant and Aristotle on this matter is simply that whereas Aristotle uses the notion of boundary, together with that of containment, to characterize place, Kant uses the idea of place, together with containment, to characterize the idea of boundary.

The relation between boundary and place is one to which we shall return, but for the moment it is important to remain with the distinction between boundary and limit, and, in particular, to note the way in which Kant's emphasis here is on the idea of boundary as opposed to that of limit. This is, of course, in keeping with Kant's geographical orientation—boundary is a geographical concept, one that calls on a notion of space and of place, in a way that limit alone does not. Admittedly, although this might be obscured by the emphatic way in which Kant (and Casey) presents the distinction, every boundary can also be construed as a limit. Yet not every limit is itself a boundary—limit implies only the idea of a certain negation, even though it is a negation that may be realized in a variety of ways.

Correlative to the distinction between boundary and limit, where limit is understood as "mere negation," is another distinction—one that obtains between two forms of ignorance. Ignorance is a limit in knowledge, but such limit can be of two kinds. It may be a limit that derives simply from the fact that there are things with which we are not yet acquainted, or it may derive from the way in which our knowledge is itself determined, and so limited, in its very nature. Ignorance that arises on the basis of the former may be termed accidental or contingent, while ignorance that arises on the basis of the latter may be termed necessary—so we can come to know things of which we are now ignorant, but we can never overcome the ignorance that follows from the nature of knowledge as such. In the discussion in "The Transcendental Doctrine of Method" Kant puts this distinction as follows: "all ignorance is either that of things or of the determination and boundaries of my cognition [*Alle Unwissenheit ist entweder die der Sachen oder der Bestimmungen und Grenzen meiner Erkenntnis*]."[19]

Yet one may well ask whether there actually are any such necessary limitations on our knowledge. Kant's response is to point out that even were we to suppose that our ignorance were merely an ignorance of things, that is to say, were merely contingent, we could never ascertain this without an investigation of the determination and possible boundaries of our cognition—which means that we could never do so without an investigation of the possibility that our ignorance is itself necessary. We may well acknowledge, on the basis of what we observe or perceive that we are ignorant of what lies beyond our past or present perceptual acquaintance—and yet we can never, on this basis alone, arrive at any understanding of whether or not or in what manner our perceptions, and the knowledge arrived at through such perception, may be necessarily constrained. Put simply we may say that the necessary boundedness of our cognition, if indeed there is such boundedness, cannot be merely observed—it is not, in other words, a matter merely of *perception*—but can only be uncovered through *science*, that is, through the critical examination of the structure of cognition.[20] The latter task is precisely what Kant attempts in the *Critique*. Moreover, since Kant associates the idea of limit with the contingent ignorance deriving from things, and the idea of bound with the necessary ignorance deriving from the structure of cognition, so the scientific task of the *Critique of Pure Reason* is indeed one of exploring the proper "bounds of sense."

The difference between an approach that looks merely to the contingent limitation on our knowledge as opposed to one that aims to uncover the necessary bounds that derive from its underlying structure is explored further in Kant's exposition in the second part of the *Critique* by means of an example that is especially relevant in the present context. Kant writes:

> If I represent the surface of the earth (in accordance with sensible appearance [*Schein*]) as a plate [*Teller*], I cannot know how far it extends. But experience teaches me this: that wherever I go, I always see a space around me in which I cannot proceed farther; thus I cognize the limits [*Schranken*] of my actual knowledge of the

> earth at any time, but not the boundaries [*Grenzen*] of all possible description of the earth. But if I have gotten as far as knowing that the earth is a sphere and its surface the surface of a sphere, then from a small part of the latter, e.g., from the magnitude of one degree, I can cognize its diameter and, by means of this, the complete boundary [*die völlige Begrenzung*].[21]

The manner in which our location on the surface of the earth appears to us can indeed be construed in terms of our location on the surface, and at the center, of a flat plate. What lies beyond the horizon of the surface, and the distance in which that surface may extend, can never be determined on the basis of our location at the center of the plate—and, of course, in terms of the way in which things appear, we are always at the center of that plate inasmuch as we are always at the center of the visual horizon. Yet if we understand the surface on which we are located to be spherical, and since to be located on the surface of the sphere is already to be located on its circumference, then from the small part of the surface that is accessible to us, on the basis of the very place in which we are located, we can determine the extent of that surface.

Not only does Kant's use of this example draw upon notions from topology and geometry, which themselves can be seen to be at work in his geographical approach to knowledge, but it also draws heavily on the character of that original situatedness on the basis of which knowledge arises. The way we find ourselves in the world is indeed within a field of appearances that has both a center and a periphery, but whose periphery as given in any particular appearance cannot be taken to be identical with the constraining boundary for all appearance. Only on the basis of an investigation of the manner in which the field of appearance is itself structured—on the basis of an investigation of the underlying character of the field that is already given in any particular portion of the field in the same way as the curvature of the surface of a sphere is itself present in any and every arc on its surface—can we possibly determine the boundaries that determine the field in its entirety.

One might argue, however, that we cannot assume the character of the field of appearance beforehand, and since its nature must therefore be viewed as uncertain, we cannot be sure that the field of appearance does not have merely the character of a flat surface rather than the curvature proper to a sphere. Kant acknowledges that "the sum total of all possible objects for our cognition [*Erkenntnis*] seems to us [*scheint uns*] to be a flat surface,"[22] yet he also asserts that "our reason [*Vernunft*] is not like an indeterminately extended plane . . . but must rather be compared with a sphere."[23] Kant does not provide us with any elucidation of the grounds that underpin the latter claim thereby making it appear almost as if it were an unsubstantiated, perhaps even "dogmatic," assertion. There is, of course, a shift in the one passage compared to the other from talk of *cognition* to talk of *reason*, and this corresponds to a shift from talk of how cognition appears to talk of how reason is (or which it is like). The key point, however, is that it is when we look to understand the *structure* of cognition, which understood in one way can be identified with reason (viewed as that which contains the a priori principles of cognition[24]), as opposed to the *content* of cognition (the term "cognition" being ambiguous between these senses), then we are forced to recognize the manner in which cognition does indeed possess an underlying character that is determinative of it, and so also determines its bounds, in just the same way that the surface of the sphere is determined by the nature that belongs to the sphere as such.

One way to express this is to say that, understood in terms of its structure, cognition, and so also reason, is indeed to be compared to a sphere. So it is not that there are two possibilities here—namely that cognition is either like a plane or like a sphere—and that we are uncertain which corresponds with cognition's true character, but rather that to understand cognition as having a determining structure is to understand it in terms that make it analogous to a sphere. Or to put matters slightly differently, one may say that cognition, understood in terms of its content, is analogous to a surface, which can itself be understood as either planar or as spherical. Understood as planar, there is no possibility of understanding its extent and boundaries on the basis of any known section of the plane (assuming, of course, that whether or not the plane has a perimeter, that perimeter is never reached), since the plane has no boundary that is given in its surface; understood as spherical, the extent and boundary of the surface can be understood on the basis of any known section of it, since, as we have already seen, from the curvature of that section, the extent of the entire surface can be determined.

If we understand the field of cognition as indeed like a flat plate, then we will never be able to demonstrate its boundaries from within it, but equally no amount of investigation will be able to rule out the possibility that it has boundaries—the most it will be able to show is that we have not so far discovered them. On the other hand, if we understand the field as possessed of its own curvature, then once we have identified that curvature, we will also be able to determine the field's boundaries. One might object here, of course, that the latter possibility depends on one other ungrounded assumption, namely, that the field is uniform. Perhaps this assumption would be ungrounded if we were operating solely in the field of empirical geography. Certainly, in the *Physische Geographie,* Kant himself notes that the curvature of the earth's surface is not completely uniform, and is constituted as not a true sphere, but a spheroid:

> Concerning the form of the earth, it is a sphere, or, as Newton has determined more precisely from the main principles and from gravity, a spheroid. This claim has been affirmed by several observations

> and measurements. Imagine the shape of the earth as if surrounded completely by water, a hydrostatic figure of it. The mountains do not make a difference as they cannot be seen in the umbra [*Erdschatten*] and the highest among them amounts to not even a 1,900th part of the diameter of the earth. Proofs of the spherical figure of the earth are as follows: 1. The sun does not rise at the same time everywhere, which would be necessary if, as was long believed by many, the earth were flat. From this, it would follow only that the earth is round from the morning until the evening. But 2. even the latitude and longitude are not the same at all spots. When we travel 15 miles further towards South, then the polar star is by about one degree lower and by about one degree higher when we travel the same further North, until it is in the vertex at the pole. From that we are right to conclude that the earth is curved from North to South. 3. At times of lunar eclipse, the umbra is, in all possible positions of the earth, always curved. 4. Even when the view on the plain sea is unrestricted, you first sight only the top of the objects and bit-by-bit the lower parts of them. 5. Man has travelled the earth by ship towards all regions, which would have been impossible if it had not been spherical in figure. The above-mentioned spheroidal shape of the earth results from the fact that, caused by laws of weight and gravity, all matter that is closer to the poles accumulates against the equator und agglomerates around it, which would happen even if the earth were completely covered by water, and because there is no movement at the pole, but most at the equator, and so the cross section through the poles, i.e. the axis of the earth, is smaller than that at the equator. Newton has proved that any body that can move freely must take this shape.[25]

Of course, what Kant's explanation here shows is that the spheroidal, rather than spherical, shape of the earth does not in fact represent a real failure of uniformity—indeed, its spheroidal character is what is to be expected of a sphere when subjected to the particular conditions that obtain in the case of the earth. There is, then, no real failure of uniformity in the case of empirical geography, and neither can there be any such failure in the epistemic case, since Kant will argue that, with respect to the field of cognition, uniformity or regularity is itself a prerequisite for the possibility of the field being constituted from the very start.

The field of cognition, the "space of reason," is by its very nature unitary and determined by a set of underlying principles. Of course this is true independently of how the field *appears*, since it is built into the idea that the field has a character that belongs to it. The very possibility of a critical investigation in fact depends on an assumption of the unity of the field to be inquired into. While we can never demonstrate that there is no necessary bounding to experience, we can demonstrate, through uncovering the principles that determine the field, that it does indeed have certain necessary bounds. But this task is identical with uncovering the character of the domain, of the place, in which we find ourselves. Recalling the connection that appears in both Kant and in Aristotle, we may say that we uncover the boundaries of cognition, and of reason, through uncovering the place, the proper *topos*, within which cognition is located. It is significant that much the same idea can also be seen at work in Kant's *Physische Geographie*. There Kant complains that in the so-called *systema naturae* of his time we find things collected and arranged, but no proper system. Kant's own physical classification instead considers things according to the place [*Stelle*] which they occupy on earth,[26] and this systemic ordering of things provides a "fore-conception" {*Vorbegriff*] that enables us to anticipate aspects of our future experience.[27]

Uncovering the place that is at issue in relation to cognition is not, as we have already noted, a matter of simply marking out a space. Crucial to the task of uncovering the proper place of cognition and of reason is the establishing of its bounds as these derive from its underlying structure. Exhibiting the nature of the place in this way can also be construed as a matter of exhibiting its underlying unity.[28] The issue of unity in Kant's account of cognition and of reason is a central one, and Kant's own characterization of the problem of the *Critique of Pure Reason* in terms of the question concerning the possibility of synthetic a priori judgment already makes this clear almost from the very start.[29] The question of the unity of a *place* is not, as such, something that Kant himself addresses—not even in his lectures on geography. But he does discuss what might be understood as an analogous question concerning the unity of a *space* (although he does not put it in that way) as that arises in terms of our grasp of that space. In his 1786 essay "What does it mean to orient oneself in thinking?" Kant argues that

> To orient oneself means to use a given direction (when we divide the horizon into four of them) in order to find the others—literally to find the *sunrise*. Now if I see the sun in the sky and know that it is now midday, then I know how to find south, west, north, and east. For this, however, I also need the feeling of a difference in my own subject, namely, the difference between my right and left hands . . . If I did not have this faculty of distinguishing, without the need of any difference in the objects, between moving from left to right and right to left and moving in the opposite direction and thereby determining a priori a difference a difference in the objects, then in describing a circle I would not know whether

> west was right or left of the southernmost point of the horizon, or whether I should complete the circle by moving north and east and thus back to south. . . .[30]

Kant takes these considerations to show that our grasp of space—our geographical orientation within it—must be based in what he calls a "subjective ground of differentiation," and this is sometimes taken as itself evidence of Kant's prioritization of the subject.[31] From the perspective of our current concerns, however, the more important point is the way in which the grasp of a space as a whole is a matter of grasping the differentiation between the different parts or regions of that space, and the ordered relation between them, but one can only do that if one is able to position oneself in that space, which means being able to relate the differentiation of the space to the differentiation in one's own body. We would add, although Kant does not emphasize this, that understanding the unity of a space therefore also means being able to move or to act within it that space in a coordinated manner (a point that does appear in Kant's exposition through his reference to the manner in which one would complete a circular movement in that space).[32]

One conclusion that can be drawn from these considerations is that the Kantian project of uncovering the proper place of cognition and of reason is necessarily a matter of understanding the way in which the cognizing subject is herself placed, and of the relation between the subject and the larger space in which she is located—finding reason's place is also to find our own place. Moreover, given the way in which Kant understands cognition and reason as necessarily constituting a single field modeled on the unified field of appearances within which we find ourselves, and given also the way in which that field is understood as constituting a certain sort of space, then we may say that the very manner in which Kant understands the unitary structure of cognition as a unified spatial field already carries within it the idea of space as itself necessarily unitary. The latter thesis has often been disputed (and Casey sees it as contributing to the problematic prioritization of space as single and homogenous extension[33]), but if that thesis is seen as itself related, even if only indirectly, to Kant's geographic or topographic understanding of cognition and of reason, then it surely appears in a somewhat different light. Indeed, Kant's insistence on there being only one space should not be understood as a claim that could possibly be subject to empirical refutation (as Anthony Quinton for one has suggested[34]), but instead as deriving from the unitary character of the place in which cognition and reason are necessarily located, and so from the unitary character of reason and cognition as such (and also, we might add, from the unity of the cognizing, and *acting*, subject).

The task of establishing the limits or bounds of cognition or knowledge, as Kant understands it, involves more than merely demarcating that which is properly known from that which is not. Although Kant designates Hume as one of the geographers of reason, the irony is that because he never attends to the proper boundaries of reason, merely identifying certain ideas as not properly within the domain of knowledge, Hume never manages to work out where or how reason is *placed.* He is thus like the explorer who wanders aimlessly across a stretch of country, enumerating the variety of features that successively present themselves, also noting that some features reported by previous travelers cannot be located within that countryside at all, but who is never able to form those features into an understanding of the country or region, of the entirety of the place, considered as a whole. In this respect, we may say that Hume is like the old-fashioned geographer whose geography consisted in little more than a series of observations of the lands through which he passes—like a series of travelers' tales—and who provides us merely with a catalogue of sights and locales, but who never provides us with any understanding of lands themselves. Such a collection of geographical observations may provide that out of which a true geography can arise, but it never becomes properly geography in its own right.

The establishing of the boundaries of cognition and of reason is thus itself dependent on the surveying of the territory in which we are already located. We cannot determine the bounds by looking beyond to what bounds (though this is not to rule out that there may be something external that bounds in this way), but must begin from the place in which we are located—in much the same way as the uncovering of the extent of a sphere can be discovered on the basis of the curvature of a part of its surface.

Yet although the geographical character of Kant's overall project is once again brought into view here through the reiteration of the connection between the boundary and place, there is also a complication in Kant's approach that needs to be further explored, and that becomes evident when we compare more closely the difference between the discussion in the "Transcendental Doctrine of Method" and that which appears in the "Conclusion" to the *Prolegomena.* Kant's distinction between contingent and necessary ignorance in the former, and the associated use there of the idea of boundary, clearly connects up with the distinction between limit and boundary in the latter. On the other hand, the plate/sphere analogy appears only in the "Transcendental Doctrine of Method," and is notably absent from the *Prolegomena* discussion. Moreover, while the way that Kant uses the idea of boundary in the discussion of contingent and necessary ignorance in the "Transcendental Doctrine of Method" depends on the idea of a boundary as involving more than a limit in the sense of a "mere negation," it does not, in contrast to the *Prolegomena* account, put any emphasis on the positive character of a boundary in the sense in which that can be taken to imply a space that lies beyond the boundary (one might well argue, in fact, that Kant's deployment of the idea of boundary in the "Transcendental Doctrine of Method" is *prima facie* inconsistent with the way he distinguishes boundary from limit in the *Prolegomena*). These differences turn out to connect with two different ways of understanding the notion of boundary between which Kant does not himself seem clearly to distinguish, but which can themselves be correlated with what might be viewed as two different modes of the "geographic."

If we look first to the account given in the *Prolegomena* we see that there Kant presents the distinction between limit and boundary in an apparently clear-cut fashion: boundaries involve something positive, so that in spatial terms, there is always a space that bounds and not merely a space that is bounded; limits are merely negative so that the assertion of a limit need involve no commitment to anything that stands outside or goes beyond that which is limited. This way of understanding the distinction between boundaries and limits is particularly important for the possibility of Kant being able to argue for the commitment to those ideas—God, the immortality of the soul, freedom—that he names "transcendental" and that are seen by him as necessary for the proper understanding of finite human being. The commitment to these ideas is itself discussed in the "Transcendental Doctrine of Method," but it comes much later in that discussion, and not in the section in which Kant makes use of the sphere/plate analogy. The "Conclusion" to the *Prolegomena*, however, is directly concerned with these ideas and their necessity—it is therefore committed to holding open a space in which these ideas can be located, and which must lie outside of the space of cognition. Consequently, the *Prolegomena* is much more concerned with the way in which the boundaries of cognition are determined, in part, by the enclosing space that surrounds cognition, rather than by the internal structure of cognition as such.[35]

This is not to say that the structure of cognition is ignored in the *Prolegomena* (and, indeed, the Transcendental Ideas may be taken to themselves play a role, or at least so Kant would argue, in the very structure of cognition), but only that the dependence of the boundaries of cognition on what lies outside cognition is emphasized over and above the way those boundaries derive from the structure of cognition as such. In the passages we have considered from the "Transcendental Doctrine of Method" the emphasis is entirely different. The way in which cognition is compared to a sphere directs attention away from the idea that the boundaries of cognition are dependent on some external enclosing space, and almost exclusively onto the way in which those boundaries derive from a structure that is, in a certain sense, "internal" to cognition.

The difference at issue here can be seen more clearly if we think about the nature of the boundedness that operates in the case of the sphere. The field of cognition is being compared here to a surface, and that surface, we discover, carries a curvature within it that determines the extent of the surface. As such, the curvature of the sphere entails a certain limit that belongs to the sphere, which is its circumference. Kant terms this limitation a boundary in order to distinguish it from the simple limit that belongs to the flat plane and which entails no such boundedness—the surface of the plane does not, as such, determine its extent (and as a result, even if the plane were finite, the perimeter of the plane could not be determined from any location interior to the surface of the plane). One might query Kant's use of the term "boundary" here, since, on the basis of his own characterization of boundary—according to which a boundary entails something positive whereas a limit involves only a negation—it might seem unclear in what sense talk of "boundary" when applied to the way in which the sphere is limited by its curvature does indeed involve something positive that bounds.

Certainly, the bounding of the sphere—which is a boundedness that applies to its *surface*—is not such that the space that is constituted by the sphere's surface is itself limited by some other space that encloses it or stands outside it. Yet the bounding of the sphere does imply something positive, namely, the character of the surface itself—the nature of the sphere—that can itself be specified and given content. The way the idea of boundary appears here sets it somewhat apart from any mere spatial boundedness—which seems to be the sort of boundedness that Kant focuses on in the *Prolegomena*—according to which a boundary is essentially a demarcation of one space from another in which the enclosed space is itself constituted through that spatial enclosure. But this does not mean that the relevant notion of boundary turns out, after all, to be set apart from any notion of the geographical, since we can say that this way of understanding boundary ties boundary directly to *topos as* understood in terms of its original meaning as *surface*—a sense that is evident both in the mathematical study of surface called topology, and in that part of geographical study known as topography.

The way in which we are brought back to the notion of *topos* at this point prompts some further consideration of the Aristotelian discussion of *topos* that we referred to above—a discussion with which Kant was surely familiar. On the one hand Aristotle emphasizes the way in which place requires the idea of some form of containment or enclosure. It is this idea that is echoed in Kant's understanding of the difference between a boundary and a limit, and which might, therefore, be said to legitimize the claim that Kant's inquiry into the bounds of cognition or of reason is also an inquiry into reason or cognition's proper place. On the other hand, the containment that is at issue here is mistakenly understood if it is treated as a containment that merely demarcates a certain limit pertaining to two otherwise similar bodies or spaces.

It is important to note that in Aristotle *topos*, while not identical with the thing placed (as it cannot be otherwise there could be no *change of place*), is nevertheless not identical with that which contains either. The *topos* or place is the inner limiting surface—the innermost boundary we might say—of the containing body, but in being such, *topos,* and the conception of boundary associated with it, is always defined from an orientation that is centered on the contained, and not the containing, body. In this respect, the notion of boundary that belongs with place, in both Aristotle and in Kant, is best understood in terms of the notion of *horizon*—itself derived from the Geek *horos* or *horismos* meaning boundary or definition—within which a certain field or locale is opened up. Such a horizon, while it does not refuse the possibility of what lies beyond it, is always turned towards the field that lies within its bounds—towards the place of which it is partly constitutive. For this reason one cannot treat the horizon as simply marking off two different spaces, one of which is within and the other without. Indeed, one can only think of the space beyond the horizon in terms of the expansion or movement of the horizon itself—strictly speaking there is no space outside the horizon at all since the horizon is just that which opens up such a space (therefore one cannot have a view of the horizon from without).

The character of the boundary as just such a horizon, and its inwardly-turned character, is something that Martin Heidegger remarks upon, in a famous passage in "Building Dwelling Thinking." There he writes,

> A space is something that has been made room for, something that is cleared and free, namely within a boundary [*Grenze*], Greek *peras*. A boundary is not that at which something stops but, as the Greeks recognized, the boundary is that from which something begins its presencing. That is why the concept is that of *horismos*, that is, the horizon, the boundary.[36]

It is worth noting that here Heidegger translates the Greek *peras*, usually taken to mean limit, as boundary, but this need not present any special problem since, as we noted above, a boundary is a limit of a certain sort, and the translation is also in accord with the reading of Aristotle's use of *peras* in his definition of *topos* to which Casey draws attention.[37]

The significance of this return to Aristotle, together with the brief foray into Heidegger, is that it indicates the way in which the notion of boundary cannot be understood, even recognizing its positive character, as directing us simply towards an enclosing space that lies outside. Equally it shows the way in which the concept of containment that appears so closely associated with the idea of boundary cannot itself be understood as always (or even primarily) a matter of the containment of one space within another such space.[38] Kant's own presentation of the matter in the *Prolegomena* misleads in its use of the example of the way the concept of boundary operates, as he specifically puts it, "in extended things," and is, in any case, not entirely in accord with the sphere/plate analogy that he employs in the "Transcendental Doctrine of Method." In the latter, the concept of boundary that is deployed (as well as the concept of containment) is much closer to the understanding of boundary that becomes evident through a closer attention to the Aristotelian account, and that is also evident in Heidegger, than to any idea of boundary as it might be supposed normally to operate in relation to spatial extendedness alone—much closer to a concept of boundary as horizon rather than boundary as a mode of mere spatial demarcation.[39]

The failure to grasp the way in which the notion of boundary must operate here has bedeviled much of the reading of Kant's critical project almost since its first appearance, and as we have seen, one of the reasons for the misunderstanding is based in Kant's own tendency to present the concept of boundary that is at stake in a way that does indeed tend towards just such a misreading.[40] To cite a relatively recent example, Graham Priest's claim that Kant is caught in a contradictory position in which he must both set bounds to reason at the same time as he oversteps those bounds[41] implicitly depends on construing the boundary at issue here as more like a form of spatial demarcation than a horizon, and on failing to recognize the way in which there can be no space beyond the horizon that is similar to the space opened up within it.

The underlying mistake that is evident here consists in confusing two different notions of boundedness. As we have already seen, one of these notions is itself close to the idea of limit as negation in that it is a notion of boundedness according to which the boundary marks off the point at which one space or territory stops and another begins. Yet this does not imply anything about the positive character of the spaces concerned, and the demarcation of those spaces need not be construed as other than arbitrary and contingent. This is the sort of boundedness that arises when we simply look to draw a line between two territories without regard for the nature of the territories so bounded. But there is another form of boundedness also and it is this latter form of boundedness—the boundedness tied to the idea of the boundary as horizon—that is so central to the understanding of the Kantian project. The latter form of boundedness arises, not on the basis of any arbitrary or contingent marking off of spaces, but is instead a function of the very nature of the spaces themselves, and enables the understanding of those spaces as constituted with a certain distinctive unity.

Earlier we suggested that the two notions of boundary that have emerged here can be correlated with two different senses or modes of the "geographic." Inasmuch as the distinction between these two conceptions of boundary are themselves tied to a notion, on the one hand, of horizontality associated with the notion of *topos* or place, and, on the other, of simple spatial demarcation, then so the first can be taken to look to the idea of place as ordered and heterogeneous, while the second gives priority to the idea of space as extended and homogenous.[42]

The space that is involved in this latter conception is one within which different features may be located and various spatial divisions—various spaces—marked out. The perspective from which that space, and the potentially multiple spaces it contains, is approached, however, does not locate itself with respect to any one space, and no space is given priority over any other. Within that space the boundaries that appear are entirely contingent, and operate, not horizonally, but rather as the means merely to delimit one portion of space from another. One might say that this is the mode of the geographical that is characteristic of much of what might broadly be understood as "physical" geography. In contrast, the focus on place that is associated with the first of the two notions of boundary at issue here is characteristic of certain key approaches associated with contemporary "cultural" geography. Such approaches do not, of course, draw upon the same weight of theoretical considerations that have been adduced in the discussion here, and yet they nevertheless draw upon notions of place and regionality that correlate with much of what also appears in Kant.

The idea of region, in particular, does not refer to some arbitrarily designated domain. Instead, a region constitutes itself *as a region* through the combination of factors—climatic, geological, cultural, and so forth—that give that region a particular unity and identity. It is, in fact, just this concept of regionality that underpins much of the tradition of place-oriented geography, particularly as influenced by the work of Paul Vidal de la Blache[43] and his successors that has been so important in twentieth-century geographical thought, as well as in certain keys strands of twentieth-century historiography.[44] On this basis, we may say that Kant is not merely one of the founders of academic geography, but that his work can be seen, in a way that Kant could never have anticipated, as pioneering the implicit theoretical underpinnings for that particular mode of geographical inquiry—particularly as realized within so-called "humanistic geography"—that takes place as a central concept.[45]

Kant's geography of reason does not constitute itself as geographic simply through its use of geographic images or metaphors—as if Kant could not help his independent interest in geographic matters from seeping through into his philosophical concerns. Kant's very conception of the critical project is determined by a set of geographic and topographic ideas to such an extent that one cannot properly understand that project without a sense of the geographic frame within which it is set. Moreover, this combination of the geographic with the critical-philosophical is not just an idiosyncrasy of Kant's own habits of thought, but instead reflects the way in which geography and philosophy are themselves connected in a certain fundamental way. Mark Johnson has argued (in a way that also draws on his work with George Lakoff) that our character as embodied, and so also as spatial, beings is fundamentally determinative of how we think and understand.[46] We would argue that geographic and topographic structures not only permeate our thinking, but that the inquiry into place is fundamental to any inquiry into the mode of human being (certainly to any critical-transcendental inquiry), such that the inquiry into the human, as well as the inquiry into being, is necessarily an inquiry into place.[47] Moreover, not only does place constitute a central focus for philosophical inquiry, but the focus on place also brings with it certain key ideas and distinctions that are fundamental to that inquiry as such. "Place" thus names a focus for philosophical inquiry just as the "geographic" (and the "topographic") can also be seen to name an essential mode of such inquiry, and in Kant's "geography of reason" we find one of the paradigmatic exemplifications of the essentially "geographic" character of philosophy. Consequently, Kant's concern with matters geographical as that is articulated in the *Physische Geographie* should not be taken to be independent of his core philosophical concerns. Indeed, Kant's own concern with understanding the essential *systematicity* of knowledge—something that, as we have seen in the discussion above, he develops using geographical concepts—itself carries over into his understanding of geography as such.[48] Not only, then, does Kant's geography inform his philosophy—to the extent that his philosophical project can rightly be termed a "philosophical geography"—but his geography is in turn shaped by his own philosophical orientation.

Notes

1. This paper was written as part of a project funded through an Australian Research Council Linkage International Fellowship at the University of Tasmania during 2008–2009. The authors gratefully acknowledge this support.

2. The comment on the title page of the original 1801 edition, "edited from his manuscript at the author's request" ("auf Verlangen des Verfassers aus seiner Handschrift herausgegeben"), might be taken to suggest that the impetus to publication of these lectures came from Kant himself. In fact, their publication was largely a result of popular demand (the 1801 comment refers to a previous unauthorised appearance of the material). See *Physische Geographie*, Akademie Edition of *Kants gesammelte Schriften* (hereafter AK, followed by volume and page number), 9: 509–11.

3. See Jeff Malpas, *Place and Experience* (Cambridge: Cambridge University Press, 1999); also *Heidegger's Topology* (Cambridge: MIT Press, 2006).

4. *Critique of Pure Reason*, Aviii—translated passages are taken from *Critique of Pure Reason*, trans. Paul Guyer and Allen Wood, The Cambridge Edition of the Works of Immanuel Kant (Cambridge: Cambridge University Press, 1998).

5. *Critique of Pure Reason*, Aix.

6. *Critique of Pure Reason*, A761/B789.

7. *Critique of Pure Reason*, Bxxiv.

8. *Critique of Pure Reason*, Bxv ff.

9. *Critique of Pure Reason*, A235/B294–A260/B315.

10. Onora O'Neill is one who does indeed recognize and draw upon the geographical and topographical in her reading of Kant—see *Constructions of Reason: Explorations in Kant's Practical Philosophy* (Cambridge: Cambridge University Press, 1989); see also the work of one of O'Neill's students, Sarah Furness, in *A Reasonable Geography: An Argument for Embodiment* (Ph.D diss., University of Essex, 1986). Malpas has also discussed some

of the issues at stake here previously in "From the transcendental to the 'topological': Heidegger on ground, unity and limit," in *From Kant to Davidson: Philosophy and the Idea of the Transcendental*, ed. Jeff Malpas, (London: Routledge, 2003), 75–99, esp. 85–6.

11. *Critique of Pure Reason*, A760/B788.

12. *Critique of Pure Reason*, A767/B795.

13. See *Critique of Pure Reason*, A707/B735—"If we regard the sum of all cognition of pure and speculative reason as an edifice for which we have in ourselves at least the idea, then I can say that in the Transcendental Doctrine of Elements we have made an estimate of the building materials and determined for what sort of edifice, with what height and strength, they would suffice. It turned out, of course, that although we had in mind a tower that would reach the heavens, the supply of materials sufficed only for a dwelling that was just roomy enough for our business on the plane of experience and high enough to survey it"—compare this with Kant's concluding comments in the *Dreams of a Spirit Seer*, trans. E. F. Goerwitz (London: Swann Sonnenschein & Co, 1900, Thoemmes Press reprint, 1992), 114 (note that this work is now considered to belong more to the early-Critical rather than pre-Critical period in Kant's œuvre).

14. *Critique of Pure Reason*, A761/B789.

15. *Prolegomena to any future metaphysics that will be able to come forth as science*, trans. Gary Hatfield, in *Theoretical Philosophy after 1781* (Cambridge: Cambridge University Press, 2002), 142 and 144 (AK 4: 352 and 354).

16. *Physics*, 212a5, the translation is from *Aristotle's Physics Books III and IV*, trans. Edward Hussey (Oxford: Clarendon Press, 1983), 28. See also *Physics*, 212a31, "A body is in place if, and only if, there is a body outside it which surrounds it."

17. Edward S. Casey, *The Fate of Place: A Philosophical History* (Berkeley: University of California Press, 1997), 63

18. Casey, *The Fate of Place*, 63.

19. *Critique of Pure Reason*, A758/B786.

20. *Critique of Pure Reason*, A758/B786.

21. *Critique of Pure Reason*, A759/B787.

22. *Critique of Pure Reason*, A759/B787.

23. *Critique of Pure Reason*, A762/B790.

24. *Critique of Pure Reason*, A11/B24.

25. *Physische Geographie*, AK 9: 166–7.

26. See *Physische Geographie*, AK 9: 160.

27. *Physische Geographie*, AK 9: 157.

28. See Malpas, "From the transcendental to the 'topological,' " 80–4.

29. See the *Critique of Pure Reason* A6/B10 ff.; see also Malpas, "The Constitution of the Mind: Kant and Davidson on the Unity of Consciousness," *International Journal of Philosophical Studies* 7 (1999): 1–2, and "From the transcendental to the 'topological,' " 80–4.

30. "What does it mean to orient oneself in thinking?," trans. Allen W. Wood, in *Religion and Rational Theology* (Cambridge: Cambridge University Press, 1996), 8–9 (AK 8:134–5).

31. Irrespective of how Kant's position on this matter is interpreted, we would argue that an adequate account of spatiality has to be understood as encompassing both subjective and objective elements—see Malpas, *Place and Experience*, 60–70.

32. For an argument to the importance of agency in making possible the unity of cognition as developed in Kant, see Malpas, "The Constitution of the Mind"; see also *Place and Experience*, 44–71.

33. See Casey, *The Fate of Place*, 187–93.

34. See Anthony Quinton, "Spaces and Times," *Philosophy* 37 (1962): 130–47. The thought experiment that Quinton employs to refute the Kantian thesis depends on the multiple spaces that are postulated nevertheless being connected in the experience of a single subject. From a Kantian perspective, the latter issue is absolutely central, and it is also an issue that Quinton seems not to address in any direct way—that a subject could be said to have experiences of multiple spaces in the way Quinton envisages seems to depend on ignoring the way in which the unity of space thesis is indeed connected with ideas about the necessary unity of cognition *and* agency.

35. For some further discussion of this matter see Karsten Thiel, *Kant und die "Eigentliche Methode der Metaphysik"* (Hildesheim: Georg Olms, 2008), 91 ff.

36. Martin Heidegger, "Building Dwelling Thinking," in *Poetry Language Thought*, trans. Albert Hofstadter (New York: Harper & Row, 1971), 154; *Vorträge und Aufsätze* (Stuttgart: Neske, 1997), 149.

37. Casey also makes use of the passage from Heidegger here (see Casey, *The Fate of Place*, 63 and 364 n.54), although he comments that Heidegger "relates 'boundary' (die Grenze) to *peras* and not to *horos* despite his simultaneous allusion to *horismos.*"

38. Here our argument can be seen to connect up, once again, with a familiar Heideggerian point. In a crucial discussion in section 12 of *Being and Time*, Heidegger argues that *Dasein* is not "in" space in the way that, for instance, water is "in" a glass, and this leads Heidegger to distinguish between two senses of being-in that could be seen to correspond to the two senses of containment or of boundedness that can also be seen to be at issue here—see Malpas, *Heidegger's Topology*, 67–83 for a more sustained discussion of the matter.

39. In "From the transcendental to the 'topological,' " Malpas argues for a distinction, one based in Aristotle, between two kinds of limit—limit as *terminus* and limit as *origin* (see 84–5). Malpas suggests that although the two sets of distinctions do not line up exactly, the distinction between these two senses of limit is closely related to Kant's distinction between limit and boundary (between *Schranke* and *Grenze*). What is evident here is that the concept of boundary itself contains an ambiguity within it that also correlates to the distinction between terminus and origin.

40. It is encouraged, in addition, by Kant's postulation of the Transcendental Ideas as themselves required by reason, and yet as not to be found within the sphere of our ordinary cognition. It is important to note, however, that the Transcendental Ideas are supposedly arrived at through a consideration of the nature of reason and cognition as such, and so their necessary postulation is a consequence, not of some examination of what lies beyond the bounds of reason, but rather of what is given within those bounds.

41. See Graham Priest, *Beyond the Limits of Thought*, 2nd ed. (New York: Oxford University Press, 2003), 100ff. Priest's overall argument is that the very attempt to speak of limits to thought, whether in Kant or elsewhere, involves us in contradictions that force the recognition of the need for a different logic—a *paraconsistent* logic.

42. For more on this distinction as it might be understood more generally, see Malpas, *Place and Experience*, 19–43.

43. See, for instance, Vidal de la Blache, *Tableau de la Geographie de la France* (Paris: Librarie Jules Tallandier, 1979 [1903]) and also *Principles of Human Geography*, ed. Emmanuel de Martonne, trans. Millicent Todd Bingham (New York, H. Holt, 1926 [1918]).

44. Of which an important foundational work, itself drawing on the ideas of Vidal de la Blache, is Lucien Febvre, *A Geographical Introduction to History*, trans. E. G. Mountford and J. H. Paxton (New York: Alfred Knopf, 1925), 20.

45. See, for instance, some of the essays in Robert David Sack, ed., *Progress: Geographical Essays* (Baltimore: Johns Hopkins University Press, 2002).

46. See Mark Johnson, *The Body in the Mind* (Chicago: Chicago University Press, 1987); see also Lakoff and Johnson, *Metaphors We Live By* (Chicago: University of Chicago Press, 1980).

47. Thus ontology, on this account, turns out to be essentially topography (or, in Heideggerian terms, topology)—see Malpas, *Heidegger's Topology*, pp.29–35.

48. See *Physische Geographie*, AK 9: 159f.

12

Orientation in Thinking

Geographical Problems, Political Solutions[1]

Onora O'Neill

Kant often uses geographical and political ideas and images to articulate philosophical questions and their resolution. There has been a lot of discussion of his use of each range of images, but the links between them are less obvious and less explored.[2] Here I shall consider some links he draws in his 1786 essay *What is Orientation in Thinking?* and their philosophical significance.[3] The shift from geographical to political imagery within this essay, I shall argue, is no superficial matter of style. The extensive geographical images in the first part of the essay articulate problems that arise in the ordinary use of human cognitive capacities, which seem inadequate for addressing many questions to which we seek answers. The political images in its closing pages are used to articulate an approach to norms of reasoning that could be used to resolve some of those questions. This interplay of geographical and political imagery recurs in many of Kant's writings, but I shall concentrate on this essay, leaving aside other writings in which he uses geographical and political imagery, as well as works in which he addresses substantive geographical and political questions.

Even my discussion of *What is Orientation in Thinking?* will have a limited focus. The essay is usually seen mainly as a discussion of the rational theology. In it Kant disputes claims put forward by Moses Mendelssohn and others that human reason can deliver *knowledge* of God.[4] Yet, profound as this theme is, it is not the fundamental issue at stake in the essay. *What is Orientation in Thinking?* is mainly of interest not because it is a notable contribution to eighteenth-century debates about knowledge of God (which it is), but because Kant approaches this debate by trying to articulate and delimit the powers of human reason. Only if this task can be addressed successfully can we find reasons for accepting—or rejecting—the theological claims advanced by Mendelssohn and others. So, with some daring, I offer an account of this essay that skirts its discussion of theological claims, including its adumbrations of Kant's own views, which allow for reasoned faith but not for knowledge of God.

The essay approaches deep questions about the powers of human reason by way of an extended comparison between terrestrial orientation, orientation in an arbitrary space, and orientation in thought. In each of these passages Kant's reliance on geographical ideas and imagery is manifest. But it less obvious, and certainly less remarked, that the later pages of the essay mainly use political ideas and imagery. As I see it, Kant's view of human reason as the capacity to orient ourselves in thought cannot be properly articulated without taking seriously his claims that this is a shared rather than a solitary task, so must meet any necessary conditions for coordination among a plurality of agents. These conditions are also necessary conditions for political coordination. So it should not surprise us that Kant articulates them using political ideas and imagery. The core of his conception of orientation is that it requires *action*, and if orientation in thought is to be possible not only for individuals taken separately but for a plurality of individuals, it must have striking parallels with the necessary conditions of *political action.*

Geographical Imagery and Philosophical Perplexity

Philosophers and others have often used geographical images to set out their most basic questions. The thought is quite intuitive: philosophical perplexity is like the experience of being lost, of not knowing the right way to go. This geographical picture of the starting point for inquiry, and specifically for philosophical inquiry, is built into our language. When we are lost we stray or wander, hence, we miss the right way, so may err morally, intellectually, or theologically.[5] Unsurprisingly, these thoughts are not unique to philosophers. They also occur again and again in religious and ethical writings, in poetry, and in other literature.[6]

Philosophers have used various geographical images to articulate this predicament. Human beings may be thought of as lost or in the dark (a cave, a deep wood, a thicket), at sea (out of their depth, drowning), imprisoned or trapped (in a fly bottle). We can find such images in Plato's great myth of the cave, and in Descartes' *Meditations*, where he likens philosophical perplexity to drowning:

> So serious are the doubts into which I have been thrown by yesterday's mediation that I can neither put them out of my mind nor see any way of resolving them. It feels as if I have fallen unexpectedly into a deep whirlpool which tumbles me around so that I can neither stand on the bottom nor swim up to the top. Nevertheless,

> I will make an effort, and once more attempt the same path which I started on yesterday. . . .[7]

Philosophical and other solutions to this predicament also reveal recurrent patterns. Often the solution is seen as a matter of finding a *guide* (a still small voice, the word of God, a knowledgeable stranger, a guiding thread), an *instruction* (Augustine hearing the words *tolle lege*), a *landmark* (the right way, the path), or a source of *illumination* (the light of reason, light dawning). These images suggest that when we are lost we can find our way only by reference to some authoritative *external* source, which providentially supplies what we need.

Orientation without Providential Assistance

The background to *What is Orientation in Thinking?* can be found in Kant's comments on the limits of human reason in the Prefaces and Introduction of the *Critique of Pure Reason*,[8] where he uses geographical images to characterize the starting point and difficulties of philosophical inquiry. In those passages he insists that human reason—here just the ordinary use of human cognitive capacities—leads us on, then lets us down. "The procedure of metaphysics," he complains, "has hitherto been a merely random groping, and what is worst of all, a groping among concepts."[9] Consequently, reason "not merely fails us, but lures us on by deceitful promises, and in the end betrays us."[10] This betrayal threatens to stymie our thinking, including our philosophical thinking; "it precipitates itself into darkness and contradiction,"[11] with the result that "ever and again we have to retrace our steps."[12] If ordinary human reason is unreliable, will not our thinking, including our attempts to address philosophical questions, be disoriented, perhaps impossible?

What is Orientation in Thinking? returns to these themes. The historical context of the essay is an extended set of disputes about the possibility of rational theology that were exercising several of Kant's contemporaries. In particular, parts of the essay comment on Moses Mendelssohn's assertion in his essay *Morgenstunden* that "healthy human reason"[13] enables us to know the existence of God. Kant rejects this position as fundamentally dogmatic. "Healthy human reason," as Kant sees it, is simply Mendelssohn's comforting term for an unconvincing *deus ex machina*.

However, Kant's essay is philosophically interesting less for its critique of Mendelssohn, or of others engaged in these debates, than for its constructive attempt to show how we can orient ourselves in thought without relying on dogmatic claims about the powers of reason. It is a swirling, suggestive, overcompressed essay, in which he often switches abruptly from discussion of rational theology to discussion of the method of reason, and many important points are made in footnotes or asides rather than integrated into the argument. Yet it is clear that the essay is *fundamentally* a discussion of the method of reason, which is to provide the context for reconsidering the debates of rational theology.

As we know from the vast demolition of rationalist metaphysics that we find in the *Transcendental Dialectic*, Kant's claims about the limitations of ordinary human reasoning are entirely serious. So is his refusal to reinstate any form of divine guarantee or providential assistance. He offers a more modest view of what it takes to find one's way—to orient oneself—than those that hold that we can or must find an authoritative external guide or source of illumination that shows us "the right way." As he sees it, we need and can have no access to a transcendent reality that can end our disorientation, and would not be helped by the illusory guidance of any *deus ex machina. What is Orientation in Thinking?* proposes a more modest account of orientation in thinking, and thereby of reason.

Varieties of Disorientation

The essay begins with extended analogies between geographical and spatial orientation, and then moves on to consider orientation in thinking. *If* we could find an authoritative method of conducting thinking and action, which we could use to overcome confusion and disorientation in thought, we would gain at least some understanding of the method of reason. Yet how can we hope to find a method of conducting thinking and action that moves beyond an account of ordinary, fallible uses of human cognitive capacities, and offers an account of the *proper* use of those capacities? Kant approaches this daunting task by comparing orientation in thought with the more manageable tasks of geographical and spatial orientation.

Near the beginning of the essay three extended passages focus on the task that must be undertaken if one is lost or disoriented. In the first passage[14] Kant offers an account of being lost and orienting oneself that is straightforwardly geographical. What must we do if we find ourselves in some unfamiliar terrestrial domain and want to work out which direction is which?

> To orientate oneself, in the proper sense of the word, means to use a given direction—and we divide the horizon into four of these—in order to find the others, and in particular that of the *sunrise*. If I see the sun in the sky and know that is now midday, I know how to find the south, west, north, and east. For this purpose, however, I must necessarily be able to feel a difference within my own *subject*, namely that between my right and left hands.[15] I call

> this a *feeling* because these two sides display no perceptible difference as far as external intuition is concerned. If I were not able, in describing a circle, to distinguish between movement from left to right and movement from right to left without reference to any differences between objects within the circle, and hence to define the different positions of such objects by *a priori* means, I would not know whether to locate west to the right or to the left of the southernmost point of the horizon in order to complete the circle through north and east and so back to south. Thus, in spite of all the objective data in the sky, I orientate myself *geographically* purely by means of a *subjective* distinction;[16] and if all the constellations, while in other respects retaining the same shape and same position in relation to one another, were one day miraculously transposed so that the former easterly direction now became west, no human eye would notice the slightest change on the next clear night, and even the astronomer, if he heeded only what he saw and not at the same time what he felt,[17] would inevitably become *disorientated.*

The most unusual feature of this account is Kant's insistence that "I orientate myself *geographically* purely by means of a *subjective* distinction." His line of thought draws on his critical accounts both of spatial and of causal knowledge.

Kant held that we cannot coherently think of space either as *absolute* (Newtonian) or as *relational.* Absolute space—as Leibniz had argued—would be inaccessible except to beings with transcendent cognitive capacities that humans lack and cannot comprehend. Relational conceptions of space cannot allow for incongruent counterparts (left hand/right hand, spirals, snail shells, etc.), where the spatial structure of distinct objects is internally the same, yet they cannot fill the same space because they differ in their orientation. Only if we inhabit a space—if, as Kant puts it, space is the form of outer sense for human knowers—can we orient ourselves within it while allowing for the possibility of incongruent counterparts.[18]

Kant had argued in the Second Analogy of the *Critique of Pure Reason* that human knowledge of events in the world depends on an ability to distinguish objective from merely subjective sequences of perceptions. Unless we can distinguish changes that are beyond our control from those that we bring about, we cannot identify events in the world that may be candidates for causal investigation. This move is echoed in the passage on terrestrial orientation when Kant states that it is by "distinguishing between movement from left to right and movement from right to left without reference to any differences between objects within the circle" that I can tell whether a sequence of perceptions is or is not a perception of change in the world. Human beings rely on their abilities to draw such distinctions in order to separate reality from dream, events from fantasies: otherwise they could not identify events in the world, support causal claims, let alone locate themselves by reference to the points of the compass.

Two further comments may be useful. First, the phrase "subjective distinction" in this passage means simply "a distinction in the subject." The intended distinction is neither merely mental, nor epistemically merely subjective as opposed to objective. My ability to distinguish which way I am facing, traveling, or turning in the world I inhabit allows me to orient myself in that world and to make my way through it, and to distinguish changes that I bring about from others that I do not. These necessary conditions of orientation in a terrestrial space constitute transcendental conditions of geographical knowledge.

Second, orientation in space need not use a terrestrial, if abstract, frame of reference, such as the one structured by the points of the compass. Often orientation is good enough for immediate purposes if I can identify directions and changes in terms of quite local landmarks, and make claims such as "I am looking towards the steeple of St. Mary's," or "the lake is on my left." In the passage on geographical orientation Kant discusses the abstract frame of reference conventionally used for terrestrial directions, which defines the points of the compass, so focuses on orientation "in the proper sense of the word";[19] but his argument also works for more local and limited frames of reference.

So the transition from the case of orientation in a terrestrial space by reference to the points of the compass to orientation in any given physical space is easy. Given an appropriate range of information and provided we can draw the necessary subjective distinction, we can orient ourselves within any sufficiently structured and differentiated space within which we find ourselves, whether or not it allows us to use the terrestrial frame of reference constituted by the points of the compass. The passage on orientation in any given space[20] reads:

> I can now extend this geographical concept of the process of orientation to signify any kind of orientation within a given space, i.e. orientation in a purely *mathematical* sense.[21] In the darkness, I can orientate myself in a familiar room so long as I can touch any one object whose position I remember. But it is obvious that the only thing which assists me here is an ability to define the position of the objects by means of a *subjective* distinction:[22] for I cannot see the objects whose position I am supposed to find; and if, for a joke, someone had shifted all the objects round in such a way that their relative positions remained the same but what was previously on the right was on the left, I would be quite unable to find my way around in a room whose walls were in other respects identical. But in fact, I can orientate myself simply by the feeling of difference between my two sides, my right and my left.[23] This

> is what happens if I have to walk, and take the correct turnings at night on streets with which I am otherwise familiar but in which I cannot at present distinguish any of the houses.

(We must remember how dark Königsberg would have been on moonless nights!)

This argument seems to me essentially the same as that in the discussion of terrestrial orientation. The difference is that the frame of reference is adjusted to the situation of an agent in a space that offers local landmarks that are not enough to identify the points of the compass. In this instance too, what is necessary is that I can tell how I am facing, turning,and moving in the space I inhabit.

Orientation in Thinking: Maxims as Subjective Distinctions

Kant extends this line of thought by abstracting entirely from geographical or spatial orientation to consider the transcendental conditions of orientation in thinking in general. The third passage on orientation[24] runs:

> Finally, I can extend this concept even further if I equate it with the ability to orientate oneself not just in space, i.e. mathematically, but also in *thought*, i.e. *logically*. It is easy to guess by analogy that this will be the means by which pure reason regulates its use when, taking leave of known objects (of experience), it seeks to extend its sphere beyond the frontiers of experience and no longer encounters any objects of intuition whatsoever, but merely a space for the latter to operate in.[25] It will then no longer be in a position, in determining its own faculty of judgement, to subsume its judgements under a specific maxim with the help of objective criteria of knowledge, but only with the help of a subjective distinction.[26] This subjective means which remains available to it is simply a **need** which is inherent in reason itself . . . if it is not just a matter of indifference whether one wishes to make a definite judgement on something or not, if this judgement is made necessary by a real *need* (in fact by a need that reason imposes on itself), and if we are at the same time limited by lack of knowledge in respect of factors essential to the judgement, we require a maxim in the light of which this judgement can be passed. . . .

The analogy with the preceding passages on terrestrial and spatial orientation lies in the insufficiency of objective criteria for organizing our thinking and action, and the consequent need to have recourse to a "subjective distinction."

However, orientation *in thought* is evidently not just a matter of being able to distinguish, face, turn, or move in one rather than another direction in the space one inhabits, or of developing a frame of reference for terrestrial or other spaces. Consequently, it is harder to see what can provide the necessary "subjective distinction." Kant asserts that the "subjective distinction" is created by agents who adopt maxims in order to make good a *deficiency* or *need* in their ordinary cognitive capacities, so that they can regulate their thought and action. Agents who do not by nature have cognitive capacities of the sort to which they might aspire "require a maxim in the light of which [this] judgement can be passed."[27] The maxim is to provide a *standard* or *norm* for a subject to use in organising her thinking and acting.

As a first step in trying to see how maxims can provide the necessary "subjective distinctions" it is useful to remember that Kant characterises maxims as *subjective* principles of volition, where once again "subjective" means neither merely mental or nor epistemically merely subjective, but rather something that belongs to, indeed is chosen by, subjects or agents as a norm to guide what they do.[28] However, this seemingly creates a further problem. Kant insists that such choices cannot be guided by an "objective" or "external" criterion: that is his objection to Mendelssohn, and to rationalist thought in general. But if choices of maxims cannot be guided by such a criterion, will they not be arbitrary? How can a maxim that is adopted in order to create a point of reference or norm for making judgments be justified?

As a first move Kant points out that although human capacities to reason are defective and unreliable, we are at least aware of their deficiency: "reason . . . perceives its own deficiency and produces a feeling of need."[29] Our ordinary cognitive capacities alert us to their own deficiencies and the threat of persistent disorientation, in which "ever and again we have to retrace our steps" in thinking. Only if we can move beyond this predicament can the "need of reason" be met. Yet how can this be done? Can recognizing a need help to meet that need? Can we perhaps choose to live with unmet needs?

Kant is alert to these problems, but aims to show that while agents are free to choose maxims arbitrarily, they need not do so. He claims that "a rational belief which is based on the need of reason for practical purposes could be described as a postulate of reason,"[30] and could provide a

> . . . signpost or compass by means of which the speculative thinker can orientate himself on his rational wanderings . . . and the man of ordinary . . . reason can use it to plan his course, for both theoretical and practical purposes.[31]

The problem is to understand how freely chosen maxims can provide a compass rather than a will-o'-the-wisp for orienting thinking.

Lawlessness and Plurality

In the final pages of *What is Orientation in Thinking?* Kant shifts his focus from the case of an individual agent who chooses a maxim to provide a subjective distinction or standard to orient his own thinking, to the case of human agents whose lives are linked to others' lives, and whose choice of maxims will affect others. This is not an arbitrary widening of focus. In many other writings Kant argues that human beings *inevitably* find themselves living in juxtaposition with others, because the world they inhabit is a finite sphere that offers only a limited arena for human life. As human numbers grow, we are bound to have contact with others.[32] Our relations with them may range from peaceful cooperation, to conflict and enmity, to deep incomprehension. However Kant rules out the possibility of human beings remaining solitary wanderers on the face of the earth, leading the life that Rousseau depicted in the first part of his essay on the origins of human inequality as "the earliest state of nature."[33] Our *geographical* situation inevitably raises *political* questions about the forms of association that a plurality of human beings can have, given that they cannot indefinitely continue to lead dispersed and solitary lives.

As Kant sees it, adequate accounts both of reason and of politics need to take account of human plurality. Both in reasoning and in politics we need to adopt principles that can be principles for all, and to reject *lawlessness*. The importance of law to any adequate polity is obvious enough: societies that lack the rule of law suffer either the imperious rule of some despot, or the uncertainties of anarchy. In either case their inhabitants will be exposed to arbitrary demands. Similarly, those who either defer to the dogmatic claims of some alleged supreme principle of reason or rely on arbitrarily chosen maxims to orient their thinking, will be exposed to one another's arbitrary claims and demands. Just as we are condemned to lawlessness in politics unless we can avoid both *despotism* and *anarchy*, so Kant thinks we are condemned to lawlessness in thinking unless we can avoid both *dogmatism* and *scepticism* about reason.

Kant's use of the term *lawless* to characterize defective conceptions of reason forms part of a rich range of political imagery in his writing on reason, in which he repeatedly likens defective conceptions of reason to defective forms of political association. *What is Orientation in Thinking?* does not use the full range of political imagery that Kant deploys elsewhere. It focuses entirely on the thought that reasoning among a plurality of agents will fail unless they shun *lawlessness in thinking*. In the first half of the essay Kant exposes *dogmatic* conceptions of reason by showing that their supposedly reasoned claims would have to be seen as arbitrary, so lawless, since they have only the illusory backing of an "argument" from authority. (We simply lack an "external" guide or source of illumination of the sort that Mendelssohn and many others (wrongly) imagined was available.) In the last pages of the essay Kant exposes *sceptical* approaches, which see what passes for reasoning only as a reflection of individual choice, so risking arbitrariness or "lawlessness." Just as *despotism* and *anarchy* are lawless and defective forms of human association, so *dogmatism* and *scepticism* about reason are lawless and defective ways of orienting thinking.

Since a reasoned proof of a supreme maxim of reason that can orient human thinking is in principle impossible, there is no way of proving that the various maxims agents may choose to orient their thinking are all unreasoned. Nevertheless, it is possible to show the calamitous costs of assuming that if there is no supreme maxim of reason—if reason is not a "dictator"[34]—anything goes. In the last pages of *What is Orientation in Thinking?* Kant does this by mounting a polemical attack on those who maintain that since no supreme principle of reason is given, we may choose whichever maxims we like to orient thought and action. In particular, he challenges those who think in this way to reconsider where their attack on reason is leading:

> Men of intellectual ability and breadth of mind! . . . have you also fully considered what you are doing and where your attacks on reason are likely to lead? . . . how much and how accurately would we *think* if we did not think, so to speak, in community with others to whom we *communicate* our thoughts and who communicate their thoughts to us![35]

Those whose lives are connected to others' lives need to realize that by viewing the choice of maxims to orient interaction and communication as arbitrary, they risk undermining not only interaction and communication but thinking itself. Such lawless views of reason lead not to emancipation, but to disorientation. Kant writes with biting sarcasm about the catastrophic results of supposing that thinking can be adequately oriented by any arbitrary choice of maxims to serve as standards for orienting thinking:

> The sequence of events is roughly as follows. The *genius* is at first delighted with its daring flights, having cast aside the thread by which reason formerly guided it. It soon captivates others in turn with its authoritative pronouncements and great expectations, and now appears to have set itself up on a throne on which slow and ponderous reason looked so out of place. It then adopts the maxim that the supreme legislation of reason is invalid, a maxim which we ordinary mortals describe as **zealotry**, but which those favourites of benevolent nature describe as *illumination*. Meanwhile a confusion of tongues must soon arise among them, for while reason alone can issue instructions that are valid for everyone, each individual now follows his own inspiration.[36]

Thinking in ways that are arbitrary does not lead to the liberation of thought: it costs us the possibility of thinking with others, and of working out or communicating what we have reason to think or to do. Kant points to protagonists of this catastrophic strategy, including purveyors of religious enthusiasm, of superstition, and of exaggerated conceptions of the powers of genius. Today he might also have pointed to postmodernists, sceptics, or deconstructionists. All of these opponents of reason disregard the disasters that will arise when agents adopt a strategy of taking just *any* maxim they choose as a basis for reason. They ignore the risk that

> . . . if reason does not wish to be subject to the laws which it imposes on itself, it must bow beneath the yoke of laws which somebody else imposes upon it: for nothing—not even the greatest absurdity—can continue to operate for long without some kind of law. Thus the inevitable result of *self-confessed* lawlessness in thinking (of emancipation from the restrictions of reason) is this: that the freedom of thought is thereby ultimately forfeited and, since the fault lies not with misfortune, but with genuine presumption, this freedom is in the true sense of the word *thrown away*.[37]

Lawless, Law-like and Lawful Thinking

But is it enough for reasoners to "operate . . . [with] some kind of law?" Adopting *law-like* principles or maxims, rather than making singular choices or edicts for particular cases is hardly demanding. Mere law-likeness sets a very minimal standard. Innumerable cognitive and practical principles (including all heteronomous principles) are formally law-like. Different agents could all adopt formally law-like maxims that varied in many ways, yet fail to light on "subjective distinctions" by which they could orient their thinking or action in ways that made it accessible others. Action on law-like principles is simply not enough to ensure that communication is possible, and will not preclude cognitive or practical disorientation, in which a Babel of voices speak past one another.[38] Agents who seek to reason with others need rather to adopt principles that are not merely *formally law-like* but *lawful*, in that they combine *law-like form* with *universal scope*.[39] Failure to adopt maxims that are *both law-like in form and universal in scope* will amount to *lawlessness*, and will prove inadequate to orient the thinking of a plurality of agents.

This demand may seem excessive. Kant is well enough aware that we often manage to communicate on the basis of principles that are formally law-like, but could not be universally adopted, so could not be principles for all, or offer reasons that all others could follow in thought or action, consider, accept, or reject. Principles that prescribe forms of epistemic or logical incoherence may be formally law-like, yet cannot be universally adopted in thinking. Principles that prescribe violence, monopolising material resources, wholesale coercion, or the destruction of agents are formally law-like, but cannot be universally adopted in action.

These realities suggest that a requirement to orient thinking by maxims that combine the form of law with universal scope may demand too much. Often we organize our thinking and action on the basis of maxims that have and can have no more than parochial scope. Just as local landmarks can be enough to orient ourselves in limited spaces for specific purposes, so parochial maxims can be used to orient aspects of thinking, acting, or communicating in specific contexts. Kant does not view all reliance on maxims that could not be adopted by all human agents as wholly unreasoned, but he sees it as *incompletely* reasoned, so prone to error, confusion, and failure. Merely adopting *some law-like maxim or other* for thought or action, while disregarding restrictions on its scope, may not take us beyond what Kant elsewhere[40] calls *private uses of reason*. Such partial reasoning is typically premised on forms of cognitive and practical deference, and provides a basis only for thinking and acting that is in certain respects arbitrary and unreasoned, and can reach only a limited audience.

Kant is emphatic that "private" uses of reason are not enough. In the last paragraphs of *What is Orientation in Thinking?* he offers some vivid illustrations of the baleful effects of relying solely on law-like principles with restricted scope in attempting to organize thought or action. If we assume that *all* that is relevant to reasoning is to adopt law-like maxims, *whether or not they meet other standards*, then thinking and acting will be inadequately oriented, and may amount to no more than "a lawless use of reason."[41] The consequence will not be a wonderful liberation of thought and action, but their subjugation "beneath the yoke of laws which somebody else imposes."[42] If we seek to reason, we must settle not for just any formally law-like principle, but rather for ways of organizing thinking and acting that are both law-like and could be followed by all, which others too could choose to accept or reject. We must beware that mere "freedom of thought if it tries to act independently even of the laws of reason, eventually destroys itself."[43]

Anything that can count as fully reasoned must therefore be fit to be given or received, exchanged or refused among all agents. Fully reasoned claims and proposals must be *followable* in thought or *adoptable* for action by others who are to consider, entertain, adopt, and reject them. Only when a plurality of agents organize their thinking and acting on maxims that they could *in principle* communicate and share, do they rely on a "subjective distinction" that sets a standard that is fully fit to orient thinking. The "subjective distinctions" that lawful maxims provide has to offer more than a "subjective distinction" for *each subject*: it must be able to serve as a common point of reference for a *plurality of subjects.* Unless agents adopt maxims that can provide distinctions that meet these standards, they will not be able to organize their thinking in ways that others can follow in thought or in action, so will not be able to offer, accept, criticise, or refuse others' specific reasons for thinking and acting.

Reasoning as a Negative Strategy

These considerations do not, of course, tell us what the "supreme" principle or other principles of reason are, nor do they identify determinate and detailed maxims for conducting thinking. Rather they make plain to anyone who needs to overcome the deficiencies of ordinary human cognition that this can be done only by adopting a *negative strategy.* The negative strategy is to reject principles that are unfit to serve as principles for all, so are lawless. But can this can be *all* that reasoning requires?

At the end of *What is Orientation in Thinking?* Kant asserts forthrightly in a remarkable (if compressed) footnote that rejecting maxims that cannot be followed by others is *all* that is basic to orienting both practical and theoretical reasoning. In both domains those who seek to offer and receive reasons must orient their thinking by choosing maxims that are lawful. He offers no more specific test for lawfulness than that of rejecting maxims that are lawless, so cannot serve for all. Reasoning, as Kant sees it, is not a matter of cleaving to some canonical list of principles of reason, but a strategy of refusing to rely on principles that do not combine *law-like form* with *universal scope*:

> To think for oneself means to look within oneself, (i.e. in one's own reason) for the supreme touchstone of truth: the maxim of thinking for oneself at all times is enlightenment. Now this requires less [note less, not more!] effort than is imagined by those who equate enlightenment with knowledge, for enlightenment consists rather in a negative principle in the use of one's own cognitive power,[44] and those who are exceedingly rich in knowledge are often least enlightened in their use of it. *To employ one's own reason means simply to ask oneself, whenever one is urged to accept something, whether one finds it possible to transform the reason for accepting it, or the rule that follows from what is accepted, into a universal principle governing the use of one's reason.*[45]

Anything that can count as reasoning must be based not solely on adopting opinions, conventions, or superstitions that may be accepted by some but disputed by others—let alone on one or another individual's idiosyncratic maxims. It must be oriented by the negative strategy of refusing to base one's thinking on lawless principles that are unfit to be principles for all.[46] This strategy provides the starting point for reasoned thinking, and can be used to pick out more determinate principles that can serve for theoretical and practical reasoning in specific domains.

This, I think, is why Kant's discussions of reason—of the most general standards or norms relevant to orienting thought and in action—shifts from geographical to political imagery. When we look more widely in the Kantian texts we find again and again that a range of political images is used to articulate the thought that reasoning must be anchored in principles that can potentially serve not just for an individual, but for all members of a plurality. Reasoning is no more than a strategy by which a plurality can keep open the *possibility* of discussing, even converging on, more specific claims and proposals. It cannot guarantee actual convergence, let alone agreement: but cutting off the possibility of convergence guarantees that there cannot be communication, agreement—or disagreement. If the possibility of communication is undermined, individual agents will not be able to engage with others' claims or proposals, and will have no response to the claims and proposals of those who fancy "lawless uses of reason." By refusing to rely on "subjective distinctions," which others with whom we seek to interact or communicate cannot follow, we keep open the possibility that those others may follow and consider our reasons for believing and for acting, and accept or refuse our claims and proposals.

Our ordinary cognitive capacities are indeed limited. They provide us (as Kant puts the matter elsewhere) with "just enough for the most pressing needs for the beginning of existence,"[47] and it is up to human agents to provide the rest. What distinguishes uses of human cognitive capacities that begin and end in disorientation from uses that can orient the thinking of a plurality of agents is just that the latter systematically reject principles for thought or action that could not be followed by all.

In a number of other works Kant uses more specific political imagery to characterize reasoning. He writes, for example, of establishing a "tribunal of reason" or a "lasting peace," and of being "governed" by "reason's law." But while his choice of institutional imagery varies, its common core is that the rejection of lawlessness is central. The task of reasoning is to adopt a negative strategy of avoiding lawlessness in thinking, as in politics.

Reason's fundamental task as Kant sees it is normative, not receptive. Reason is not given by any external source: it is a standard that thinkers can adopt by rejecting maxims that would undermine the possibility of a plurality of free agents articulating and communicating their beliefs and proposals for action. That task parallels the political task of establishing institutions (whatever specific features they may need in actual historical circumstances) that constitute a polity in which a plurality of otherwise uncoordinated agents can seek to enact common laws.

Kant does not argue for a unique, substantive "supreme" principle of reason. Rather he argues for a second-order strategy that enjoins would-be reasoners to reject maxims for thinking or acting that cannot be adopted by all, and to rely on maxims for thinking and acting that can be adopted by all. This limited principle of reason is a far cry from the supposedly powerful principles that rationalists commend, and also from those vaguer standards that Mendelssohn (and others) invoked. However, unlike grander proposals that cannot be justified, it can be vindicated as a transcendental condition for orienting thought and action among a plurality of agents.

Notes

1. I would like to thank Stuart Elden, Eduardo Mendieta, Jens Timmermann, Ken Westphal, Garrath Williams, and anonymous referees for helpful comments.

2. For some discussion of Kant's use of geographical imagery beyond his writings on geography see David W. Tarbet, "The Fabric of Metaphor in Kant's Critique of Pure Reason," *Journal of the History of Philosophy* 6 (1968): 257–70. For some discussion of his uses of political imagery beyond his writings on politics see Hans Saner, *Kants Weg vom Krieg zum Frieden*, vol. 1: *Widerstreit und Einheit* (München: Piper, 1967), translated as *Kant's Political Thought: Its Origins and Development*, trans. E. B. Ashton (Chicago: University of Chicago Press, 1973); Ronald Beiner, *Political Judgment* (Chicago: University of Chicago Press, 1983); Hannah Arendt, *Lectures on Kant's Political Philosophy*, ed. Ronald Beiner (Chicago: University of Chicago Press, 1982); Onora O'Neill, "Reason and Politics in the Kantian Enterprise," in *Constructions of Reason: Explorations of Kant's Practical Philosophy* (Cambridge: Cambridge University Press, 1989), 3–27; Natalie Brender, "What is disorientation in thought?," in *New Essays on the History of Autonomy: a Collection Honouring J. B. Schneewind*, ed. Natalie Brender and Larry Krasnoff (Cambridge: Cambridge University Press, 2004), 154–80.

3. This is the conventional English translation of the title, but a more accurate one would run "*What is It to Orient Oneself in Thought?*": the reference to agents is central to Kant's argument. I shall quote from *What is Orientation in Thought?*, in *Kant's Political Writings*, 2nd ed., ed. Hans Reiss, trans. H. Barry Nisbet (Cambridge: Cambridge University Press, 1991, 237–49), and give the conventional Academy page and line reference, with the page in the translation in parentheses. In citing passages from Kant's other works I shall where available use the Cambridge Edition of the works of Immanuel Kant, and give the page references of the Prussian Academy edition. Where it may be helpful I will give the German in the footnotes.

4. Moses Mendelssohn (1785), *Morgenstunden oder Vorlesungen über das Daysein Gottes* [Morning Hours or Lectures on the Existence of God*]* (Berlin: Christian Friedrich Boß, 1786).

5. The Anglican General Confession compresses several of these images: "Almighty and most merciful Father, We have erred and strayed from your ways like lost sheep . . ." *Book of Common Prayer*.

6. Two famous and striking twentieth-century uses of this range of imagery: "We shall not cease from exploration / And the end of all our exploring / Will be to arrive where we started / And know the place for the first time," T. S. Eliot, *Little Gidding, in Four Quartets* (London: Faber, 1944); "When we first learn to believe anything, what we believe is not a single proposition, it is a whole system of propositions, (Light dawns gradually over the whole.)" Ludwig Wittgenstein, *On Certainty*, ed. G. E. M. Anscombe and G. H. von Wright, trans. Denis Paul and G. E. M. Anscombe (Oxford: Basil Blackwell, 1969), §141.

7. René Descartes, *Meditations on First Philosophy*, trans. John Cottingham (Cambridge: Cambridge University Press, 1996), 16 (*Meditation* II, §1).

8. *Critique of Pure Reason*, trans. Paul Guyer and Allen W. Wood (Cambridge: Cambridge University Press, 1999). This is now the standard English translation, but where a passage is well-known in the Kemp Smith translation, (London: Macmillan, 1929), I shall quote that version.

9. *Critique of Pure Reason*, Bxv. I quote these four short passages using the very well-known and more resonant words of the Kemp Smith translation.

10. *Critique of Pure Reason*, Bxv.

11. *Critique of Pure Reason*, Aviii.

12. *Critique of Pure Reason*, Bxiv.

13. Kant points out that Mendelssohn used various equally vague terms for this idea in different writings: ". . . certain guideline which he sometimes described as *common sense* (in his *Morgenstunden*), sometimes as *healthy reason*, and sometimes as *plain understanding* (in *An Lessings Freunde*)." *What is Orientation in Thought?*, AK 8: 133, 25–7 (237).

14. *What is Orientation in Thought?*, AK 8: 134, 27–135, 22 (238–9).

15. "Zu diesem Behuf bedarf ich aber durchaus das Gefühl eines Unterschiedes an meinem eigenen *Subject*. . . ." *What is Orientation in Thought?*, AK 8: 134, 31–2 (238).

16. ". . . ur durch einen *subjectiven* Unterscheidungsgrund. . . ." *What is Orientation in Thought?*, AK 8: 135, 10–1 (239). The claim that I orientate myself *only* by reference to a subjective distinction is too strong. Kant in fact argues that orientation requires a subjective distinction, but not that a subjective distinction is sufficient. He evidently thinks that terrestrial (in his example terrestrial orientation in mid-latitudes) also requires perceptual data, remembered information and bodily knowledge. But it also requires movement by which the subject creates the "subjective distinction" needed for orientation.

17. ". . . wenn er bloß auf das, was er sieht, und nicht zugleich was er fühlt, Acht gäbe." *What is Orientation in Thought?*, AK 8: 135, 16–7 (239). Here Kant clearly makes the more plausible claim that orientation requires both a subjective distinction and other information.

18. We can, of course, distinguish locations in spaces we do not inhabit—for example, within imagined spaces—provided that we also imagine a frame of reference that allows us to do so.

19. *What is Orientation in Thought?*, AK 8: 134, 27 (238).

20. *What is Orientation in Thought?*, AK 8: 135, 22–37 (239).

21. ". . . sich in einem gegebenen Raum überhaupt, mithin bloß *mathematisch* orientiren. . . ." *What is Orientation in Thought?*, AK 8: 135, 26 (239).

22. ". . . nach einem *subjectiven* Unterscheidungsgrunde. . . ." *What is Orientation in Thought?*, AK 8: 135, 29 (239). Once again the subjective distinction is not the only thing that is needed; but it is needed.

23. ". . . das bloße Gefühl eines Unterschiedes meiner zwei Seiten, der rechten und der linken. . . ." *What is Orientation in Thought?*, AK 8: 135, 35 (239). Once again Kant can be read as if he held that the "subjective distinction" were sufficient for orientation, but a moment's thought shows that his claim is that it is necessary rather than sufficient.

24. *What is Orientation in Thought?*, AK 8: 136, 1–23 (239–40).

25. In speaking of "a sphere beyond the frontiers of experience" Kant indicates only that it is a "sphere" within which spatial, *a fortiori* geographical, orientation is not possible. That "sphere" would include not only any domain of transcendent objects (to which Kant thinks we have no access), but also the domains of transcendental presuppositions and normative claims.

26. Kant's footnote to this sentence encapsulates his central point: "Thus to *orientate* oneself in thought means to be guided, in one's conviction of truth [literally: "in holding anything to be true"], by a subjective principle where objective principles of reason are inadequate."

27. *What is Orientation in Thought?*, AK 8: 136, 22–3 (240).

28. *Groundwork of the Metaphysics of Morals*, trans. Mary J. Gregor (Cambridge: Cambridge University Press, 1997), AK 4: 421 n.

29. Cf. "Reason does not feel. It perceives its own deficiency and produces a feeling of need through the *cognitive impulse*. The same applies in this case as in the case of moral feeling, which is not the source of the moral law, for this is entirely a product of reason; on the contrary, moral feeling is itself produced or occasioned by the moral laws and hence by reason, because the active yet free will needs specific grounds [on which to act]." *What is Orientation in Thought?*, AK 8: 139 n. and 140 n. (243). Compare the discussion of moral feeling—"reverence"—at *Groundwork of the Metaphysics of Morals*, AK 4: 401 n.

30. *What is Orientation in Thought?*, AK 8: 140, 29–31 (245).

31. *What is Orientation in Thought?*, AK 8: 142, 1–6 (245). "Ein reiner Vernunftglaube ist also der Wegweiser oder Compaß, wodurch der speculative Denker sich auf seinen Vernunftstreifereien im Felde übersinnlicher Gegenstände orientiren, der Mensch von gemeiner, doch (moralisch) gesunder Vernunft aber seinen Weg sowohl in theoretischer als praktischer Absicht dem ganzen Zwecke seiner Bestimmung völlig angemessen vorzeichnen kann."

32. For discussion of the theme see Katrin Flikschuh, *Kant and Modern Political Philosophy* (Cambridge: Cambridge University Press, 2000).

33. Jean-Jacques Rousseau, (1754) *What is the Origin of Inequality among Men, and is it Authorised by the Natural Law?*, in *'The Discourses' and Other Early Political Writings*, ed. and trans. Victor Gourevitch (Cambridge: Cambridge University Press, 1997).

34. Cf. "Reason has no dictatorial authority," *Critique of Pure Reason*, A738/B767.

35. *What is Orientation in Thought?*, AK 8: 144, 6–22 (247).

36. *What is Orientation in Thought?*, AK 8: 145, 10–8 (248).

37. *What is Orientation in Thought?*, AK 8: 145, 19–30 (248).

38. Cf. *Critique of Pure Reason*, A707/B735.

39. For present proposes I shall not say anything about the scope of these law-like principles. This is an important and difficult topic, to which one can gesture by indicating that their scope must include all relevant agents, but that differing understandings of relevance may be needed in different contexts.

40. See especially Immanuel Kant, *What is Enlightenment?*, in *Practical Philosophy*, ed. and trans. Mary J. Gregor (Cambridge: Cambridge University Press, 1996), AK 8: 35–42. For more detailed textual analysis than I can provide here see Onora O'Neill, "Vindicating Reason," in *The Cambridge Companion to Kant*, ed. Paul Guyer, (Cambridge: Cambridge University Press, 1992), 280–308 and "Kant's Conception of Public Reason," in *Kant und die Berliner Aufklärung* (Akten des IX. Internationalen Kant-Kongresses), vol. 1, ed. Volker Gerhardt, Rolf-Peter Horstmann and Ralph Schumacher, (Berlin: Walter de Gruyter, 2001), 35–47.

41. ". . . eines gesetzlosen Gebrauchs der Vernunft." *What is Orientation in Thought?*, AK 8: 145, 8 (247).

42. *What is Orientation in Thought?*, AK 8: 146, 11–3 (248).

43. *What is Orientation in Thought?*, AK 8: 146, 21–2 (249).

44. *What is Orientation in Thought?*, AK 8: 146 n., "ein negativer Grundsatz im Gebrauche seines Erkenntnißvermögens."

45. *What is Orientation in Thought?*, AK 8: 146 n., continued on 147 (249 n.), my italics. The italicized passage runs: "Sich seiner eigenen Vernunft bedienen, will nichts weiter sagen, als bei allem dem, was man annehmen soll, sich selbst fragen: ob man es wohl thunlich finde, den Grund, warum man etwas annimmt, oder auch die Regel, die aus dem, was man annimmt, folgt, zum allgemeinen Grundsatze seines Vernunftgebrauchs zu machen."

46. Some may worry that this inserts the Kant's supposedly exigent account of duty into the heart of cognitive and practical life. This is an unnecessary worry. In refusing to base thought or action on principles others who are to follow what is claimed or proposed cannot follow we indeed discard certain ways of thinking or acting, but are generally left with a wide area of freedom.

47. *Idea for a Universal History from a Cosmopolitan Point of View*, in *Kant's Political Writings*, AK 8: 19–20 (43).

13

"The Unity of All Places on the Face of the Earth"

Original Community, Acquisition, and Universal Will in Kant's Doctrine of Right[1]

Jeffrey Edwards

We encounter a description of political geography in the introduction to the *Physical Geography* that is published in the Prussian Academy edition of Kant's works. Kant evidently holds that political geography has to be understood above all in relation to the universal law that furnishes "the first principle of a civil society."[2] But his substantive determination of political geography's field of investigation is couched in terms of the laws that "are related to the character [*Beschaffenheit*] of the land and of the inhabitants,"[3] and he clearly holds that laws of this type have to do with the subject matter of physical geography. Thus, despite the initial reference to a universal law of civil society, Kant's description seems to imply that political geography is fundamentally concerned with special laws that are conditioned by geographical variations and the varying anthropological characteristics corresponding to them.[4] If this is in fact the case, then we can see that Kant has sufficiently staked out the topical area that political geography is supposed to investigate. Yet what exactly are we to say about the relationship between political geography and the universal law of civil society that Kant is reported as mentioning? Or more generally, how are we to understand the relationship between (1) the particular set of empirically conditioned laws involved in the physical geographic discipline that grounds political geography, and (2) the a priori principles and laws that form the groundwork of the specifically juridical theory to which a *universal* law of civil society belongs?

The topics furnished by these questions are far too broad for adequate discussion in a chapter-length study. Thus, I propose to concentrate on the ramifications of attributing an objective property to the earth—a property that figures in both Kant's treatment of the preliminary concepts of physical geography and his mature account of the foundations of his juridical doctrine. The property in question is the spherical shape of the earth's surface. It is in view of this property that Kant, in the context of his juridical theory of right, introduces the concept of the unity of all places on the face of the earth and employs it in combination with the correlative concepts of original community and universal will in order to establish the foundations of property law. By following the logical connections that underlie this conceptual combination in conjunction with an examination of its historical background, we will ultimately be confronted with a conclusion concerning Kant's foundational theory of property that is as historically unexpected as it is remarkable from a strictly systematic point of view. But this is already getting well ahead of ourselves at the outset, so let me conclude my introductory remarks by sketching out the general framework of Kant's fully developed theory of property law.

Kant's theory of private law (or private right: *Privatrecht*) is presented in Part I of the Metaphysical Foundations of the Doctrine of Right (= *DR*) of 1796. The theory treats the conditions under which external objects of the power of choice (*Willkür*) can be rightfully "mine and yours." It is divided into three main parts, all of which concern the principles that ought to govern the power of choice in relation to anything that can be acquired as externally mine or yours.[5] The first part, the doctrine of possession (*DR* §§1–9), determines the general conditions under which one can have something external as "one's own [*das Seine*]." The second part, the doctrine of acquisition (*DR* §§10–35), establishes the way in which something external may be rightfully acquired. The third part (Of Subjectively Conditioned Acquisition through Decision by a Public Court of Justice: *DR* §§36–42) gives an a priori account of basic contractual relations and other normative factors in keeping with which the public administration of justice should take place. In this chapter, I discuss the doctrine of possession in connection with the portion of the doctrine of acquisition in which Kant establishes the principles in accordance with which corporeal objects can become property (*Eigentum* or *dominium*). I therefore focus on the concepts of possession and acquisition as they apply to the law of things (*Sachenrecht*).[6] The primary point of interest is the role that the idea of an original community of possession plays in Kant's theory of acquisition when combined with the concepts of global geographic unity and universal will mentioned in the preceding paragraph. I treat this role against part of its historical background in early modern natural-law theory, and I bring out its implications for modern theories of the foundations of property law.

Intelligible Possession and the Juridical Postulate of Practical Reason

The doctrine of possession revolves around two systematic factors: the exposition of the concept of intelligible possession; and the argument by which Kant establishes the universal validity of the juridical postulate of practical reason. In this section, I will clarify these factors in the order just mentioned.

Kant defines what is "rightfully mine" (*meum iuris*) as something with which "I am so connected that another's use of it without my consent would wrong me."[7] In view of this definition, he stipulates that the subjective condition under which I can make use of anything is possession; and he holds that the possibility of my possessing (and hence using) something as rightfully mine presupposes that a clear distinction can be drawn between two basic meanings of possession.[8] It is necessary to distinguish between sensible or physical possession and intelligible or merely rightful possession (*bloß rechtlicher Besitz*). These different meanings correspond to a distinction that underlies our thinking of objects, a distinction implicit in the expression "external to me." It is the distinction between an object conceived as something "merely distinct from me (the subject)" and an object regarded as something located in another position in space or existing in another time. The type of possession that corresponds to the first sense of externality—intelligible possession—is possession taken without limiting reference to the conditions under which something exists apart from me in space or time.[9] Since (according to Kant) these conditions are all sensible, intelligible possession leads to the understanding of rightful possession as purely rational possession (*Vernunftbesitz*).

Interpreted as intelligible or rational possession, rightful possession is not merely non-sensible. It is also non-empirical. Possession is empirical when a subject's relation to objects depends on the spatial and temporal limitations that are characteristic of physically having some object in hand or at hand.[10] As Kant puts this point: "Intelligible possession (if such is possible) is possession without *holding* [*Inhabung*] (*detentio*)."[11] Accordingly, the possibility of intelligible possession, as rational possession, has fundamentally to do with our capacity to abstract from the conditions of physical possession. As rational beings, we must be able to disregard the set of sensible conditions under which "empirical possession" denotes a subject's phenomenal relation to certain objects of choice (i.e., *possessio phaenomenon*). Intelligible possession must therefore denote a noumenal possessive relation to such objects (*possessio noumenon*), and the practical proposition that grounds the juridical possibility of intelligible (or non-physical) possession is one which presupposes that *all* conditions of empirical possession in space and time can be set aside. For only in this way can possession without holding, i.e., intelligible possession, be affirmed as "necessary for the concept of something externally mine and yours."[12]

The practical proposition just referred to is the juridical postulate of practical reason. This postulate is given its primary formulation in *DR* §2.[13] The postulate states:

> It is possible for me to have any [*einen jeden*] object of my power of choice as mine; that is, a maxim by which, if it were to become a law, an object of choice would *in itself* (objectively) have to become something belonging to no one (*res nullius*) is contrary to right.[14]

Kant argues that the juridical postulate of practical reason provides us with a permissive law (*lex permissiva*), a law that authorizes each of us to put all other rational agents under an obligation that they would not otherwise have, namely, the obligation to "refrain from the use of certain objects of our power of choice because we were the *first* to take them into our possession."[15] He maintains further that the possibility of intelligible (i.e., non-physical) possession must be "inferred" from the juridical postulate as an *immediate* consequence of the latter.[16] As we will see, the combination of these two factors—the interpretation of the juridical postulate as a permissive law that grounds obligation through first possession; and the view that intelligible possession is grounded as a direct implication of the same postulate—determines the character of Kant's theory of the foundations of property law. But first we need to understand these factors in the context of the doctrine of possession in which they are introduced.

Since the juridical postulate asserts that *any* object of my power of choice is something that is possibly mine, we can see that physical possession cannot be the defining mark of the rightful possession that correlates with the property of being rightfully mine. Thus, if rightful possession is to be possible, the spatio-temporal conditions that limit physical (or empirical) possession must be set aside. And if this is true, we can readily understand how the possibility of rightful possession (as intelligible possession) is anchored directly in the categorical requirement of practical reason expressed by the juridical postulate. What is not so readily apparent, however, is how this grounding relation fits together with the interpretation of the juridical postulate as a law that authorizes me (or anyone else) to obligate others to refrain from making use of some object of my power of choice because I was the first to take it into my possession. To better understand this problematic, let us turn to *DR* §§8–9, where Kant treats the concept of provisional rightful possession in connection with his account of the juridical and civil condition (*Zustand*).

In the sections just cited, Kant establishes that a subject can be in the state of having something external as her own only if she exists in a juridical condition (*rechtlicher Zustand*) with respect to other subjects, that is, in a civil condition (*bürgerlicher Zustand*) involving a publicly legislative power. Such a power requires a civil constitution that conforms with the universal principle and a priori laws of right set forth as principles of natural law in the Doctrine of Right.[17] This civil constitution is what represents the juridical condition in which everything that belongs to each person is secured for every single person respectively as his own. Kant thus takes the civil constitution to be the fundamental guarantee of right with respect to everything that is externally mine and yours. At the same time, though, he insists that this guarantee *presupposes* a normatively valid possessive relation, and he links this presupposition to the assumption that something externally mine and yours must be possible even apart from the condition under which it is guaranteed:

> Any guarantee, then, already presupposes that something belongs to someone (to whom it secures it) [*setzt . . . das Seine von jemanden voraus (dem es gesichert wird)*]. Thus, prior to the civil constitution (or in *abstraction* from it), something externally mine or yours must be assumed as possible, and with it also a right to compel everyone with whom we could have any dealings to enter with us into a constitution in which what is externally mine and yours can be secured.[18]

We thus arrive at the interpretation of rightful possession that furnishes the centerpiece of the description of the state of nature relevant to the theory of private right. Apart from the civil constitution, anything that can be thought of as externally mine and yours must be understood as a merely provisional rightful possession.[19] In a provisional sense, possession can be rightful even for agents who are thought of as existing in a state of nature. Possession in this natural condition can be rightful in the sense that it permits anyone to resist those who are unwilling to enter into the civil condition and who would interfere with what one has come to possess. But this way of portraying what it is to have something as one's own in the state of nature must still be characterized in terms of *physical* (or *empirical*) possession. It qualifies as rightful possession only in a derivative sense: it has in its favor "the rightful presumption" that it will be *made* rightful possession "through unification with the will of all in a public lawgiving."[20] Thus, by treating the relationship between the different meanings of rightful possession, Kant brings to light the necessary connection between the concepts of rightful possession and universal will. But he also acknowledges, as a crucial component of his theory of private right, a "prerogative of right arising from empirical possession."[21]

Now consider again the problem addressed above, which concerns the interpretation of the juridical postulate as a permissive law by which the obligation-founding character of first possession is determined. In keeping with Kant's considerations thus far, we can understand first possession-taking (*Bestiznehmung*) as the type of act that gives rise to empirical possession. We can also accept that the empirical possession achieved through first possession-taking has to be consistent with the deontic requirements of right which are taken from Kant's concept of universal will. On these assumptions, we are presumably in a position to grasp how possession-taking can establish a prerogative of right that necessarily correlates with the obligation of others (or more precisely: with the obligation to be placed on others) to refrain from using objects of our choice *because* we were, each of us respectively, the first to take them into our possession. To see whether these conjectures bear out, however, we must move beyond Kant's treatment of rightful possession to his account of the possibility conditions for rightful acquisition. We thus turn to the theory of original acquisition.

Acquisition, Occupation, and Original Community

It will not be possible here to provide fully detailed analysis of Kant's arguments on rightful acquisition as they pertain to the law of things (*Sachenrecht*). My considerations on the doctrine of acquisition will therefore have to presuppose that Kant effectively substantiates the following points in the course of his reflections on the principles of private right. *First,* whoever wants to claim that he has a thing (*Sache*) as his own must be *in* possession of an object; for otherwise he could not be wronged by another's use of any object without his consent, and such use would for that reason not be contrary to right.[22] *Second,* the fundamental principle of external acquisition in general is fully consistent with the universal law of right, the juridical postulate of practical reason, and the juridical requirements of practical reason that are contained in the concept of universal will.[23] *Third,* only corporeal things can be originally acquired.[24] *Fourth,* the basic argument presented in the second paragraph of *DR* §12 is sound. That is to say, Kant successfully establishes that first acquisition of a thing must be thought of in terms of the acquisition of land if the idea of first acquisition is to have foundational significance for the law of things.[25] *Finally,* the possibility of originally acquiring an external object of the power of choice is the immediate consequence of the juridical postulate of practical reason.[26]

How exactly are we to understand the concept of original acquisition in accordance with the assumptions just stated, especially given Kant's position that the possibility of original acquisition is the *immediate consequence* of the juridical postulate? If, in keeping with this postulate, I am to acquire anything originally, then what I acquire must be something external that is not originally mine. For nothing external to me can be originally mine; and whatever I can acquire originally *as* mine must be something that does not belong to anyone else as her own, since all acquisition from another is derivative, and so cannot be original.[27] Consequently, if original acquisition is to be possible as a rightful act, then two things must obtain. I must not only be able to think of myself as a subject that can relate itself to something external. I must also be able to think of this external something as a thing that can be neither originally mine nor possessed by another subject as something that is her own. According to Kant, thinking jointly in these ways is possible only if the following conditions are satisfied. Acquisition must take place through the act of bringing something corporeal under my control. The control-seizing act (*Bemächtigung*) by which things are brought under my control must be the result of *unilateral* choice. Yet such a unilaterally performed act of seizure, or occupation (*occupatio*), must be able to take place consistently with the understanding of universal will as an *omnilateral* will—that is, it must be compatible with the idea of a will that is "united not contingently but a priori and therefore necessarily, and because of this is the only will that is lawgiving."[28]

It is in view of the relation between action proceeding from unilateral choice and action performed in conformity with the idea of an omnilateral will that Kant specifies three aspects or "moments" (*Momente*) of original acquisition through seizure or occupation: (1) the apprehension (*apprehensio*) of an object that belongs to no one, which is effected by a subject's taking possession of an object of choice in space and time, and which therefore results in *possessio phaenomenon*; (2) the designation (*declaratio*) of the possession of this object as mine and of my act of choice to exclude everyone else from it; and (3) appropriation (*appropriatio*) conceived as an act performed in conformity with the idea of an externally and universally lawgiving will, whereby everyone is bound to agree with my choice.[29] Kant understands the connections between these different aspects of original acquisition in terms of the stages of syllogistic inference. The descriptions given for apprehension and designation provide, respectively, the contents of the major and minor premises. The conclusion asserts that what I possess in virtue of the acts described under (1) and (2) is merely rightful possession (or *possessio noumenon*).[30] Thus, the movement of thought from the first to the last conceptual moments of original acquisition makes the transition from empirical possession (*possessio phaenomenon*) to intelligible possession by an inference of reason. This inference is meant to demonstrate how, in accordance with the universality requirements of practical reason, "abstraction can be made from the empirical conditions of possession, so that the conclusion, 'the external object is mine,' is correctly drawn from sensible to intelligible possession."[31]

But here, of course, we must ask how anything that I come to possess empirically as the result of unilateral choice could be something that I *can* possess in conformity with the idea of an omnilateral will. Kant formulates this specific problem of original acquisition as follows:

> Original acquisition of an external object of the power of choice is called *seizing control* [*Bemächtigung*] (*occupatio*), and it can take

> place only with respect to corporeal things [*Dinge*] (substances). When it takes place, what it requires as the condition of empirical possession is priority in time to anyone else who wills to seize control of a thing [*Sache*]. . . . As original, it is only the result of unilateral choice, . . .—Still, if an acquisition is *first*, it is not yet for that reason *original*. For the acquisition of a public juridical condition by the union of the will of all for universal lawgiving would be an acquisition such that none could precede it, yet it would be derived from the particular wills of each and would be *omnilateral*, whereas original acquisition can proceed only from a unilateral will.[32]

Thus, the specifically Kantian problem of original acquisition has to be understood in view of the following requirements of intelligibility. First, original acquisition has to be thought of as taking place only in connection with a condition of empirical possession, namely, the temporal priority of the act by which one wills to seize control of—occupy—corporeal entities. Second, original acquisition is conceivable only as the result of unilateral choice in keeping with this temporal condition. Third, because of its conceptual link to such a condition of empirical possession, the act by which original acquisition takes place must somehow be made consistent with the act to which it may seem to be most radically opposed. We must be in a position to think of the act of unilateral choice by which a particular acquisitive agent aims to seize control of (or occupy) things as compatible with the omnilaterally unifying and universally lawgiving act by which a public juridical condition is to be established.

By this point, then, we can see how Kant clarifies the concept of original acquisition by explicating the notion of *occupatio* (*Bemächtigung*). We can also see how this explicative work enables him to formulate the problem of original acquisition so as to make it the systematic centerpiece of the law of things. But the considerations on original acquisition summarized thus far do not yet provide an adequate platform from which to address the problem of obligation that emerges from the interpretation of the juridical postulate as a permissive law, i.e., from the interpretation of this postulate as the law that authorizes me (or anyone) to hold that everyone else *ought* to refrain from using some object of my power of choice *because* I was the first to take possession of it. Those considerations do, of course, confirm one of the conjectural assumptions mentioned above. If taking possession of an object of choice represents the initial aspect (or moment) of original acquisition, and if priority in time with respect to the occupying will of others provides the condition for empirical possession, then we are surely justified in regarding first possession-taking as the act that gives rise to empirical possession. Consequently, as long as it can be demonstrated that the achievement of empirical possession through first possession-taking is necessarily consistent with the categorical prescriptions of right involved in the concept of universal will, we should be able to grasp how this kind of acquisitive action establishes a prerogative of right corresponding to the obligation placed on others by the juridical postulate. The ground of this purported consistency, however, is precisely what is not yet evident. Indeed, the contrast that Kant brings out between the necessarily unilateral character of the initial moment of the procedure of original acquisition and the union of the will of all for universal lawgiving serves to highlight the problem of obligation that issues from the interpretation of the juridical postulate as a use-excluding permissive law. Just how is this problem to be solved in the context of a doctrine of acquisition that takes the possibility of original acquisition to be the immediate consequence of the juridical postulate of practical reason?

Kant holds that original acquisition, regarded as the result of unilateral choice, is possible as a rightful act because the original possession of an external object can *only* be possession in common.[33] Accordingly, he maintains that first possession-taking can be thought of as an act performed in agreement with the law of the external freedom of everyone only if it linked to an "original possession in common [*ursprünglicher Gesamtbesitz*] (*communio possessionis originaria*) whose concept is non-empirical and independent of temporal conditions."[34] If one accepts, with Kant, that first acquisition must be thought of as the acquisition of land, it follows that the relationship between the subjects who possess something in common must be represented in terms of what Kant calls the "original community of land in general."[35] Moreover, because the earth's surface is spherical, this community must extend to "the possession of the land of the entire earth."[36] This conception of the universal scope of original community is yielded when we take account of a fundamental material factor that conditions human existence on earth: the unending finitude, as it were, of the terrestrial globe's surface makes it physically impossible for human beings to be dispersed to the extent that they exist beyond the limits of possible community with one another.[37]

By specifying the notion of original possession in common in terms of the universal community of land, Kant introduces "a practical rational concept that contains a priori the principle in accordance with which alone human beings can use a place on earth in accordance with laws of right."[38] He thus bases his theory of rightful acquisition on the idea of "the possession of all human beings on earth that precedes every rightful act of theirs," a possession that is "constituted by nature itself."[39] By employing the idea of the original community of land, Kant seeks to explain how it is possible to think of first possession-taking as an act by which I (or anyone else) can come to use, in accordance with laws of external freedom, things that I *already* possess in common with all others. By means of his conception of original community, then, Kant wants to show that the act of physically taking possession of something (*apprehensio physica*) furnishes an "empirical title" of acquisition that correlates directly, and necessarily, with "the title to an intellectual possession-taking [*intellektuelle Besitznehmung*] (setting aside all empirical conditions of space and time)."[40] The latter serves as a principle of entitlement that authorizes my (or anyone's) control-seizing activity with respect to objects that can only be thought of as belonging to *everyone* once all *empirical* conditions pertaining to the spatial and temporal features of the possession-taking act are set aside. As such, it provides for "the rational title of acquisition" that "can lie only in the idea of the will of all united a priori."[41]

The analysis of this concept of universal will thus yields the warranting principle by which external acquisition is authorized. Accordingly, we can say that the idea of universal will must itself be presupposed as an indispensable condition for *all* external acquisition. For all external acquisition must accord with the requirements of reason developed from the concept of the unified will of everyone if the control-seizing activity that proceeds from a unilateral will is to place upon others the obligation to refrain from using the things which one has taken into one's possession and designated as one's own.[42] Otherwise, the rightful quality of derivative acquisition could have no strictly rational ground; and the possession-taking act required by original acquisition could not be consistent with the formal requirements set by practical reason's universal lawgiving in the domain of external freedom of action.[43] With reference to this domain, the idea of original community is a conceptual representation that picks out the objective correlate of the idea of universal will. The employment of this idea of will in conjunction with its correlate leads to the theoretical insight that something external can be rightfully acquired only if its way of acquisition conforms to the specific requirements of the juridical or civil condition: *anything* external that I can have as mine must be acquired "in conformity with the idea of a civil condition."[44] For Kant, this means that whatever one chooses to seize (i.e., occupy) as one's own from the originally common possession of humankind must be something acquired *with a view to* the realization of the civil condition but prior to its establishment. Original acquisition is therefore necessarily provisional; and anything originally acquired in keeping with laws of external freedom must be thought of in terms of the provisional acquisition of land through occupation.[45]

In Kant's doctrine of acquisition, then, reference to the global unity of the earth's surface—i.e., reference to the physically insurmountable unity exhibited by the universal object and material substrate of all possible original acquisition—is what puts us in a position to regard possession-taking as the type of act that can be performed in conformity with the idea of an a priori united will. Kant introduces the notion of "the unity of all places on the face of the earth as a spherical surface"[46] as an essential feature of the rational concept of an original community of possession. He employs this idea to show how every subject's possession-taking acts can be represented as ultimately grounded in an authorization to occupy that fulfills the universality requirements of practical reason's external lawgiving. The idea of an original community of land thus furnishes an a priori conceptual representation that structurally corresponds to, and combines with, the idea of universal will as an omnilateral will. Although this rational concept of original community is both non-empirical and applicable independently of temporal conditions, its correlation with the idea of universal will allows for the determination of certain sensible conditions—specifically, the spatial conditions—for everyone's original external possession as well as the conditions under which empirical possession is established through a unilateral will to occupy:

> We have found the *title* of acquisition in an original community of land, and thus among the spatial conditions of an external possession. The way of acquisition [*Erwerbungsart*], however, we have found in the empirical conditions of an external possession-taking (*apprehensio*), joined with the will to have the external object as one's own.[47]

There will be a good deal more to say below about the connection between original community and the spatial conditions for external possession. At this juncture, the major item of interest lies in the connection that Kant discerns between the empirical conditions of external possession-taking and the naturally given unity exhibited by the universal object of possible acquisition.

Original Community and Universal Possession

To shed further light on the notion of the community of possession that the latter connection presupposes, I refer here to several passages from Kant's preliminary work on private right, as published in volume 23 of the Academy Edition.[48] In these passages, Kant is concerned with two different views of universality involved in the idea of original possession in common. He distinguishes between disjunctively (or distributively) universal possession and collectively universal possession. We must think of the earth as the disjunctive-universal possession (*disjunktiv-allgemeiner Besitz*) of humankind when we take account of the physical fact that the earth's spherical surface, and hence its finite magnitude, sets all of the earth's inhabitants "in a relation of thoroughgoing reciprocal possible influence." In virtue of this relation of possible causal reciprocity, every human being is potentially in possession of any place on the earth's surface. Each human agent is therefore "in a potential but only disjunctive-universal possession of all places on the earth's land [*Erdboden*]."[49] Regarded from this distributive viewpoint, i.e., from the point of view by which each human inhabitant of the earth is considered in relation to what any given agent can individually possess, the original community of possession is something more than a merely spatial relation of human beings. It has to be understood as a spatial relation of possible causal community between all agents, that is, as a universal relation of possible reciprocal influence. We must therefore think of it in terms of the dynamical community—the *commercium*—of all originally possessive agents, and not just in terms of the spatial juxtaposition (*communio spatii*) of separate human entities.[50] Now, because the possible reciprocal causal relation between these agents extends to their possessive relations to all places on the earth's surface, the relation of community between the earth's inhabitants must itself be thought of in terms of a universal community of possession. Moreover, the form and scope of the relation of possessors is such that this community of possession must be regarded from a collective point of view as well as from the distributive point of view just characterized. Specifically, it must be thought of as the

> possession in common which, as collective-universal possession through resistance in occupying the space that everyone on earth requires, makes a rightful act possible and practically, i.e., objectively, necessary—an act by which the possession of each is *determined* distributively.[51]

Thus, the universal community of possession, the form of which consists in the relation of possible reciprocal influence between all of the earth's inhabitants, puts anyone in a position to possess any place on earth, which is what it means to say that the earth is originally possessed in common. But this is precisely why

> this possession must also be regarded as collectively universal, i.e., as the common possession of the human species to which corresponds an objectively united will or will that is to be united; for without a principle of distribution (which can only be found in the united will as law) the right of human beings to be anywhere at all [*irgend wo zu sein*] would be entirely without effect and would be destroyed by universal conflict.[52]

Taken from the standpoint of its disjunctive universality, humankind's original community of possession represents a state of nature. Since the juridically salient feature of this natural condition is its potential for universal conflict, the specification of the properties of the disjunctive-universal possession makes evident practical reason's demand for a principle of distribution by which the possession of each can determined as rightful possession. According to Kant, the practical rational ground of this principle can lie only in the idea of an a priori united will that corresponds to the collective view of the originally common possession of all humanity.

Let us return to the published text of the Doctrine of Right. The considerations on disjunctively universal and collectively universal possession just treated are fully congruent with what Kant argues in his metaphysical exposition of the concept of the original acquisition of land.[53] Given that all human beings are originally in common possession of the land of the entire earth, and assuming that each has the will to use this land, it follows that the "naturally unavoidable opposition of the power of choice of one in relation to another" would "do away with all use of it [the land] if that will to use it did not also contain the law by which a *particular possession* can be determined."[54] This law—the distributive law of mine and yours with respect to land (*das Austeilende Gesetz des Mein und Dein eines jeden am Boden*)—can issue only from the a priori united will that normatively underlies, and is realized in, the civil condition. Still, the account of the juridical state of nature as a condition of unavoidable opposition between human agents' powers of choice must include the "rightful capacity [*Vermögen*] of the will of everyone to recognize the act of possession-taking and appropriation as valid, even though it is only unilateral."[55] That is because original acquisition, understood as the provisional acquisition of land, "requires and has the favor of the law (*lex permissiva*) in view of the determination of the limits of rightful acquisition."[56] This favor (*Gunst*) of the law in question (that is, the favor of the permissive law by which the obligation-founding character of first possession-taking is determined) does not reach beyond the point at which the juridical condition is established. In other words, it obtains only with a *view to* the achievement of the condition in which the limits of rightful acquisition on the part of each agent are determined in accordance with the principle of distribution that proceeds from an a priori united will of all. Nonetheless, that favor of law carries with it "all the effects of acquisition in conformity with right."[57]

It is essential to take note here that the favor extended by *lex permissiva* with respect to external acquisition cannot derive from any consideration of need or equity that may pertain to the distribution of some good. This type of material consideration can play no grounding role in a doctrine of acquisition that is consistent with the general account of strict right (*ius strictum*) at issue in the Doctrine of Right as a whole. This account concerns *only* the conditions under which the power of choice of one can be united with the power of choice of another in accordance with a universal law of freedom. Within the systematic framework of the theory of these conditions, the concept of right "has to do solely with the external and practical relations of one person to another, insofar as their actions . . . can have (direct or indirect) influence on each other."[58] It concerns the "reciprocal relation of powers of choice" in which "no account at all is taken of the *matter* of choice, that is, of the end that each has in mind with the objects he wants."[59] The theory of the metaphysical foundations of right therefore treats only the formal conditions under which the relation of possible influence between subjects whose powers of choice are reciprocally related can be in accordance with practical reason's universally prescriptive lawgiving role.[60]

The strict formalism of Kant's foundational juridical position entails that this lawgiving role of reason, as clarified by means of the concept an a priori united will, furnishes the only possible basis for limiting the scope of original acquisition by occupation. To clarify this thought, Kant employs the idea of an original contract in conformity with which the juridical or civil condition is to be established. By introducing this conceptual device, Kant wants to show how a given agent's authorization unilaterally to occupy land is necessarily subject to restriction, given the deontic requirements developed from the concept of the united will of all. This restriction on the scope of original acquisition, however, is itself subject to a crucial qualification. According to Kant, it is not possible to specify, either quantitatively or qualitatively, the limits of what *can* be originally acquired unless the conditions that feature in the idea of the original contract actually apply to the whole human species:

> The indeterminacy with regard to quantity as well as quality of the external object that can be acquired makes this problem (of the sole, original acquisition) the hardest of all to solve. Still, there must be some original acquisition or other of what is external since not all acquisition can be derived. So this problem cannot be abandoned as insoluble and intrinsically impossible. But even if it is solved through the original contract, such acquisition will always remain provisional unless this contract extends to the whole human race.[61]

What follows from the qualification just stated?

It is essential to bear in mind here that the contractualist idea at issue in this passage must in any event apply normatively to the entire human species, whether or not all acquisition is merely provisional. So the implication of Kant's qualifying claim must be this: no definite limits can be placed upon the unilateral act of occupation by which original acquisition occurs as long as the juridical condition of humanity as a whole is not *realized*. Apart from the universal realization of a juridical condition in accordance with the idea of an original contract that extends to the entire human species, original acquisition through unilateral occupation cannot be understood as the type of act that is subject to specific distributive restriction.

This interpretation follows from Kant's understanding of the relationship between the juridical state of nature and the civil condition. It is a required interpretation because the laws of juridico-practical reason prescribe the realization of the juridical condition of humanity, and because these laws therefore maintain their prescriptive force for human beings, as rational beings, no matter what the condition may be in which their powers of choice are in fact reciprocally related. At the same time, though, that interpretation sheds light on a fundamental problem for Kant's theory of private right. Let me formulate this problem with reference to the disjunctively universal quality of the originally common possession of humankind and the dynamical nature of the original community of possession. According to Kant, *all* acquisition must satisfy one of two conditions. Either it must take place in view of the determination of the limits of rightful possession (that is, it must be thought of as taking place with a view to the establishment of the juridical condition), or else it must occur in accordance with the distributive law of mine and yours that is actually operative in a universally realized juridical condition. But if this is true, why should original acquisition receive the favor of *any* law of right if we have to think of it as following from a unilaterally appropriative act of possession-taking—that is, if we have to think of it as something achieved by an act of occupation that *can* generate a condition of universal conflict by virtue of the nature-given form of the causal community of all human possessors? As we have seen, this is a condition characterized by the unavoidable opposition of the powers of choice of all appropriating agents; and this opposition is such that it would do away with *all use* of things already possessed in common if the will of each to use such things did not "contain" (in other words, were not capable of conforming with) a law by which the particular possession of each can be determined. We have also seen that the quantitative and qualitative indeterminacy of the object of external acquisition is what makes original acquisition a problem to be solved by means of the idea of an original contract. Why, then, should we accept that the type of act which gives rise to such indeterminacy with respect to particular possession is compatible with the categorical requirements of right that demand, in keeping with the contractualist idea, the realization of a universal juridical condition in which all indeterminacy of possession is overcome?

Dominium, Right, and the Reach of Cannons

In view of the problem just posed, let us consider several passages from the two segments of the doctrine of acquisition in which Kant concretely treats empirical conditions of external possession-taking. At the beginning of the remarks appended to the main argument *of DR* §15, which establishes the provisional status of all acquisition in the state of nature, Kant writes:

> The question arises, how far does the authorization to take possession of land [*die Befügnis zur Besitznehmung eines Bodens*] extend?

> As far as the capacity for having it under one's control extends, that is, as far as whoever wills to appropriate it can defend it—as if the land were to say, if you cannot protect me, you cannot command me. This is also how the dispute over whether the sea is *free* or *closed* would have to be decided; for example, within the range of cannon fire no one may fish, haul up amber from the ocean floor, and so forth along the coast of a territory that already belongs to a certain state.[62]

And in a similar vein, Kant puts forward the following claim as part of his remarks on the deduction of the concept of original acquisition in *DR* §17:

> My *possession* extends as far as I have the mechanical capacity, from where I reside, to secure my land against encroachment by others (e.g., as far as cannon reach from the shore) and up to this limit the sea is closed (*mare clausum*). But since it is not possible to *reside* on the high seas themselves, possession also cannot extend to them and the open seas are free (*mare liberum*).[63]

Both of these passages refer to a dispute in modern international law concerning the permissibility of acts and policies that would establish exclusive *dominium* in the seas. Hugo Grotius and John Selden set the terms of the dispute in the early seventeenth century. In order to determine with precision how Kant relates to these particular figures, it would be necessary to investigate various references (direct or indirect) to the *mare liberum* vs. *mare clausum* controversy found in the eighteenth-century compendia on natural jurisprudence that Kant used in connection with his lectures on natural law. But we need not be concerned here with the philological details. For the dispute between Grotius and Selden was of fundamental importance for the whole development of modern natural-law theories of property, and Kant makes it quite clear that he addresses an issue of international maritime law only because it pertains to a basic question of his general theory of original acquisition. In keeping with Kant's broader concern with this question about the limits of occupation, I will discuss Grotius and Selden in view of the problem of original acquisition that underlies the *mare liberum/mare clausum* controversy.[64]

Selden's treatise against Grotius—*Mare clausum*—is intended to establish that the high seas are just as much subject to exclusive appropriation and use as is landed territory. Selden's argument is based on an account of the principally unrestricted scope of the unilateral act of occupation. While Selden directly opposes Grotius on the question of *dominium* in the seas, his account of occupation makes use of the key elements of the theory of original acquisition that Grotius constructs in his *Mare liberum* and in *De iure belli ac pacis*.[65] Given our focus on Kant, Selden's critical employment of these elements will be the primary item of interest in this section. But before turning to this employment, a few comments on the Grotian theory of property are in order.

Grotius' account of the origins of property employs the notion of an originally common *dominium*. Since, according to Grotius, all human beings share this *dominium* in the earth and its products, every man may arbitrarily seize for his own use whatever may offer itself for consumption.[66] The result of this universally permissible seizure for use—physical possession—generally takes the place of property in the state of nature. The need for definite property relations arises only when primitive use is replaced by productive activity requiring the distribution of livestock and land.[67] Property as such is established by universal agreement:

> At the same time we learn how things came to be property. It was not by an act of the mind alone; for some could not know what others wanted to have as their own so as to abstain from it, and many might want one and the same thing. Rather, it was by some agreement [*pactum*]—an agreement either expressed, as by division, or tacit, as by occupation. For it must be supposed that as soon as community was no longer satisfactory, and before any division was instituted, all agreed to this: that which anyone had occupied he should have as his own.[68]

Notice that the postulated agreement of all to dissolve the original community of things takes no account of any distinction that can be drawn between the primitive use of things for the purpose of immediate consumption and the more developed forms of relationship between human beings and the natural world that emerge from productive activity and requires the establishment of property. Grotius' postulate of universal agreement pertains *generally* to the universally (even if tacitly) accepted principle that whatever one *had* occupied should be something that one possesses as one's own. Selden's account of the normative grounds of acquisition hinges on this general point.

Selden's portrayal of the origins of property, as found in the fourth chapter of *Mare clausum,* revolves around the ideas of common *dominium*, occupation, and universal agreement. Despite the similarities with Grotius' corresponding portrayal, Selden employs these ideas in order to draw conclusions about the permissible scope of unilateral occupation that undermine Grotius' systematic intentions in the controversy over freedom of the seas. Selden holds that a viable theory of property must begin by supposing that all things were at first given to all human beings in common, and that the earth and its products therefore originally constituted humanity's common *dominium.* Given this theoretical requirement, he uses the idea of universal agreement (*consensus humani generis corporis seu universitatis*) to explain how people could relinquish their original common right to things (*ius commune pristinum*) in such a way that exclusive individual *dominia* could be established in accordance with the law of nature.[69] Building on this explanation, he offers an explicitly contractualist interpretation of the Grotius passage quoted above. Selden thereby wants to show that the theory of the normative grounds of acquisition cannot place any restriction on the scope of unilateral occupation with respect to unoccupied things. The thrust of Selden's argument is as follows.[70]

If all people were originally owners (*domini*) of what was given to everyone in common, then it seems that they must have retained their community of ownership with respect to anything not subject to distribution in the first partition of things through universal consent. Yet, although it is true that all humans must be regarded as joint owners of what was given to everyone in common prior to the first partition, we must conclude that the original common title and right to things given to everyone in common was renounced in a way that any individual might become the owner of what remained vacant (i.e., undistributed) after that consent-based partition. This accords with a parity of reason (*parilis sane ratio*). For if we accept, with Grotius, that each should retain as his own whatever he *had* come to possess before this partition, then we also have reason to conclude that anyone may occupy any thing that was afterwards left vacant as *res nullius*.

Consider carefully the implication of Selden's "parity of reason" argument: *all* components of the originally common stock of things are subject to exclusive appropriation by way of first seizure or occupation. Exactly this is the lesson that Selden is concerned to draw from the link that Grotius forges between the concepts of universal agreement and *occupatio*. For Selden, the unilateral act of occupation performed subsequent to the partition of humanity's common *dominium* is (qua act) conceptually indistinguishable from such an act performed beforehand. Thus, there is no reason to employ the principle of universal agreement in order to restrict the quantitative or qualitative scope of occupation after the dissolution of the common *dominium* while not placing any explicit restriction on it prior to this dissolution. For the same universality principle that allows an agent to retain what he had seized prior to dissolution must apply to all things not owned by particular subjects afterwards. (This is the underlying point of Selden's parity of reason claim.) And, according to Selden, there is no basis whatsoever for restricting the scope of anyone's unilateral act of occupation prior to everyone's agreement that whatever anyone had occupied should be something that one has as one's own.

Selden holds, then, that every agent is entitled unilaterally to occupy any given thing from the common stock of things available to humankind. The *ius commune pristinum*, as Selden designates it, determines the character of the human agent's possessive relation to things seized before the first partition. It also determines this agent's possible appropriative relation to everything afterwards left vacant. Thus, with regard to both the originally common *dominium* and all things that can qualify as *res nullius* subsequent to the dissolution of the original community, Selden's *ius commune* serves as the normative ground for unrestricted appropriation. It is, in effect, *ius commune in omnia*: a right of all to all things–at least to those things not previously subject to exclusive occupation and particularized distribution.[71]

It is on the basis of his doctrine of acquisition that Selden seeks to refute Grotius' arguments against the permissibility of exclusive appropriation of the world's oceans. In his most telling counterargument, Selden maintains that Grotius fails to comply with what is entailed by his own crucial claim, i.e., the claim that the seas do not present a proper object of acquisition.[72] Grotius insists that the seas are not, and never have been, subject to exclusive appropriation. Yet he also concedes that it is possible to establish *dominium* in certain parts of the maritime world (stretches near the shore, bays, and the like). As Selden points out, this concession implicitly reduces the whole problem of the permissibility or impermissibility of the oceans' exclusive appropriation to an empirical question about the pragmatically feasible reach of occupation—which is in fact to miss the philosophic import of the point in question. This point, after all, concerns the juridical possibility of unilateral occupation; and a coherent doctrine of acquisition—even one that takes its key elements from Grotius' own theoretical arsenal—must allow for the possibility of the unilateral and *global* occupation of the oceans. Selden grants, of course, that this global occupation would be an exceedingly difficult act to carry out unilaterally. But if such an act of occupation *could* be successfully performed, then the object of acquisition—the oceans and, indeed, the entire globe for that matter—would become part of the occupant's *dominium*;[73] and this would be entirely consistent with the requirements of natural law in its permissive sense.[74]

So much for the historical background to Kant's references to the *mare liberum* vs. *mare clausum* debate. Having explored a bit of this background, we can see how Kant's claims regarding the extent of the seas' occupation are in keeping with the position that Grotius wanted to defend. It might be an interesting exercise to investigate the extent to which Kant actually understood his doctrine of acquisition as offering support for Grotius' attempt to set limits on the appropriation of the world's oceans. Our interest in this particular issue in international maritime law, however, extends only as far as it sheds light on a basic problem in Kant's theory of original acquisition. The problem, again, is this: why should acquisition through unilateral occupation receive the favor of any law of right if it represents the type of control-seizing act that can generate a condition of universal conflict by virtue of the dynamical community of all possessive agents? In view of this question, the focus of our interest in Grotius' and Selden's doctrines of acquisition has been the connection between the concepts of occupation and community of possession. Thus, in assessing the significance of those doctrines for Kant's theory of original acquisition, we need to bear in mind the following considerations.

The conceptions of common *dominium* found in Grotius' and Selden's doctrines fall under the Kantian heading of "primitive community [*uranfängliche Gemeinschaft*]."[75] According to Kant, the concept of primitive community, or primitive possession in common, underlies the quasi-historical and unavoidably fictional (*gedichtet*) descriptions of humankind's natural condition that Grotius and others put forward in order to depict the beginnings of property and juridical relations among human beings.[76] It is an empirical concept that depends on temporal conditions.[77] The theory of original acquisition, however, is not concerned with temporal conditions that underlie the description of supposed historical (or prehistorical) states of affairs. Rather, this theory aims to determine a priori the necessary conditions for all rightful acquisition. To achieve this aim, it requires a concept of the community of possession, but this must be a rational practical concept whose application in no way depends on empirically specifiable temporal conditions. No empirical concept of the community of possession can qualify as such a concept of reason, and the narrative portrayals of the beginnings of property that rely on this kind of empirical concept cannot meet the conceptual demands of the theory of original acquisition.

This Kantian criticism of traditional doctrines of acquisition certainly applies to Grotius's portrayal of the origins of property relations.[78] It also applies to Selden, at least to the extent that he follows Grotius in giving a historicizing depiction of the state of nature and the dissolution of primitive community. But we should notice as well an interesting implication of the "parity of reason" argument that Selden directs against the Grotian doctrine of acquisition. Selden contends that the contractualist principle of universal agreement can furnish no ground for restricting the quantitative or qualitative scope of unilateral occupation after the dissolution of primitive community if it does not place any explicit restriction on this occupation beforehand. But if this is true, then (*parilis sane ratio*) it must also be the case that there can be no such restriction on the scope of occupation prior to that dissolution either. This follows for two reasons: first, because there is no basis for restricting the scope of anyone's unilateral act of occupation apart from some principle of agreement to do so; and second, because the whole point of employing the principle of universal agreement is to establish that each occupying agent should have *as his own* whatever he had occupied prior to everyone's (tacit or expressed) agreement to restrict each agent's unilaterally occupying acts to those things not already occupied by another.

Selden himself does not bring out this implication regarding the unrestricted scope of occupation prior to universal agreement. His main line of attack against Grotius in *Mare clausum* led him to concentrate instead on the temporal conditions under which the dissolution of primitive community leaves a world full of objects (including all the seas) that have *yet* to be occupied. But if the further inference step is taken, then the general upshot of Selden's parity of reason argument against Grotius becomes clear: apart from the determination of an occupying act's priority in the *order* of time, the specification of temporal conditions is not relevant to a theory of acquisition that employs a contractualist principle of universal agreement. As long as an agent's unilaterally occupying act is thought of as being performed prior to the corresponding act of another, it makes no difference whether occupation occurs before or after the point in time at which universal agreement is supposed (imagined or stipulated) to take place. For the relevant conditions to account for in a theory of *original* acquisition are the conditions by which all occupying agents are set in community with one another in virtue of the relation of each of these agents to all of the outer objects that are subject to unilateral occupation.

It is (logically, if not historically) but a short further step to link the conceptual representation of the sum and the material substrate of these spatial conditions to the potential for universal conflict and the naturally unavoidable opposition of occupying agents' powers of choice. But it is precisely in view of this potential and this opposition that we must again consider the passages from *DR* §15 and §17 quoted at the outset of this section. The general question that Kant poses is the question of the extent to which one is authorized to take possession of unoccupied land. His response, which is keyed to the *mare liberum/mare clausum* controversy, is in keeping with his standpoint that the theory of original acquisition must not rely on the substantive specification of particular conditions of empirical possession apart from the condition of temporal priority in possession-taking. But notice that his treatment of the permissible extent of occupation under actual empirical conditions of possession-taking is entirely unilluminating as far as the pivotal task of that theory is concerned. To put the matter in terms as blunt as Kant's own: settling the permissible spatial scope of possession-taking (hence occupation) by calculating the range of one's weapons is unhelpful if the theoretical task is to clarify how the will of each to use humanity's originally common possession can lead to unilaterally possession-taking acts that are necessarily in keeping with the universality requirements of juridico-practical reason. That is because the insurmountable global unity of this object of possible use is such that the weapons of each *can* be within the range of everyone else's weapons, in which case the nature-given will of each to use the originally common object can give rise to a condition of universal conflict in which no one's use of that object is possible.

It will hardly do to think that an indispensable systematic component of Kant's doctrine of right depends on empirical assumptions about the limitations on human beings' technical capacity for the use-prohibiting mutual infliction of physical damage. There must be something still missing from our treatment of original acquisition. It is certainly not the possible existence of the cannon, which nowadays are real enough in any event.

Dynamical Community and the Formal Principle of Material Equality in Distribution

DR §17 gives Kant's formal "Deduction of the Concept of Original Acquisition."[79] The overall argument of the doctrine of acquisition prior to this deduction leads to the following results, which Kant summarizes in the passage from the first paragraph of §17 translated at the end of the section titled *Acquisition, Occupation, and Original Community*.[80] The title of acquisition (*Title der Erwerbung*) is found in the original community of land, and thus "among the spatial conditions of external possession." The way of acquisition (*Erwerbungsart*), however, is found by taking account of empirical conditions of external possession-taking, and by conceptualizing how these conditions connect up with the will to have external objects as one's own. The remaining task—the actual deduction of the concept of original acquisition—is to "develop *acquisition* itself, i.e., the external mine and yours . . . from the principles of pure practical reason."[81] Perhaps, then, we can find the way to complete the pivotal task of the theory of original acquisition by carefully examining this deduction.

In the second paragraph of *DR* §17, Kant explicates the concept of intelligible possession (*possessio noumenon*), showing that it has as one of its features a representation that denotes "the connection of an external object with me insofar as this connection is the subjective condition of the possibility of the object's use."[82] Since it features in a concept of pure practical reason, this representation must itself be both non-empirical and intellectual. Specifically, it must be the a priori conceptual representation that signifies the property of "*having* an external object *under my control* [des in *meiner Gewalt Habens* . . . des äußeren Gegenstandes]."[83] Even as applied to sensible objects, the concept of having here at issue is one that abstracts from the sensible conditions by which possession appears as "a relation of person to objects that have no obligation."[84] Now when these sensible conditions of concept application are disregarded, possession proves to be "nothing other than the relation of a person to persons, all of whom are subject to the first person's *will* to *bind*"[85] to the extent that this first person's will accords with the basic principles of juridico-practical reason and the universally lawgiving function of the united will of all. In sum, the deduction of the concept of original acquisition involves the specification of the concept of intelligible possession in terms of having something external to me under my control; and this is the particular concept of having that we can employ to explain how one person's will to bind can actually place an obligation upon all other persons with respect to certain objects of the power of choice.

By means of that specification, then, Kant wants to show how the concept of having that plays so central a role in the doctrine of possession also provides the key to understanding the justificatory basis of original acquisition.

Is Kant successful in this endeavor? The analytic procedure of *DR* §17 clearly does show the way to think of the possession of objects as a relation of obligation between persons, instead of merely as a physical relation of persons to objects. It therefore sheds considerable light above all on the culminating aspect (or "moment") of original acquisition mentioned above. (In particular, it allows for a proper understanding of what Kant means in when he asserts, in *DR* §10, that appropriation is "the act of an externally and universally lawgiving will (in idea) through which everyone is bound to agreement with my choice."[86]) Given its focus on an a priori concept of *having*, however, it is far from obvious that Kant's procedure in *DR* §17 can even address what is fundamentally at issue in the required account of original acquisition as long as acquisition as such must be understood as an act of *acquiring* things in order to come to be in the state of having them. With regard to the use of these things, Kant's aim in §17 is to show how all persons can be bound by the will of a first person, i.e., by the will of a unilaterally control-seizing agent who was the first to take possession of an object. The argument is that all persons are so bound *insofar* as that first person's will to have something under her control accords with the universality requirements of juridico-practical reason, and is therefore consistent with "the universal *lawgiving* of the will that is thought as united a priori."[87] Yet it remains entirely unclear how we are to go about determining the extent to which the way of acquisition (*Erwerbungsart*) can be in conformity with this universally prescriptive role of the a priori united will of all as long as that way of acquiring things is to be found in the empirical conditions of possession-taking that are linked to a person's will to have objects under her control.

The thrust of this line of questioning becomes evident when we consider again three things that Kant establishes in the sections that are supposed to prepare the terrain for the deduction of the concept of original acquisition in *DR* §17. As we have seen, Kant argues in §15 and §16 that (a) taking possession of objects from the originally common possession of all human beings gives rise to quantitative and qualitative indeterminacy with regard to particular possession (*besonderer Besitz*); (b) each human being's nature-given will to use that common possession is unavoidably faced with the opposition of human powers of choice; and (c) this opposition would do away with all use of the common possession unless that will to use contained the law by which the particular possession of each can be determined. What, then, is the determining ground of the will that this law expresses? Kant makes it quite clear, of course, that the law itself must derive from the *idea* of an a priori united will of all. Consequently, the ground in question obviously cannot be found in empirical conditions of possession-taking. But neither Kant's expository work on the concept of original acquisition prior to the Deduction of the Concept of Original Acquisition nor the Deduction itself provides for the explanation of any *a priori* ground (or condition) by which the particular possession of each can be *determined.*

There is but one remaining path to take if we are to discover this kind of ground. It is the path that leads from the title of acquisition, not from the empirical conditions of an external possession-taking. Let us therefore examine more closely the "spatial conditions"[88] that feature in the a priori concept of the original community of land which furnishes the title of acquisition. This is best accomplished with reference to the considerations on disjunctively universal and collectively universal possession found in Kant's preliminary work on the Doctrine of Right. As we have seen,[89] the thought that the earth furnishes the disjunctive-universal possession of humankind takes account of a physical factor that conditions human existence: the spatial properties of the earth's surface are such that all of the earth's inhabitants stand in a relation of possible reciprocal influence or possible dynamical community (*commercium*). In virtue of this relation, each human being is *always* potentially in possession of any place among *all* places on the face of the earth, which is what requires the collective view of the universal dynamical community of possession. The explication of this collective view shows both the possibility and the practical necessity of a rightful act "by which the possession of each is determined distributively."[90] Regarded from the collective point of view, the common possession of the human species must therefore be thought of as something subject to a form of particularized distribution that *can* coincide with "an objectively united will or will that is to be united." Apart from the consideration of this possible structural overlap, there is no way to ground the principle of distribution without which, as Kant says, "the right of [particular] human beings to be anywhere at all would be entirely without effect and would be destroyed by universal conflict."[91]

All this is in keeping with the argument presented in the published version of Doctrine of Right. Yet Kant's concern to highlight the causal features of the universal spatial community of possession allows us to discern the import of a key requirement of practical reason that is not readily apparent in the published text. On its collective view, the original community of possession may not be understood *simply* in terms of the reciprocal causal relation of all persons that can emerge from their possessive relation to usable objects of the power of choice. For besides representing a universal relation of persons to persons, it must also be thought of as representing the relation of all persons to all possible objects that are subject to original acquisition in virtue of the fact that all such objects are constitutive components of the common possession of all human beings. It is this collective causal relation of *persons to objects*—that is, the possessive relation that persons have to usable objects of the power of choice on the basis of their possible reciprocal causal relations to the powers of choice of all other persons—that must be constituted in conformity with a principle of distribution that derives from the idea of a united and universally lawgiving will of all persons.

Now as regards acquired rights, lawgiving in accordance with this idea of practical reason is subject to a fundamental requirement that is established in the Introduction to the Doctrine of Right. It has to be consistent with a basic principle of juridico-practical reason involved in Kant's understanding of freedom as "the only original right belonging to every human being by virtue of their humanity."[92] The relevant principle is that of innate equality, i.e., "independence from being bound by others to more than one can also reciprocally bind them."[93] Kant's theory of private right, and hence his account of original acquisition, presupposes the practical objective validity of this principle of independence. The principle of distribution that applies to original possession in common, i.e., to the original community of land, must therefore accord with the constraint against the non-reciprocal imposition of obligation in terms of which the innate equality of human beings is defined.[94] With respect to the conditions of original acquisition, then, what Kant calls the "rightful capacity of the will to bind everyone"[95] must be governed by a principle which recognizes that *no* person's unilateral act of possession-taking and appropriation can bind other persons to more than that first person can be bound by them. With regard to all things that can be acquired for use, this principle allows for indefiniteness or indeterminacy from the point of view of disjunctive-universal possession, i.e., from the point of view according to which every human being is potentially in possession of any arbitrarily chosen place on earth because of the possible dynamical community—the *commercium*—of all the planet's inhabitants. Yet Kant's understanding of original possession in common is such that the principle of distribution contained in the idea of the united will of all must be a principle of *equality* in distribution. For even if I am (originally) the first to *take* possession of something that I already possess in common with all others, I would violate the constraint against the non-reciprocal placement of obligation unless I limited the extent of my possession-taking in accordance with the following precept: we may place upon all others an obligation to refrain from using certain objects of our powers of choice because we were (each of us respectively) the first to take possession of these objects—but *only insofar* as this does not bind all others, in their use of objects, to refrain from *more* than we are reciprocally bound by all of them to refrain from using.

However quantitatively or qualitatively indeterminate the object of acquisition may turn out to be under empirical conditions of possession-taking, the consideration of the collective relation of all persons to the universal object and material substrate of possible acquisition gives rise to the following demand. The principle of distribution at issue in the conceptual determination of original acquisition must be a principle of equality. It must be the principle (or law) through which the a priori determinable condition of possession-taking is *specified as* the ground of equality in the distribution of things. This is the only way to establish that the will of each to use the originally common possession of humankind is necessarily consistent with the universal lawgiving at issue in the idea of an a priori united will of all. The normative basis for all derivative acquisition therefore lies in a distributive principle of original equality in use. This principle must retain its objective validity and practical necessity with respect to all conditions of possession-taking. That is to say: it must retain its universally prescriptive force independently of the extent to which any given outcome of the unilaterally control-seizing activity of human beings in fact satisfies the imperative of reason that such a principle of equality yields.

We should bear in mind that the principle in question furnishes a law of practical reason that belongs to the theory of *ius strictum*. This is a law of right that concerns merely the reciprocal relations of powers of choice. Its derivation does not depend in any way on considerations of equity or human need;[96] and it takes no account of the matter of choice, i.e., the end that one may have in mind with whatever object one might want to have. The principle of distribution in question is therefore a strictly formal principle of equality. Nonetheless, it is a principle of *material* equality as far as the acquisition of objects is concerned. Only if it incorporates a formal principle of material equality can Kant's theory of original acquisition provide a coherent normative basis for property law.

Needless to say, this conclusion is hardly what one would expect to get from a classic modern theory of the foundations of property, and I hasten to emphasize that Kant himself by no means draws it in the Doctrine of Right. Yet there seems to be no other defensible Kantian way of solving the problem of original acquisition—unless, of course, we are prepared to accept that a person's will to use and rightfully to have objects need not be in keeping with practical reason's *exeundum* prescription, that is, with its commandment to go forth from the juridical state of nature and leave it behind.

Notes

1. Kant's works are cited by volume, page, and line number of *Kants gesammelte Schriften*, Königlich Preußische Akademie der Wissenschaften (Berlin: G. Reimer [now De Gruyter], 1902– (hereafter cited as AK). My renderings of passages from Kant's *Metaphysik der Sitten* conform with Mary Gregor's English translation, which is available in the Cambridge Edition of the Works of Immanuel Kant (see Immanuel Kant, *Practical Philosophy*, ed. and trans. M. Gregor [Cambridge: Cambridge University Press, 1996]), 363–615. I have, however, substantially altered the Gregor translation whenever I have judged it appropriate to do so.

2. *Physical Geography*, AK 9: 164, 24–5.

3. *Physical Geography*, AK 9: 164, 27.

4. This, I take it, is the meaning of Kant's claim (if he in fact stated it in the course of his lecturing activities) that political geography "is entirely based on physical geography" (*Physical Geography*, AK 9: 164, 29). Support for the interpretation here offered can be gathered from pages 5–6 of the Powalski manuscript on Kant's geography (available in Werner Stark's transcription at http://kant.bbaw.de/base.htm/geo_pow.htm). Pay special attention to Kant's reported references to Montesquieu and Hume.

5. See the *Metaphysics of Morals*, AK 6: 245, 5. (The Metaphysical Foundations of the Doctrine of Right—or simply Doctrine of Right [= *DR*]—comprises the first main part of the *Metaphysics of Morals*. All references to the Doctrine of Right are therefore keyed to this text.) For general discussion of the structure and systematic character of the theory of private right, see Hans Friedrich Fulda, "Zur Systematik des Privatrechts in Kants 'Metaphysik der Sitten,' " in *Recht, Staat, und Völkerrecht bei Immanuel Kant*, ed. Dieter Hüning and Burkhard Tuschling (Berlin: Duncker & Humblot, 1998), 141–56; Wolfgang Kersting, *Wohlgeordnete Freiheit: Immanuel Kants Rechts- und Staatsphilosophie* (Frankfurt a.M.: Suhrkamp, 1993), 225–321. See also Paul Guyer, "Kant's Deductions of the Principles of Right," in *Kant's Metaphysics of Morals: Interpretive Essays*, ed. Mark Timmons (Oxford: Oxford University Press, 2002), 23–64.

6. On the relationship between *Sachenrecht* and the theory of property in Kant, see Rainer Friedrich, *Eigentum und Staatsbegründung in Kants 'Metaphysik der Sitten'* (Berlin: de Gruyter, 2004), 89–95. On modern theories of property and their developmental background, see Reinhard Brandt, *Eigentumstheorien von Grotius bis Kant* (Stuttgart: Frommann-Holzboog, 1974); Manfred Brocker, *Arbeit und Eigentum: Der Paradigmenwechsel in der neuzeitlichen Eigentumstheorie* (Darmstadt: Wissenschaftliche Buchgesellschft, 1992); Stephen Buckle, *Natural Law and the Theory of Property: Grotius to Hume* (Oxford: Oxford University Press, 1991); Jeffrey Edwards, "Property and *communitas rerum*: Ockham, Suarez, Grotius, Hobbes," in *Societas rationis: Festschrift für Burkhard Tuschling zum 65. Geburtstag* (Berlin: Duncker & Humblot, 2002), 41–60; Brian Tierney, *The Idea of Natural Rights: Studies on Natural Rights, Natural Law and Church Law 1150–1625* (Atlanta: Scholars Press/Emory University, 1997); Brian Tierney, "Permissive Natural Law and Property: Gratian to Kant," *Journal of the History of Ideas* 62 (2001): 381–99.

7. *Metaphysics of Morals*, AK 6: 245, 9–11.

8. See *Metaphysics of Morals*, AK 6: 245, 13–21.

9. See *Metaphysics of Morals*, AK 6: 245, 23–27.

10. See *Metaphysics of Morals*, AK 6: 245, 16–27.

11. *Metaphysics of Morals*, AK 6: 245, 27–246, 2.

12. *Metaphysics of Morals*, AK 6: 250, 13–4.

13. As has long been widely acknowledged, *DR* §6 (Deduction of the Concept of the Merely Rightful Possession of an External Object [*possessio noumenon*]) was flawed by the insertion of a piece of text the content of which belongs to Kant's reflections on original acquisition. Bernd Ludgwig, in his *Philosophische Bibliotek* edition of Kant's *Rechtslehre* (Hamburg: Felix Meiner, 1998), has maintained that the faulty insert should be replaced by transferring the entire text of §2 to §6. This is not the place to examine the extensive controversy in the secondary literature that has arisen in the wake of Ludwig's editorial work and other publications relating to the textual reconstruction of

the Doctrine of Right. It must suffice here for me to state that I reject these reconstructive efforts because I see no plausible reason to advocate the destruction of Kant's logical line of argument in the Doctrine of Right.

14. *Metaphysics of Morals*, AK 6: 246, 5–8.

15. *Metaphysics of Morals*, AK 6: 247, 2–6, italics mine.

16. See *Metaphysics of Morals*, AK 6: 252, 17–21.

17. On the signficance of Kant's account of private right for natural-law theory, see, e.g., *Metaphysics of Morals*, AK 6: 242, 3–19; 256, 22–35; 257, 14–9.

18. *Metaphysics of Morals*, AK 6:256, 30–5. The analytic connections between the concepts of right and compulsion or constraint [*Zwang*] are established in §§D and E of the introduction to *DR* (see *Metaphysics of Morals*, AK 6: 231, 22–233, 25).

19. See *Metaphysics of Morals*, AK 6: 256, 35–257, 6.

20. *Metaphysics of Morals*, AK 6: 257, 14–8.

21. *Metaphysics of Morals*, AK 6: 257, 20.

22. See *Metaphysics of Morals*, AK 6: 247, 10-3.

23. "The principle of external acquisition is as follows: that is mine which I bring under my control [*Gewalt*] (in accordance with the law of outer *freedom*); which, as an object of my power of choice, is something that I have the capacity [*Vermögen*] to use (in accordance with the postulate of practical reason); and which, finally, I *will* that it is to be mine (in conformity with the idea of a possible united will)" (*Metaphysics of Morals*, AK 6: 258, 22–7).

24. See *Metaphysics of Morals*, AK 6: 258, 22–5.

25. See *Metaphysics of Morals*, AK 6: 262, 1–10; cf. Karl Marx, *Grundrisse der Kritik der politischen Ökonomie* (Berlin: Dietz, 1974), 391–2.

26. See *Metaphysics of Morals*, AK 6: 263, 12–9.

27. See *Metaphysics of Morals*, AK 6: 256, 1–11; 259, 17–20.

28. *Metaphysics of Morals*, AK 6: 263, 26–7.

29. *Metaphysics of Morals*, AK 6: 258, 28–259, 4. For discussion, see Kersting, *Wohlgeordnete Freiheit*, 264-7, Ludwig, *Kants Rechtslehre*, 127–33.

30. *Metaphysics of Morals*, AK 6: 259, 4–7.

31. *Metaphysics of Morals*, AK 6: 259, 8–10.

32. *Metaphysics of Morals*, AK 6: 259, 12–8.

33. See *Metaphysics of Morals*, AK 6: 258, 9–14.

34. *Metaphysics of Morals*, AK 6: 262, 28–30.

35. *Metaphysics of Morals*, AK 6: 262, 13–4. See also 268, 3–4. It is natural to think of "*Ursprünglicher Gesamtbesitz*" as the term to use when referring to the object possessed in common (that is, when referring to the relation of possessors *to* their common object). It is also natural to think of "*Gemeinschaft des Bodens*" as the expression for the mutual relations that possessors have by virtue of their relations to that common object. Kant, however, seems to use the two terms interchangeably. In any event, nothing in his arguments on original acquisition hinges on the distinction just mentioned.

36. *Metaphysics of Morals*, AK 6: 267, 4–5.

37. See *Metaphysics of Morals*, AK 6: 262, 20–6.

38. *Metaphysics of Morals*, AK 6: 262, 32–4.

39. *Metaphysics of Morals*, AK 6: 262, 2–28.

40. *Metaphysics of Morals*, AK 6: 264, 9–14.

41. *Metaphysics of Morals*, AK 6: 264, 17–8.

42. See *Metaphysics of Morals*, AK 6: 264, 17–22; 265, 19–30.

43. See *Metaphysics of Morals*, AK 6: 259, 17–20; 260, 33–261, 14.

44. *Metaphysics of Morals*, AK 6: 264, 24.

45. See *Metaphysics of Morals*, AK 6: 264, 29–35.

46. *Metaphysics of Morals*, AK 6: 262, 22–3.

47. *Metaphysics of Morals*, AK 6: 268, 3–7. See also 263, 4–12.

48. See Loose Leaf 58, especially the following passages: AK 23: 320, 19–321, 8; 323, 26–324, 2.

49. Loose Leaf 58, AK 23: 320, 22–3.

50. These terms are taken from the *Critique of Pure Reason's* Third Analogy of Experience (see AK 3: 182, 27–31; cf. *Metaphysics of Morals*, AK 6: 352, 6–22). For discussion, see Jeffrey Edwards, *Substance, Force, and the Possibility of Knowledge: On Kant's Philosophy of Material Nature* (Berkeley: University of California Press, 2000), 20–1, 37–43.

51. Loose Leaf 58, AK 23: 320, 30–1. Italics mine.

52. Loose Leaf 58, AK 23: 323, 30–324, 2.

53. *Metaphysics of Morals*, §16.

54. *Metaphysics of Morals*, AK 6: 267, 7–11.

55. *Metaphysics of Morals*, AK 6: 267, 19–21.

56. *Metaphysics of Morals*, AK 6: 267, 24—6.

57. *Metaphysics of Morals*, AK 6: 267, 31.

58. *Metaphysics of Morals*, AK 6: 230, 9-11.

59. *Metaphysics of Morals*, AK 6: 230, 16–7.

60. Which is *not* to say that this theory disregards the force of moral requirements derived in view of need, or that it assigns no place to the determination of equity in the public administration of justice. See *Metaphysics of Morals*, AK 6: 235, 6–11; 235, 16–236, 7.

61. *Metaphysics of Morals*, AK 6: 266, 28–37.

62. *Metaphysics of Morals*, AK 6: 265, 1–10.

63. *Metaphysics of Morals*, AK 6: 269, 31–6.

64. For more extensive treatment of both Grotius and Selden, see Jeffrey Edwards, "Natural Right and Acquisition in Grotius, Selden, and Hobbes," in *Der lange Schatten des Leviathan. Hobbes' politische Philosophie nach 350 Jahren*, ed. Dieter Hüning (Berlin: Duncker & Humblot, 2005), 153–78.

65. For a succinct account of the exceptionally interesting (not to mention complicated) publication histories of the works mentioned in this paragraph, see Tuck, Richard, *Philosophy and Government: 1572–1651* (Cambridge: Cambridge University Press, 1993), 169–70, 190–2, 211–14.

66. On Grotius' conception of originally common *dominium*, and for discussion of its broader historical context, see Edwards, "Property and *communitas rerum*," 52–6; Tierney, *The Idea of Natural Rights*, 316–42.

67. See Hugo Grotius, *The Freedom of the Seas*, trans. Ralph Van Deman Magoffin (Oxford: Oxford University Press, 1916 [1609]), 25. Reference is to the pagination of the Latin text of Grotius' *Mare liberum*, which is published in conjunction with the Magoffin translation. Also see Hugo Grotius, *De iure belli ac pacis*, ed. James Brown

Scott (Washington, DC: Carnegie Institute, 1913 [1625, 2nd ed. 1631]), II.2 §2. Photomechanical reproduction of 1746 Latin text. In keeping with standard citation practice, my references to Grotius' *De iure belli ac pacis* are to the relevant book, chapter, and section numbers.

68. Grotius, *De iure belli ac pacis*, II.2 §2.

69. See John Selden, *Opera omnia* (London: S. Palmer, 1726 [1636]), II col. 1197. My references to Selden are to the volume and column numbers of the eighteenth-century edition of his collected works.

70. See Selden, *Opera omnia*, II cols. 1197–8.

71. For recent discussion of Selden in connection with Thomas Hobbes' account of natural right and the state of nature, see Richard Tuck, *The Rights of War and Peace: Political Thought and the International Order from Grotius to Kant* (Oxford: Oxford University Press, 1999), 16–50.

72. See Selden, *Opera omnia*, II col. 1274.

73. See Selden, *Opera omnia*, II col. 1274.

74. Selden lays out his conception of universal permissive natural law (*ius permissivum*) in chapters 5–7 of *Mare clausum*.

75. See *Metaphysics of Morals*, AK 6: 258, 14–21; 262, 26–34.

76. See *Metaphysics of Morals*, AK 6: 251, 1–13. Regarding this passage's underlying references to Grotius and Pufendorf, see Jeffrey Edwards, "Disjunktiv- und kollektiv-allgemeiner Besitz: Überlegungen zu Kants Theorie der ursprünglichen Erwerbung," in *Recht, Staat, und Völkerrecht bei Immanuel Kant*, 121–40, here 127 n. 17.

77. See *Metaphysics of Morals*, AK 6: 262, 28–31.

78. For extended discussion of the medieval and modern background relevant to pre-Kantian doctrines of acquisition and theories of property, see Annabel Brett, *Liberty, Right, and Nature: Individual Rights in Later Scholastic Thought* (Cambridge: Cambridge University Press, 1997); Brocker, *Arbeit und Eigentum*; Edwards, "Property and *communitas rerum*;" Tierney, *The Idea of Natural Rights*; Tierney, "Permissive Natural Law and Property."

79. *Metaphysics of Morals*, AK 6: 268, 2.

80. Page 15 above.

81. *Metaphysics of Morals*, AK 6: 268, 8–11.

82. *Metaphysics of Morals*, AK 6: 268, 18–9.

83. *Metaphysics of Morals*, AK 6: 268, 18–20.

84. *Metaphysics of Morals*, AK 6: 268, 22–3.

85. *Metaphysics of Morals*, AK 6: 268, 23–8.

86. *Metaphysics of Morals*, AK 6: 259, 2–4.

87. *Metaphysics of Morals*, AK 6: 268, 25–6.

88. *Metaphysics of Morals*, AK 6: 268, 3.

89. See section titled *Original Community and Universal Possession.*

90. Loose Leaf 58, AK 23: 320, 36.

91. Loose Leaf 58, AK 23, 323, 31–324, 2.

92. *Metaphysics of Morals*, AK 6: 237, 32–3.

93. *Metaphysics of Morals*, AK 6: 237, 34.

94. Which is not to say that this principle can be *derived* from Kant's account of the innate account of freedom.

95. *Metaphysics of Morals*, AK 6: 267, 19–20.

96. In this regard, the position that I take on what is required by Kant's doctrine of original acquisition may well differ significantly from Guyer's approach to the same set of issues. See Guyer, *Kant on Freedom, Law, and Happiness* (Cambridge, Cambridge University Press, 2000), 253–8; cf. Guyer, "Life, Liberty, and Property: Rawls and the Reconstruction of Kant's Political Philosophy," in *Recht, Staat, und Völkerrecht bei Immanuel Kant*, 273–92, here 281–4.

V

Gender, Race, History, and Geography

14

Cosmopolitanism in the *Anthropology* and *Geography*

David Harvey

Kant's inspiration for the contemporary approach to cosmopolitanism is impossible to ignore. I cite his most celebrated passage from "Perpetual Peace":

> The peoples of the earth have entered in varying degrees into a universal community, and it is developed to the point where a violation of laws in *one* part of the world is felt *everywhere*. The idea of a cosmopolitan law is therefore not fantastic and overstrained; it is a necessary complement to the unwritten code of political and international law, transforming it into a universal law of humanity.[1]

Kant's conception of cosmopolitan law arises in the context of a certain kind of geographical structure. The finite quality of the globe defines limits within which human beings, by virtue of their common possession of the surface of the earth, are forced to accommodate (sometimes violently) with each other. Human beings have the inherent right, if they so desire, to range across the surface of the earth and to associate with each other (through trade and commerce, for example). Means of transport (Kant mentions the ship and the camel) facilitate increasing contacts over space. But in Kant's schema, the earth's surface is presumed to be territorially divided into sovereign states. These will tend in the long run to become both democratic and republican. Inhabitants will then possess distinctive rights of citizenship within their states. Relations between states will be regulated by a growing requirement to establish perpetual peace because of increasing interdependence through trade and commerce. War between states becomes less likely for two reasons. In a democratic state it will be necessary to gain the consent of a public that would have to bear the brunt of the costs. The habit of sovereigns, emperors, and the nobility of waging war for reasons of personal prestige or aggrandizement will be constrained. Second, trade disruptions from war will inflict greater and greater losses as the levels of economic interdependence between states increase. The cosmopolitan ethic requires that individuals (presumed citizens of one state) would have the right to hospitality when they cross borders (particularly for purposes of trade):

> Hospitality means the right of a stranger not to be treated as an enemy when he arrives in the land of another. One may refuse to receive him when this can be done without causing his destruction; but so long as he peacefully occupies his place, one may not treat him with hostility. It is not the right to be a permanent visitor that one may demand. A special contract of benificence would be needed in order to give an outsider a right to become a fellow inhabitant for a certain length of time. It is only the right of temporary sojourn, a right to associate, which all men have. They have it by virtue of their common possession of the surface of the earth, where, as a globe, they cannot infinitely disperse and hence must finally tolerate the presence of each other.[2]

Cosmopolitan right is, therefore, circumscribed. "The right of hospitality," Benhabib notes, "occupies that space between human rights and civil rights, between the right of humanity in our person and the rights that accrue to us insofar as we are members of specific republics."[3] The presumption of a sovereign (preferably democratic and republican) state authority defined by its distinctive territoriality lies at the basis of this formulation. For purposes of citizenship the territoriality of the state is regarded as an absolute space (i.e., it is fixed and immovable and has a clear boundary). But it is the universal (i.e., deracinated) right to hospitality that opens the absolute spaces of all states to others under very specific conditions.

Kant's formulation of the cosmopolitan ethic has been the subject of considerable analysis and debate. But no one has cared to explore the implications of Kant's assumptions about geographical structure for the cosmopolitanism he derives. The only substantive discussion I can find concerns the role that the common possession of a finite globe plays in Kant's justification of cosmopolitan right. The consensus seems to be that "the spherical surface of the earth constitutes a circumstance of justice but does not function as a moral justificatory premise to ground cosmopolitan right."[4] This conclusion is understandable. To conclude otherwise would be to commit the naturalistic fallacy or, worse still, fall into a crude environmental determinism (the idea that spatial structure—the sphericity of the globe—has direct causative powers). But relegation of the geographical circumstances to the status of a mere "circumstance of justice" is not the end of the issue. It makes it seem as if the nature of the geographical space has no bearing in relation to principles applied to it. Though the material (historical and geographical) circumstances may be contingent, this does not mean that the characterization of those circumstances in the form of anthropological and geographical knowledges is irrelevant to the formulation of a cosmopolitan ethic. Nussbaum and, as we shall see, Kant himself, clearly think the circumstances matter. And so, it turns out, does Foucault. So how and why does it matter?

Kant's philosophical teaching concentrated on logic, metaphysics, and ethics. But he also taught geography and anthropology on a regular basis. Is there any relation between these teachings? His writings on anthropology and particularly geography are generally ignored or relegated to a zone of insignificance in relation to his three major critiques. The anthropology has, however, been translated into several languages and subjected to some commentary. Foucault, for example, translated the *Anthropology* into French in 1970, promising a deeper analysis of it in a subsequent publication. He never made good on this promise (though he did leave behind an extended unpublished commentary that is now available to us).[5] Kant's geography is known hardly at all (Foucault, interestingly, totally ignores it). Whenever I have questioned Kantian scholars about it, their response has almost always been the same. It is "irrelevant," "not to be taken seriously," or "there is nothing of interest in it." There is no published English edition (though there is a translation of Part I as a Master's Thesis by Bolin).[6] A French version finally appeared in 1999.[7] There is no serious study of Kant's *Geography* in the English language other than May's[8] coupled with occasional forays by geographers into understanding Kant's role in the history of geographical thought (without any attempt to link this to his metaphysics or ethics). The introduction to the French edition of the geography does attempt an evaluation.

This neglect of the geography does not accord with Kant's own assessment. He went out of his way to gain an exemption from university regulations in order to teach geography in place of cosmology. He taught geography forty-nine times, compared to the fifty-four occasions he taught logic and metaphysics and the forty-six and twenty-eight times he taught ethics and anthropology, respectively. He explicitly argued that geography and anthropology defined the "conditions of possibility" of all knowledge. He considered these knowledges a necessary preparation—a "propaedeutic" as he termed it—for everything else.[9] While, therefore, both anthropology and geography were in a "precritical" or "prescientific" state, their foundational role required that they be paid close attention. How else can we interpret the fact that he taught geography and anthropology so persistently alongside his metaphysics and ethics? Though he signally failed in his mission, he plainly thought it important to bring anthropology and geography into a more critical and scientific condition. The question is: why did he think so?

Van de Pitte, in his introduction to the *Anthropology*, provides one answer to this question. As Kant increasingly recognized that "metaphysics could not follow the method of pure mathematics" then, as Kant himself put it, "the true method of metaphysics is basically the same as that introduced by Newton into natural science." Metaphysics must rest, therefore, upon a scientific understanding of human experience:

"But if metaphysics now was to begin in experience, where would it find the fixed principles in terms of which it could build with assurance? As Kant himself expressed it, the variations in taste and different aspects of man give to the flow of experience an uncertain and delusive character. 'Where shall I find fixed points of nature which man can never shift and which can give him indications of the shore on which he must bring himself to rest?' " Kant, according to Van de Pitte, turned to Rousseau's writings to find an answer. There he discovered that "because man can consider an array of possibilities, and which among them is more desirable, he can strive to make himself and his world in a realization of his ideals." This could be so because human beings possessed powers of rational thought (though mere possession of these powers did not guarantee their appropriate use). But this meant in turn that metaphysics need no longer be purely speculative. It must proceed "in terms of clearly defined absolute principles derived from man's potential."[10] By what means, then, could man's potential be established, if God and traditional cosmology could not provide the answer?

At several points in his articulation of the cosmopolitan ethic, Kant expresses the view that the ethic arises out of nature or out of human nature (he sometimes seems to conflate the two). The cosmopolitan ethic is therefore based on something other than pure speculation or idealism. Kant (unlike Former President George W. Bush) refuses to invoke any notion of God's design. The attention Kant pays to both geography and anthropology then makes more sense. If theology and cosmology could no longer provide adequate answers to the question "what is man?" (hence Kant's determination to eliminate cosmology from the curriculum and replace it with geography) then something more scientific was needed. Where was that "science of man" to come from if not from anthropology and geography? The distinction between geography and anthropology rested, in Kant's view, on a difference of perspective between the "outer knowledge" given by observation of "man's" place in nature and the "inner knowledge" of subjectivities (which sometimes comes close to psychology in practice). The fact that he began teaching the geography first (in 1756) and that much of what he there examines concerns the physical processes that affect the earth's surface and human life upon it, suggests a certain initial attraction to an underlying theory of environmental determinism as providing a potentially secure scientific basis for metaphysical reflection (and, as we shall see, many of the examples he evokes in his geography reflect that tendency). His later turn to anthropology (which he began teaching in 1772), and the fact that he paid far greater attention to elaborating upon it (even preparing it for publication) in his later years, suggests that he increasingly found the inner knowledge of subjectivities more relevant to his philosophical project. It is significant that the final passages of the *Anthropology* address the whole question of cosmopolitan law directly, whereas there is no mention of this topic in the geography.

Consider first, then, the implications of his anthropology. The work amounts to a detailed enquiry into our species being (it foreshadows, therefore, Marx's examination of the concept in the *Economic and Philosophic Manuscripts of 1844*). The purpose is plain enough. Not only must we understand what we have been about and how we now are as a human species, we must also understand what we can become by virtue of our particular capacities and powers. Human nature is not fixed but evolving and by studying that evolution we can say something about the destiny of the human race. Foucault, in his commentary, is as profoundly admiring of Kant's capacity to ask these questions as he is critical of Kant's actual answers. "Man is not simply 'what he is,' but 'what he makes of himself.' And is this not precisely the field that *Anthropology* defines for its investigation?" Foucault asks. The *Anthropology* is, therefore, in Foucault's view, a central rather than marginal text in relation to the three major philosophical critiques that Kant contributed.[11] Amy Allen summarizes Foucault's argument this way:

> Thus, Foucault suggests, the *Anthropology* (perhaps unwittingly) breaks open the framework of the critical philosophy, revealing the historical specificity of our a priori categories, their rootedness in historically variable social and linguistic practices and institutions. Foucault's reading of Kant's *Anthropology* thus suggests that Kant's system contains the seeds of its own radical transformation, a transformation that Foucault will take up in his own work: namely the transformation from the conception of the a priori as universal and necessary to the historical a priori; and the related transformation from the transcendental subject that serves as the condition of possibility of all experience to the subject that is conditioned by its rootedness in specific historical, social and cultural circumstances.[12]

It is this transition in thinking from a disembodied to a rooted human subject that is critical, and the vehicle is in the first instance supplied by the anthropology. Kant's views on our species being are not confined to his text on anthropology, so on this point some contextualization is needed. Kant generally rejects any notion of the inherent goodness of humanity. He does not appeal to any figure of the noble savage or of Godly innocence. "Everything," he says, "is made up of folly and childish vanity, and often of childish malice and destructiveness."[13] Enlightenment, he says in his celebrated essay on that subject, depends upon "man's emergence from his self-incurred immaturity," defined as the inability to use understanding "without the guidance of another."[14] Only a few, Kant suggests, "have succeeded in freeing themselves from immaturity and in continuing boldly on their way" while all manner of prejudices (even the new ones created in the course of revolution) "will serve as a leash to control the great unthinking mass." For enlightenment to progress depends on "the most innocuous form" of freedom—"the freedom to make *public use* of one's reason in all matters."[15] While we live in an age of enlightenment, we do not live in an enlightened age. This way of thinking enters into the final passages of the *Anthropology*. Human beings, he says:

> cannot be without peaceful coexistence, and yet they cannot avoid continuous disagreement with one another. Consequently, they feel destined by nature to develop, through mutual compulsion and laws written by them, into a cosmopolitan society which is constantly threatened by dissension but generally progressing toward a coalition. The cosmopolitan society is in itself an unreachable idea, but it is not a constitutive principle . . . It is only a regulative principle demanding that we yield generously to the cosmopolitan society as the destiny of the human race; and this not without reasonable grounds for supposition that there is a natural inclination in this direction . . . (W)e tend to present the human species not as evil, but as a species of rational beings, striving among obstacles to advance constantly from the evil to the good. In this respect our intention in general is good, but achievement is difficult because we cannot expect to reach our goal by the free consent of individuals, but only through progressive organization of the citizens of the earth within and toward the species as a system which is united by cosmopolitical bonds.[16]

The mission of Kant's anthropology, written, as he insists, from "a pragmatic point of view" is, therefore, to define "the conditions of possibility" for that "regulative principle" that can lead us from a condition of folly and childish vanity to "our destiny" of a cosmopolitan society. This entails an analysis of our cognitive faculties, feelings (of pleasure and displeasure), and of desire (the influence on Foucault's work is obvious). It also entails reflection on how and why natural endowments ("temperaments") are transformed by human practices into "character." Kant writes: "what nature makes of man belongs to temperament (wherein the subject is for the most part passive) and only what man makes of himself reveals whether he has character."[17] Pheng Cheah summarizes Kant's argument as follows:

> As natural creatures with passions and sensuous inclinations, we are, like things and animals, creatures of a world merely give to us

> and are bound by the same arational mechanical laws of causality governing all natural objects. However, as moral subjects we are self-legislating rational agents. We belong to a transcendent realm of freedom we create for ourselves, a world that encompasses all rational beings governed by universal laws we prescribe through our reason. The moral world is supersensible and infinite because it is not subject to the blind chance of meaningless contingency that characterizes finite human existence . . . culture provides a bridge to the transcendent world of freedom because it minimizes our natural bondage by enhancing the human aptitude for purposive self-determination . . . (it) liberates the human will from the despotism of natural desires and redirects human skill toward rational purposes by forming the will in accordance with a rational image.[18]

The general proposition that "man makes himself" carries over very strongly, of course, into the Marxist tradition. Echoes of Kant's transcendent definition of freedom can also be heard in Marx's pronouncement that "the realm of freedom actually begins only where labour which is determined by necessity and of mundane considerations ceases" and that this "lies beyond the realm of material production."[19] Kant reflects on how far we have progressed in re-shaping temperaments into character through the making of culture by examining differentials in national character and cultures. The text is lighthearted, anecdotal, and on occasion deliberately amusing in the national stereotypes it evokes (rather like Jim Jarmusch's film "Night Over the World"). But this should not detract from the seriousness of Kant's purpose. Human beings have made themselves differently in different places and produced different cultures. Our task, and on this point Kant's arguments are surely powerfully to the point, is to exercise both judgment and intelligence with respect to this process:

> Just as the faculty of discovering the particular for the universal (the rule) is called judgment, so the faculty of discovering the universal for the particular is called intelligence. Judgment concentrates on detecting the differences within the manifold as to partial identities; intelligence concentrates on marking the identity within the manifold as to partial differences. The superior talent of both lies in noticing either the smallest similarity or dissimilarity. The faculty to do this is acuteness, and observations of this sort are called subtleties, which, if they do not advance knowledge, are either called empty sophistries or conceited prattlings. . . .[20]

Judged against this high-sounding standard, Kant's own formulations often appear unduly crude (if not empty sophistries and conceited prattlings). Throughout the *Anthropology* we are assaulted by all manner of seemingly prejudicial statements about race, class, gender, and nation. His statements on the nature of woman and the feminine character will likely outrage even the mildest feminist (although they will probably delight some ardent evangelicals). In seeking to understand the differentiations that plainly occurred within our species being, Kant initiated the idea (that later had a very unfortunate history) that the question of race should be put upon a purely scientific footing. And his consideration of the roots of national identity are problematic:

> By the word people (populus) we mean the number of inhabitants living together in a certain district, so far as these inhabitants constitute a unit. Those inhabitants, or even a part of them, which recognize themselves as being united into a civil society through common descent, are called a nation (gens); the part which segregates itself from these laws (the unruly group among these people) is called rabble (vulgus), and their illegal union is called a mob (agere per turbas), a behavior which excludes them from the privileges of citizen.[21]

Not all residents, by this account, qualify as citizens. It all depends, according to Kant, upon the "maturity" of the individual, a normative concept of rational behavior that Foucault challenges head-on in his essay "What is Enlightenment?"[22] Is a character like Baudelaire to be excluded from citizenship, Foucault asks, by virtue of what Kant would almost certainly consider his irrational immaturity?

The definition and significance to humanity of nationhood by common descent, however, leads Kant to one of his most important conclusions. A singular world government could only exist as monarchical despotism because it would have to erase and suppress national differences based upon common descent within territorial configurations. A world government of that sort would, in short, go against nature and human nature. The only form of cosmopolitan government that will work is that based on a federation of independent (preferably democratic and republican) nation states. This may or may not be a good idea, but it is important to recognize that Kant's derivation of it arises out of his highly questionable anthropological conception of the nation state as a civil society based on common descent (to say nothing of the exclusion of "troublesome elements"—however defined—from rights of citizenship). This presumption also helps explain why the cosmopolitan right to cross borders is so circumscribed and why the right to hospitality must be temporary. Permanent residence for foreigners is inconsistent with the requirement of common descent. Those, like Benhabib, who want to extend the rights of migrants in meaningful ways have therefore to struggle mightily with the restrictions of the Kantian cosmopolitan frame. The real problem lies, however, in the questionable anthropological foundations for Kant's arguments. But if Kant's specific anthropological foundation is rejected, as I think it must be, then the question arises as what is or what might be an adequate anthropological foundation for understanding the territorial structures of human association? Indeed, questions might reasonably be asked as to whether the cosmopolitan ethic requires any kind of anthropological foundation whatsoever. Plainly, both Nussbaum and Foucault, as well as Kant, believe it does.

An examination of Kant's geography raises even deeper problems. The lack of interest in it on the part of Kant scholars is understandable since its content is nothing short of an intellectual and political embarrassment. As Droit remarks, reading it "comes as a real shock" because it appears as "an unbelievable hodge-podge of heterogeneous remarks, of knowledges without system, of disconnected curiosities."[23] The thought that this might provide a secure foundation for metaphysical reflection is just absurd. To be sure, Kant seeks to sift the sillier and obviously false tales from those that have some factual credibility, but we are still left with an incredible mix of materials more likely to generate hilarity than scientific credibility. However, there is a more sinister side to it. While most of the text is given over to often bizarre facts of physical geography (indeed that was the title of his lectures) his remarks on "man" within the system of nature are deeply troubling. Kant repeats without critical examination all manner of prejudicial remarks concerning the customs and habits of different populations. Thus we find:

> In hot countries men mature more quickly in every respect but they do not attain the perfection of the temperate zones. Humanity achieves its greatest perfection with the White race. The yellow Indians have somewhat less talent. The Negroes are much inferior and some of the peoples of the Americas are well below them.[24]

> All inhabitants of hot lands are exceptionally lazy; they are also timid and the same two traits characterize also folk living in the far north. Timidity engenders superstition and in lands ruled by Kings leads to slavery. Ostoyaks, Samoyeds, Lapps, Greenlanders, etc. resemble people of hot lands in their timidity, laziness, superstition and desire for strong drink, but lack the jealousy characteristic of the latter since their climate does not stimulate their passion greatly.[25]

> Too little and also too much perspiration makes the blood thick and viscous . . . In mountain lands men are persevering, merry, brave, lovers of freedom and of their country. Animals and men which migrate to another country are gradually changed by their environment . . . The northern folk who moved southward to Spain

> have left progeny neither so big nor so strong as they, and which is also dissimilar to Norwegians and Danes in temperament.[26]

Burmese women wear indecent clothing and take pride in getting pregnant by Europeans, the Hottentots are dirty and you can smell them from far away, the Javanese are thieving, conniving and servile, sometimes full of rage and at other times craven with fear. It is difficult to attribute any notion of rationality or maturity to such populations.

When projected into a world of sovereign democratic and republic states, this conjures up a threatening image of unwashed Hottentots, drunken Samoyeds, conniving and thieving Javanese, and hordes of Burmese women lusting to get pregnant by Europeans, all clamoring for the right to cross borders and not be treated with hostility. It is precisely in such geographical "circumstances" that we can better understand why Kant included in his cosmopolitan ethic and in his notion of justice the right to refuse entry (provided it does not result in the destruction of the other), the temporary nature of the right to hospitality (provided the entrant does not create any trouble) and the condition that permanent residency depends entirely on an act of beneficence on the part of a sovereign state that in any case always has the right to deny rights of citizenship to those that create trouble. Only those who exhibit maturity, presumably, will be granted the right to stay permanently. Again, those like Benhabib who struggle mightily to loosen the constraints of Kantian cosmopolitan law as it relates to the rights of migrants, in effect have to undo the hidden trace of these geographical preconceptions upon Kant's formulation of cosmopolitan law.

None of this has gone away. Acrimonious debates about the rights of minorities and of migrants to be received "without hostility" even on a temporary basis abound in our contemporary world. All manner of prejudicial and stereotypical conceptions about "others" and "strangers" exist, even among highly educated political elites. Denial of rights of citizenship to strangers on the grounds that they are immature and not like us is all too familiar. And while Kant's excursus into the idea of national character may be barely acceptable, the long tradition of writing on the "peculiarities" of the English, the particularities of the French, the Spanish or the Italians, and the like by eminent and much respected writers (such as Anderson,[27] Zeldin,[28] and Barzini[29] with tacit support from epic works such as that of E. P. Thompson on *The Making of the English Working Class*[30]) suggests that Kant was onto something. Furthermore, when political philosophers of the stature of John Rawls (most particularly in *The Law of Peoples*[31]) and Michael Walzer (particularly in *Spheres of Justice*[32]) ground their arguments in something akin to Kant's original idea with respect to national character and culture, then we have to take the whole question of the anthropological and geographical rootedness of political philosophy if not of politics itself far more seriously than is our wont.

Kant's geographical depictions can, of course, be excused as mere quotes from or echoes of Montesquieu and other scholars with environmental determinist and racist leanings, such as Hume and Buffon (to say nothing of the lore that Kant picked up from merchants, missionaries, and sailors passing through Königsburg). Many of the fervent defenders of universal reason and of universal rights at that time, Droit notes, cheerfully peddled all manner of similarly prejudicial materials, making it seem as if racial superiorities and ethnic cleansings might easily be reconciled with universal rights and ethics (though Kant, to his credit, did go out of his way to condemn colonialism on the grounds that this was occupation without permission and therefore a violation of cosmopolitan law). And all manner of other extenuating circumstances can be evoked: Kant's geographical information was limited, his course in geography was introductory, meant to inform and raise issues rather than solve them, and Kant never revised the materials for publication (the text that comes down to us was compiled from Kant's notes supplemented by those of students).[33] And his later shift in emphasis towards the *Anthropology* certainly suggests a gradual shift away from some of the grosser forms of environmental determinism that are featured in his *Geography*.

But the fact that Kant's *Geography* is such an embarrassment is no justification for ignoring it. Indeed, this is precisely what makes it so interesting, particularly when set against his much-vaunted universal ethics and cosmopolitanism. Dismissal of his geography does not accord with Kant's own positioning of it as a "condition of possibility" and as a "propaedeutic" for all other forms of reasoning (including his metaphysics and his ethics). The problem is that Kant failed entirely to bring geographical knowledge out of its "precritical state" and place it on a rational scientific basis. He later hinted as to why. He simply could not make his ideas about final causes work on the terrain of geographical knowledge. "Strictly speaking," he wrote (in a passage that Glacken regards as key[34]), "the organization of nature has nothing analogous to any causality known to us" and this problem blocked his attempt to construct geographical understandings in a style akin to Newtonian natural science. If this is so, then his metaphysics and his ethics lack the solid scientific foundations he considered essential to their formulation. They revert to the sphere of mere speculation as to "man's species being."

The problem that Nussbaum poses of how anthropological and geographical knowledges might be better constructed and positioned in relation to the "proper" formulation of a cosmopolitan ethic is left open by Kant, and hardly anyone has cared since to investigate it. Most contemporary commentators either ignore this question or, as in the cases of Benhabib[35] or Brennan,[36] attempt to deal with some of the issues that Kant left dangling through ad hoc adjustments to his concept of cosmopolitan law. Laudable though such adjustments may be, such writers deal with symptoms rather than underlying structural problems not only with Kant's original formulations but with almost all subsequent work on the subject. So what would it take, then, to reconstruct anthropological and geographical knowledges in a way that could better inform struggles over the proper conception of cosmopolitan law? How, in short, should we attempt to answer Nussbaum's foundational demand for a proper set of geographical and anthropological understandings? And what, under contemporary conditions, could "proper" possibly mean? While such questions may appear daunting if not unanswerable in any simple sense, this should not deter us from investigating them.

Foucault's position on this is interesting. He seems to have been profoundly affected by his reading of Kant's *Anthropology* and clearly saw this as a propaedeutic to Kant's ethics. And his own writings bear the trace of that influence throughout. But Foucault never apparently read Kant's texts on geography. He did, however, make frequent use of spatial concepts. This was particularly evident in the relatively early and long-unpublished essay on heterotopia and in his careful delineation of spatial forms (such as his celebrated use of the panoptican) in his enquiries into prisons and hospitals in texts like *Madness and Civilization* and *Discipline and Punish*. He accepted that spatiality was a key concept and initially at least seems to have accepted Kant's views on how space should be understood. But later in life, he also openly worried, perhaps with Kant as well as his own formulations of heterotopia in mind, at the way "space was treated as the dead, the fixed, the undialectical, the immobile" while "time, on the contrary, was richness fecundity, life, dialectic."[37] If "space is fundamental in any form of communal life" then space must also be "fundamental in any exercise of power," he argued. More surprisingly, when asked in 1976 by the editors of the newly founded radical geography journal Herodote to clarify his arguments on space and geography, Foucault gives evasive and seemingly incomprehending answers to what, on the whole, were quite reasonable probing questions. By refusing again and again to elaborate or even speculate on the material grounding for his incredible arsenal of spatial metaphors, he evades the issue of a geographical knowledge proper to his or anyone else's understandings (even in the face of his use of actual spatial forms such as the panoptican to establish his themes). He fails, furthermore, to give tangible material meaning to the way space is "fundamental to the exercise of power." And his final admission that a proper understanding of geography is "a condition of possibility" for his arguments—the Kantian propaedeutic once more—reads more like a tactic to get his geographer interlocutors off his back than a genuine thought. Yet this is what he eventually did say by way of conclusion:

> I have enjoyed this discussion with you because I've changed my mind since we started . . . Now I can see that the problems you put to me about geography are crucial ones for me. Geography

> acted as the support, the condition of possibility for the passage between a series of factors I tried to relate. Where geography itself was concerned, I either left the question hanging or established a series of arbitrary connections. . . . Geography must indeed necessarily lie at the heart of my concerns.[38]

Foucault here accords, albeit somewhat reluctantly, a parallel status for geography to that earlier assigned to Kant's anthropology. It is, therefore, the geographical rootedness in specific historical, social, and cultural circumstances—the historical and geographical a priori if you will—that now must be taken into account. So how then are we to understand this relation between the geographical as opposed to the historical a priori, and what role does this play in relation to Foucauldian ethics (to say nothing of politics)?

On this point, critical engagement with Kant provides some useful pointers for, as May argues, it is possible to reconstruct some of his putative principles of geographical knowledge from the general corpus of his writings. Geography was not only a precursor, as we have seen, but also, together with anthropology, a synthetic end-point of all of our knowledge of the world (understood as the surface of the earth as "man's" habitation). And Kant saw this end point as more than simply a posteriori knowledge of the world. It is in some sense constructive of our "destiny" for actually living in the world. In other words, we need to examine not only what our geography and anthropology have been and are, but consider what they might become.

Geography, however, looks at "man" as a "natural object within the system of nature." In the eighteenth century this meant that geographical knowledge was prone to those forms of environmental determinism that could all too easily lurch over (as we have already seen in Kant's case) into blatant racism. But the general question Kant poses, of how to understand the metabolic relation between human evolution and environmental transformations, is as vital now as ever. Just because Kant plainly got it wrong is no excuse for ignoring the question. And environmentalism has by no means disappeared. In the contemporary work of Jared Diamond and Jeffrey Sachs, as we will later see, it even acquires a seeming scientific respectability. How to conceptualize "man" as a "natural object within the system of nature" remains a core question and how we answer it will affect not only the technicalities of application of cosmopolitan law but the whole destiny of humanity.

But Kant excluded environmental history from the definition of geography per se. Geographical knowledge concerns the study of spatial order, he argued. History is considered distinctive because it provides a narration in time. Space and time are therefore considered quite distinct from each other in the Kantian scheme of things. This positions Kant firmly in the Newtonian (rather than in, say, the Liebnizian) camp in conceptualizing space and time and, as we shall later see, this seriously inhibits the Kantian perspective.[39] Kant's geography is then defined as an empirical form of knowledge about spatial ordering and spatial structures. Kant's definition of the subject as a "synoptic discipline synthesizing findings of other sciences through the concept of Raum (area or space)" became very influential in the discipline with very unfortunate results.[40] The particularity of spatial positioning is marked by contingency and under this restrictive definition geography can only be concerned with the unique and the particular. This contrasts radically with the universality that attaches to the concept of a unidirectional time that points us teleologically towards our destiny of cosmopolitan governance. In Kant's scheme of things, spatial ordering therefore produces regional and local truths and laws as opposed to universals. These local laws are derived territorially by way of the specific rules of citizenship derived within the history of nation states defined in terms of common descent. This "absolute" Newtonian conception of space (and of time) then frames Kant's territorial anthropological approach to cosmopolitan law, much as it also frames Rawls' and Walzer's approach to questions of local justice. It also seems to underpin Foucault's conclusion that metatheory is inadmissible and that the politics of the contingent and the local is all that matters.[41]

May does not tell us how Kant proposed to relate local truths and laws (e.g., national character) to the universals of reason (humanity in the abstract). But if May's account is right, then Kant's geographical and anthropological knowledges appear potentially in conflict with his universal ethics. What happens, for example, when universal ethical ideals get applied to issues of global governance in a world in which nation states set up their own distinctive rules consistent with their national character? Worse still, how do we apply a universal ethic to a world in which some people are considered immature or inferior and others are thought indolent, smelly, or just plain untrustworthy? Either the smelly Hottentots, the lazy Samoyeds, the thieving Javanese, and the indecent Burmese women have to reform themselves for consideration under the universal ethical code (thereby flattening out all kinds of geographical and cultural differences in favor of some normative definition of maturity), or the universal principles operate across different geographical conditions as an intensely discriminatory code masquerading as the universal good. There are reasonable grounds for inferring that Kant actually thought the former, since in his famous essay on "What is Enlightenment?" he made much ado about human "maturity" as a necessary condition for proper engagement in a public realm where certain freedoms were institutionally established and politically guaranteed. His rules for the exclusion of troublesome elements from citizenship within a sovereign democratic republic support that view. But, as we have also already seen, the supposedly universal principles laid out in his specification of cosmopolitan law entail all manner of hidden concessions to a certain version of anthropological and geographical realities. One suspects that it is precisely the attraction of Kant's cosmopolitanism that it can somehow sustain a veneer of attachment to some theory of universal goodness while allowing, even justifying, innumerable concessions to prejudicial exclusions on the ground.

This is, as many have recognized, a fundamental and unresolved difficulty in Kant's whole approach to knowledge. Arendt puts the dilemma this way:

> The chief difficulty in judgment is that it is "the faculty of thinking the particular"; but to think means to generalize, hence it is the faculty of mysteriously combining the particular and the general. This is reasonably easy if the general is given—as a rule, a principle, a law—so that the judgment merely subsumes the particular under it. The difficulty becomes great if only the particular be given for which the general has to be found. For the standard cannot be borrowed from experience and cannot be derived from outside. I cannot judge one particular by another particular; in order to determine its worth.[42]

Kant's answer, as we have seen, is to invoke that acute intelligence that acknowledges subtleties. But then the danger is the production of "empty sophistries and conceited prattlings." Kant's actual geography amounts to an incoherent bunch of anecdotic particulars for which the general has yet to be found. To this day, geographical knowledge continues to lie very much in this state, in spite of some of the best efforts of geographers and others to reform its ways. Kant's anthropology, though more systematic, is also deeply flawed. And contemporary anthropology, in spite of the efforts of its best practitioners, has hardly eliminated "empty sophistries and conceited prattlings" (particularly in its so-called "postmodern" guise). All of this would not be a problem were it not for the fact that the political consequences (as in Iraq, Ruanda, and Darfur) can on occasion be nothing short of catastrophic.

What appears so dramatically with Kant has widespread ramifications for politics. Popular geographical and anthropological knowledges in the public domain (in the United States in particular) are either entirely lacking or of a similar prejudicial quality to that which Kant portrayed. Stereotypes about geographical "others" abound and prejudicial commentary can be heard daily in casual conversations even in elite circles (listen in to any conversation about Mexicans, sub-Saharan Africans, or Arabs in university common rooms let alone upon the street, and see how quickly stereotypes are invoked only to usually pass unchallenged).

It then becomes all too easy for the U.S. government to portray itself as the bearer of universal principles of justice, democracy, liberty, freedom, and goodness (as Former President George W. Bush did on such a persistent basis) while in practice he operated in an intensely discriminatory way against others judged different, unfamiliar or in some sense lacking in proper qualifications or human qualities. Bush propounded his version of the Kantian cosmopolitan ethic while shock and awe over Baghdad and the horrors of thousands of Iraqi deaths plus the sordid sights of Abu Ghraib brought us back to what in the technical language of remote sensing is referred to as the "ground-truthing" of abstract concepts in relation to anthropological and geographical realities.

This contrast between the universality of Kant's cosmopolitanism and his ethics and the awkward and intractable particularities of his anthropology and geography is therefore of critical importance. If knowledge of the latter defines (as Kant himself held) the "conditions of possibility" of all other forms of practical knowledge of the world, then on what grounds can we trust Kant's cosmopolitanism if his anthropological and geographical groundings are so suspect? Yet there is a way to see this as a fruitful starting point for discussion. For while it is possible to complain endlessly at "the damage done by faction and intense local loyalties to our political lives,"[43] it is also important to recognize how "human passions" (which Kant believed to be inherently aggressive and capable of evil) so often acquire a local and disruptive expression. In the face of this, it will take a tremendous effort to even approach that cosmopolitan state of which we are, at least in Kant's judgment, potentially capable. Is there, then, some way in which we can facilitate that effort by answering Nussbaum's call for a radical overhaul in our curricula in the teaching of geography, ecology, and anthropology? Kant identifies the questions but fails to provide adequate answers. He may have lived in an age of enlightenment but it was most certainly not, as he himself understood, an enlightened age. We are then faced with an interesting choice. We can either reject the whole enlightenment project along with all of Bush's rhetoric about freedom and liberty as a sordid and hypocritical justification for imperial rule and global domination or accept the basic thrust of what the enlightenment (and its U.S. off-shoot) was about with the clear understanding that that particular stab at enlightenment was not enlightened enough. And one of the prime areas of knowledge that remains to be reconstructed is that of "appropriate" anthropological and geographical understandings that can illuminate the way to a genuinely cosmopolitan future. But behind this there lies a certain imperative that pushed the whole question of cosmpolitanism and a federated republicanism into the forefront of Kant's concerns. Why, for example, do we need to cross borders anyway? On this point it is clear that the needs of trade are paramount and that implies that the legal requirements of merchant capital in particular and perhaps even of capital in general play a highly significant closet role in Kant's formulations.

Notes

1. Immanuel Kant, "Perpetual Peace: A Philosophical Sketch," in *Kant: Political Writings*, ed. Hans S. Reiss, trans. H. Barry Nisbet (Cambridge: Cambridge University Press, 1991), 107–8.

2. Cited in Seyla Benhabib, *The Rights of Others: Aliens, Residents and Citizens* (Cambridge: Cambridge University Press, 2004), 27.

3. Benhabib, *The Rights of Others*, 27.

4. Benhabib, *The Rights of Others*, 33.

5. Michel Foucault, "Introduction to Kant's *Anthropology from a pragmatic point of view*," trans. Arianna Bove, http://www.generation-online.org/p/fpfoucault1.htm.

6. Ronald L. Bolin, trans., *Immanuel Kant's Physical Geography* (MA diss., University of Indiana, 1968).

7. Michelle Cohen-Halimi, Max Marcuzzi, and Valérie Seroussi, trans., *Géographie (Physische Geographie)* (Paris: Bibliothéque Philosophique, 1999).

8. Joseph A. May, *Kant's Concept of Geography and its Relation to Recent Geographical Thought* (Toronto: Toronto University Press, 1970).

9. May, *Kant's Concept of Geography*, 132–6.

10. Fred Van de Pitte, Introduction to *Anthropology from a Pragmatic Point of View* (The Hague: Martinus Nijhoff, 1974), xiii. See also John Zammito, *Kant, Herder, and the Birth of Anthropology* (Chicago: Chicago University Press, 2000).

11. Foucault, "Commentary of Kant's *Anthropology from a pragmatic point of view*."

12. Amy Allen, "Foucault and Enlightenment: A Critical Reappraisal," *Constellations* 10 (2003): 186.

13. Cited in Martha Nussbaum, "Kant and Stoic Cosmopolitanism," *Journal of Political Philosophy* 5 (March 1997): 10.

14. "An Answer to the Question: 'What is Enlightenment?,' " in *Kant: Political Writings*, 59, 54.

15. "An Answer to the Question: 'What is Enlightenment?,' " 55.

16. *Anthropology from a Pragmatic Point of View* (The Hague: Martinus Nijhoff, 1974), 249–51.

17. *Anthropology from a Pragmatic Point of View*, 203.

18. Pheng Cheah, "Given Culture: Rethinking Cosmopolitan Freedom in Transnationalism," in *Cosmopolitics: Thinking and Feeling Beyond the Nation*, eds. Pheng Cheah and Bruce Robins (Minneapolis: University of Minnesota Press, 1998), 304–5.

19. Karl Marx, *Capital, Vol. 3* (New York: Penguin, 1981), 958–9.

20. *Anthropology from a Pragmatic Point of View*, 96.

21. *Anthropology from a Pragmatic Point of View*, 225.

22. Michel Foucault, "What is Enlightement?," in *The Foucault Reader*, ed. Paul Rabinow (Harmondsworth: Penguin, 1984).

23. Roger-Paul Droit, "Kant et les Fournis du Congo," *Le Monde*, February 5, 1999.

24. *Géographie*, 223, my translation from the French version.

25. *Physical Geography*, cited in May, *Kant's Concept of Geography*, 66.

26. *Physical Geography*, cited in May, *Kant's Concept of Geography*, 66.

27. Perry Anderson, *English Questions* (London: Verso, 1992).

28. Theodore Zeldin, *The French* (New York: Vintage Books, 1984).

29. Luigi Barzini, *The Italians* (New York: Athaneum, 1967).

30. E. P. Thompson, *The Making of the English Working Class* (Harmondsworth: Penguin, 1968).

31. John Rawls, *The Law of Peoples* (Cambridge: Harvard University Press, 1999).

32. Michael Walzer, *Spheres of Justice* (Oxford: Basil Blackwell, 1983).

33. Stuart Elden, "Reasessing Kant's geography" *Journal of Historical Geography*, 35 (2009), 3–25.

34. Clarence J. Glacken, *Traces on the Rhodian Shore: Nature and Culture in Western Thought from Ancient Times to the End of the Eighteenth Century* (Berkeley: University of California Press, 1967), 532.

35. Benhabib, *The Rights of Others.*

36. Timothy Brennan, *At Home in the World: Cosmopolitanism Now* (Cambridge: Harvard University Press, 1997).

37. Foucault, "What is Enlightement?," 70.

38. Michel Foucault, "Questions on Geography," in *Power/Knowledge*, ed. Colin Gordon (London: Harvester, 1980), 77.

39. The peculiar way in which Kant compromised between Newton and Leibniz makes my assertions as to the absolute outcome of Kant's arguments controversial. The important point is his strict separation of history from geography and the only view of space and time consistent with this is absolute. That Kant in effect sealed in the absolute view is supported, among others, by Edward S. Casey, *The Fate of Place* (Berkeley: University of California Press, 1996), 14.

40. May, *Kant's Concept of Geography.*

41. See Chris Philo, " 'Bellicose History' and 'Local Discursivities': An Archaeological Reading of Michel Foucault's *Society Must be Defended*," in *Space, Knowledge and Power: Foucault and Geography*, eds. Jeremy W. Crampton and Stuart Elden (Aldershot: Ashgate Publishing, 2007).

42. Hannah Arendt, *Lectures on Kant's Political Philosophy* (Chicago: University of Chicago Press, 1989), 76.

43. Nussbaum, "Kant and Stoic Cosmopolitanism": 8.

15

Is there Still Room for Freedom?

A Commentary on David Harvey's "Cosmopolitanism in the *Anthropology* and *Geography*"

Edward S. Casey

David Harvey's paper has the singular merit of bringing Kant's lectures on geography and anthropology into the scene of contemporary discourse by taking up such insistent issues as immigration, hospitality, and even the Bush administration's pseudo-cosmopolitan rhetoric of democracy. Thus it complements earlier contributions that focused on questions of the translation, pedagogical purpose, and historical influence of these same lectures (especially those in geography). As was adumbrated in Robert Bernasconi's oral presentation at the Manhattan meeting that led to this book, we are moving from issues of the aim, exact content, and literary form of the lectures to concrete matters of the greatest current concern. The sequence is at once apt and timely.

In my brief remarks, I want to focus on several closely related conceptual clusters in Professor Harvey's essay. These are neither philological nor political but bear on claims he makes that call for further philosophical analysis. I find four such clusters: space vs. time, reflective judgment, environmental determinism, and the historical-geographical a priori.

Space vs. Time

Toward the end of his paper, Harvey remarks that for Kant "spatial positioning is marked by contingency and under this restrictive definition geography can only be concerned with the unique and the particular. This contrasts radically with the universality that attaches to the concept of a unidirectional time that points us teleologically towards our destiny of cosmopolitan goverance." (p. 280 above) Plausible as this claim seems to be, I wonder if it is true. On the one hand, spatial positioning arises in geography only by reference to the unique surface of the earth, a surface that acts as a kind of sheer container for all particular positions—and thus as the geographical equivalent of what Kant calls in the Critique of Pure Reason "one all-embracing space," arguing there that "if we speak of diverse spaces, we mean thereby only parts of one and the same unique space" (CPR, A 27–29; N. K. Smith translation). Just as space, understood transcendentally (that is, as a universal and necessary form of intuition) is "essentially one" (ibid.), so the earth that is the basis of all geography (and anthropology) is also essentially one. On the other hand, to say that time is "teleological" in its movement toward "cosmopolitan governance" is not to make a universal claim but one that is historically and politically quite specific. In particular, the unidirectionality of successive time is not to be confused with the teleology of historically realized time. Or, if we do wish to interpret an eventual cosmopolitan world-governance as "universal" in the sense that this is what world-history must (or at least should) eventually come to, this involves a sense of universal that is very different from that at stake in Kant's claim in the transcendental aesthetic that "time is the formal a priori condition of all appearances whatsoever" (CPR, A 34). As Kant himself avers, in a passage from the Anthropology cited by Harvey, a "cosmopolitan society" is at most a "regulative idea": this sense of universality does not belong to time as such, since it is a matter of the end or limit of measurable time, i.e., it is something on the far side of sheerly successive time.

Reflective Judgment

Late in his paper, Harvey cites Hannah Arendt: "The chief difficulty in judgment . . . [occurs] if only the particular be given for which the general has to be found" (p. 281)—in contrast with judgments in which a particular is given as subsumed or subsumable under a known concept. Arendt's point in her lectures on Kant is that his idea of reflective judgment is promising and productive for understanding history and politics as well as art and natural process, even if not for grasping the judgments operative in natural science at the level of pure theory. If so, this means that the idea of deriving geographical and (above all) anthropological knowledge from the facts on the ground is dubious—if it is the case that human beings are not merely factual beings for whom determinate judgments are definitive. Such judgments do obtain for the strictly physical dimensions of any beings, but just insofar as human being on earth entails making history—a crucial form of their more general ability to "make themselves" in the Enlightenment sense—then the judgments that are appropriate for understanding them in such history-making (as for the historizing actors themselves) must be reflective. As such, they are uniquely well equipped for dealing with the uncertainties of free decision and with the indeterminate outcome which historical actions bring with them. No merely factual anthropology or geography—no matter how extensive or accurate they may be, regarded as sciences—can be an adequate basis for such uncertainty and indeterminacy, which call for reflective rather than determinate judgments.

Environmental Determinism

Similarly, Kant escapes any charge of "environmental determinism" (in Harvey's phrase) when we take into account his later thinking about reflective judgment as well as his doctrine of human freedom. Harvey allows that this escape may well be the case, but he is troubled by Kant's "initial attraction to an underlying theory of environmental determinism" (p. 270) and by "the grosser forms of environmental determinism that are featured in his Geography" (p. 277) but that are mitigated in his Anthropology. Here I would question if Kant ever explicitly claims, even early on, an outright geographical determinism at any level above that of acquired characteristics. My own reading of the lectures on physical geography points in another direction: namely, that living in particular parts of the world does affect certain physical traits such as skin color in humans, hair patterns of dogs, etc.—as well as certain character traits, as when he says that "in the mountains, people are long-lived, vivacious, brave, patriotic, and freedom-loving,"[1] not to mention the all too many derogatory character traits Kant attributes to peoples of southern climes. But to claim climactic or environmental influence of these two sorts is not the same as saying that the world's peoples are determined tout court by their regions of upbringing. In the very midst of the blatant racism and chauvinism of the observations he makes (and cites others as making), Kant keeps open a margin of free self-determination that is consonant with his mature doctrine of ethical freedom and reflective judgment, even though such freedom and judgment are not invoked as such in his lecture courses.

Historical-Geographical A Priori

But let us leave aside these mitigating factors, which (at least in their explicit formulation) come from Kant's later thought, and come to the heart of the matter. Harvey may not insist on Kant's environmental determinism, but he does point to something even more pernicious—something subtler and more insidious on Kant's part. This is the implicit notion of the "trace" of, or "hidden concession" to, the prejudiced observations that infect and infest the lectures on anthropology as well as on geography. Harvey proposes this notion in the wake of Foucault's reading of Kant (especially as interpreted by Amy Allen) as someone who provides certain "historical a priori" structures that act as "conditions of possibility" for the life of human beings on earth. Such historical a priori are certainly oxymoronic in Kantian transcendentalism considered as positing necessary and universal structures of human experience and thus they "break open the framework of the critical philosophy" (Allen, cited on p. 271 of Harvey's text). Nevertheless, such historical a priori are deeply determinative of this experience. We here enter the domain of what Max Scheler, one of the most trenchant critics of Kantian formalism in ethics, called the "material a priori" and that was subsequently extended by Merleau-Ponty to specifically corporeal a priori in the Phenomenology of Perception, social and poetic a priori structures by Dufrenne in his *Notion of the A Priori*, and to certain a priori structures of history in Sartre's *Critique of Dialectical Reason*. In this regard, Foucault was only building on something opened up by these immediate predecessors.

The most challenging thesis in Harvey's challenging and rewarding paper is that Kant's overtly prejudicial observations at the level of constitution and character cannot be confined to the empirical level of factual error or personal bias, but that they seep into his more advanced notions of hospitality, right of entry, and permanent residency. All these latter are inherently limited in scope despite their inclusion in Kant's cosmopolitan ethic and model of justice. On Harvey's reading, they represent "the hidden trace of [earlier] geographical [and anthropological] preconceptions upon Kant's formulation of cosmopolitan law" (Harvey, p. 276). They are in effect "hidden concessions to a certain version of anthropological and geographical realities . . . [allowing] Kant's cosmopolitanism [to] somehow sustain a veneer of attachment to some theory of universal goodness while allowing, even justifying innumerable concessions to prejudicial exclusions on the ground" (Harvey, p. 280). The result is that Kant's anthropology and geography are not just "potentially in conflict with his universal ethics" (p. 280)—as Harvey puts it rather tamely—but that they are actually in such conflict. For an unacknowledged hidden trace or concession has sufficient subversive force to taint any claim to a pure cosmopolitan universalism.

This leaves us with a special dilemma. Either we have to do merely with a "contrast between the universality of Kant's cosmopolitanism and his ethics and [the] awkward and intractable particularities of his anthropology and geography" (Harvey, p. 282)—as Harvey states near the close of his paper—or we are confronted with a more promising albeit more complex circumstance. Harvey's way of putting it points to a radical difference of levels of discourse that, as such, might take Kant off the hook of Harvey's explicit critique, since it suggests that we have to do with sheer disparity of level rather than with a model that is tainted throughout. But what if we were to invoke the very historical a priori structures Harvey endorses after Foucault (and others) and propose the following paradigm: namely, that such anthropological and geographical structures, though certainly not necessary or universal, are also not merely factual or contingent (as are implied in the phrase "intractable particularities")? Instead, these structures would be seen as possessing a certain latitude that allows for free choice and reflective judgment even in the midst of discernible patterns of conduct and character. This would admit a factor of self-determination in the very midst of empirically specifiable traits and trends.

The resulting spectrum of description would allow for diverse interventions in the geographical or anthropological field—ranging from the most highly biased "observations" (which I put into parentheses to emphasize that these may not be observational in any strict sense but, rather, imputational if not outright imaginary) to more nuanced descriptions that acknowledge the free play of the faculties (and thus the freely chosen actions) belonging to peoples of the most disparate genealogies and locations. The fact that Kant endorsed the former end of the spectrum most often, guilelessly or not, and that he can do so short of any appeal to cosmopolitanism (which would itself be yet another, admittedly complex instance of the historical a priori possibilities of human beings on planet Earth), does not mean that he left no room for the latter.

Even if this expanded model of descriptive integrity obtains, however, this does not redeem Kant's lectures on physical geography and human anthropology from their rank racism, speciecism, and other ethical opacities. But it does complicate the philosophical drama into which they draw us more than two centuries after they were delivered in the icy winters of eighteenth-century Königsberg.

Note

1. Rink ed., p. 5.

16

Kant's Third Thoughts on Race

Robert Bernasconi

Is Kantian Cosmopolitanism Compatible with Racism?

Pauline Kleingeld's recent essay "Kant's Second Thoughts on Race" has catapulted Kant's *Lectures on Physical Geography* to the center of the current debate about whether he remained committed both to the concept of race and to a certain racism throughout his life or whether he abandoned one or both of these toward the end.[1] In her essay Kleingeld acknowledges that student notes of his lectures on physical geography that have recently been transcribed by Werner Stark show that Kant continued to maintain a hierarchical approach to the races as late as 1792, but her overall argument is that Kant subsequently abandoned this position. On the basis of a criticism of the slave trade that can be found in a draft of Kant's essay *Toward Perpetual Peace*, presumably written some time in 1794 or 1795, she argues that Kant must have dropped his hierarchical conception of races during the intervening period. In this essay I show that although his comments on the slave trade are significant, the texts she cites, when taken in their entirety, do not support her overall reconstruction of his trajectory.

Kleingeld is only the most recent of a number of scholars who have attempted to show that toward the end of his life Kant modified his racism and perhaps even renounced the concept of race itself. Versions of this argument were made earlier by Susan Shell, Peter Fenves, and Sankar Muthu.[2] There is a ready audience for such efforts because it is hard not to be shocked and disappointed when one learns of Kant's racist statements that seem to grow in intensity, not diminish, during the 1780s. However, one has to wonder whether we are just being served up a new secular version of the old story of a deathbed conversion, which is retold more because it is comforting than because it meets the basic standards of historical scholarship.

Kleingeld sets out from the fact that "Kant's race theory and its implications for global migration cast his cosmopolitanism in a disconcerting light."[3]

Even though Kleingeld is the author of an important essay "Six Varieties of Cosmopolitanism in Late Eighteenth Century Germany," where she identifies international, legal, cultural, market, moral, and romantic forms of cosmopolitanism,[4] I suspect that she has been misled by those who, like Allen Wood, believe that Kantian cosmopolitanism works as an antidote to racism. In the context of a discussion of cosmopolitan right Wood writes: "A more civilized race has no right to enslave, to dispossess, or to impose its civilization on a less civilized race. And Kant never wavers from his severe condemnation of Europeans for behaving as if it were otherwise."[5] Wood fails to cite any such condemnation by Kant of European involvement in the enslavement of Africans: I will show that on this issue Kant's record is a great deal more troubling than he suggests. However, the usual defense of Kant has been to draw a wedge between Kant's racist statements and his moral philosophy.[6] More recently Todd Hendrick challenged this view that Kant's views on race can "easily be decoupled from his cosmopolitanism and moral philosophy."[7]

To be sure, when we today read Kant's mature moral philosophy in isolation from what we know about his views both on race itself and on the different races, we find it hard to imagine that it could coexist with racism. To us in the twenty-first century there is no room for hesitation: the contradiction is so blatant that we judge that it would have been impossible for anyone, let alone one of the greatest of philosophers, not to recognize it. And yet, tempting though it is to imagine that Kant must eventually have seen the light and resolved this issue in favor of a radical egalitarianism, there is no explicit acknowledgment on his part that he saw the problem. Kleingeld concedes that Kant "gives no indication of when or why he changed his views" [on race], and she cannot do so because there is no evidence that he did renounce his views either about the scientific character of race as such or about the hierarchy of the races, although he did appear to modify his views on the slave trade.[8] We will find that Kant's particular brand of cosmopolitanism does have implications for how he saw slavery, but it does not mark a radical transformation in his view of race as such. In fact a similar contradiction arises in relation to Kant's sexism, and it is striking that Kleingeld, who had already written eloquently on this theme, refuses the idea that Kant had finally resolved the problem by abandoning his sexism.[9] In a footnote Kleingeld insists that Kant's move to egalitarianism was limited: "His views on women, for example, did not undergo a similar development."[10] It is, of course, perfectly possible to be sexist without being racist, and vice versa, but to the extent that Kleingeld allows the contradiction between the universal moral theory and the racism to provide the motivation for his alleged abandonment of his hierarchical conception of race this interpretation merely highlights the puzzle without resolving it.

Both contradictions are intolerable to us, but that does not mean that historians of philosophy should project a solution to the contradiction onto Kant's text where evidence is lacking. This is what Charles Mills does when he imports the loaded term *Untermenschen* to make sense of what he calls "the *actual* (as against officially represented) logic of his writings."[11] But the whole thrust of Kant's theory of race was to establish the unity of the human species against the polygenists who believed that the different human varieties were the products of distinct local creations. If Mill's attempt at a "solution" obliges him to deny this, then I would agree with Kleingeld that it is not sustainable.[12] However, the greater problem, which Kleingeld is as guilty of as Mills, is succumbing to that feeling that we must have a reading of Kant that imposes consistency on him from outside when it seems that there is none. Historians of philosophy cannot operate in that way and still claim to be writing history.

Kleingeld herself acknowledges that Kant managed to live with the inconsistency throughout the 1780s and at least until 1792. She explicitly concedes that in the 1780s "Kant simultaneously defended a universalist moral theory and a racial hierarchy."[13] If the contradiction between general statements of human equality and the existence of slavery was as striking then as it is now, how did Kant tolerate it for so long? But if it is not, then we cannot use general statements of equality as incontrovertible evidence that Kant had renounced his earlier belief in a racial hierarchy. One can always appeal to this contradiction for an explanation of why Kant might have changed his mind, if we know for sure that he did, but in the absence of further evidence it does not show that he must have changed his mind. And yet, in the final analysis, that is precisely what Kleingeld among others do when they appeal to consistency in order to push their interpretations through. Kleingeld herself acknowledges that for at least a decade Kant had already lived with the inconsistency of a universalist moral theory and a racial hierarchy, as if he were oblivious to it, so however impressive the development of this universalist moral theory might be in his later years, the mere fact that it "contradicts his earlier views on a racial hierarchy" does not of itself constitute evidence on its own that he must have changed his mind.[14]

Kleingeld asks whether on the question of race Kant promotes "a consistent inegalitarianism," a position she attributes to Charles Mills, or "an inconsistent universalism."[15] She believes that in the 1780s he was an "inconsistent universalist" and that he became in the course of the 1790s "more egalitarian with regard to race."[16] Perhaps we need to start asking ourselves whether these categories are sufficiently rich to throw light on a process of such complexity. Although I cannot go into this question now and must reserve it for another occasion, I am convinced that philosophers will never succeed in deepening their understanding of moral insight unless they take cases such as this and appreciate how these failures take place by viewing them in their historical context.

That Black men and women were indeed human did not mean to Kant that they were equal in their capacities with members of the White race. Nor does cosmopolitanism as Kant understood the term call that into question. Indeed, Kant did not think that cosmopolitanism was about giving equal weight to all people. He understood it to mean moving outwards from one's own circle to embrace others by degrees. This is made clear in his lectures on *The Metaphysics of Morals* from 1793–94 as recorded by Johann Friedrich Vigilantius. In the course of a discussion of what it means to talk about a general love for everybody, Kant highlights patriotism and cosmopolitanism as two forms of the love of others resting on common descent, but as the first is local, the second is directed to our common world ancestry. This focus on common inheritance locates Kant's discussion in the context of the human species and its races. Given that he clearly indicated that the divisions between peoples represented a natural breakpoint on this graduated scale, it seems likely that the divisions between races operated in the same way, not at this point according to a hierarchy in nature, but simply as a natural psychological tendency.

Our love for others is most intense when directed to a particular group. Kant offered as examples the order of freemasons and sects like the Protestant Herrenhuter, but he specifically criticized these groups because he considered that their love for each other was detrimental to the general love of mankind: focus on the smaller group promotes indifference based on separation. Kant similarly criticized the Greeks and the Jews for their failure to reach out beyond their own. He wrote that among the Jews "all estimation for other men, who are not Jews, is totally lost, and goodwill is reduced merely to love of their own tribe."[17] But one could also go too far the other way. Kant also warned that the friend of all humanity is as open to censure as the freemason, albeit from the other direction: the cosmopolite in this sense lacks adherence to individuals.[18] Nevertheless, Kant did not even conceive here of someone who in the name of cosmopolitanism rejected patriotism. The danger he foresaw was that one might love people in general but not individual people. Cosmopolitanism was not a correction of nationalism but an extension of it: the cosmopolite in the good sense is someone "who in fealty to his country must have an inclination to promote the well-being of the entire world."[19] Kant recognized that love for others embraces the whole human race, but that this love for others extends outward from the groups with which one is most closely involved, to one's fatherland and finally to everyone because of our common descent.[20] It is by loving one's fellow countrymen more than one loves foreigners that one contributes to the love for all humanity.

Kant was not alone in thinking that love of fatherland is naturally extended to embrace everyone in this way so that cosmopolitanism is genuinely patriotic. This is a commonplace of late nineteenth and early twentieth century German intellectual history. Friedrich Meineke cited Kuno Fischer as saying, "The cosmopolitanism of the *Wissenschaftslehre* and the patriotism of the *Addresses* [to the German Nation] are one and the same concept."[21] This same view was supported by Wilhelm Windelband, who wrote, "This patriotism of the *Addresses to the German Nation* resembles Fichte's cosmopolitanism as one twin brother does the other."[22] On the basis of the 1793–94 lectures Kant's idea of cosmopolitanism would seem to be much closer to Fichte's than to our own.

Even while granting that patriotism and racism are very different, it is possible to see Kant's relation to the different races in a similar light or, more precisely, as part of the same process of extending one's love outwards. This might help to explain why he did not see his cosmopolitanism and his racism as directly in conflict. I am not insisting on this explanation. My main point here is to insist on a truly historical approach, one that acknowledges that people in Kant's time might have thought differently from the way we tend to think today. Cosmopolitanism is not Kant's answer to racism, at least in the sense of a prejudice in favor of one's own race over other races, its antidote, but its natural accompaniment in the sense that in Kant a certain cosmopolitanism and a certain racism do not seem to conflict with each other. They are part of the same love of others, which starts with those most like himself. I highlight Kant's discussion of cosmopolitanism from 1794, not because I believe it holds the key to understanding the character of Kant's racism, but because it reminds us that Kant saw these issues very differently from the way we see them. He had a different set of concerns and priorities to those that tend to motivate philosophers today.

The fact of the matter is, of course, that the coexistence of the advocacy of a moral universalism together with assertions of racism was by no means unique to Kant at that time. Indeed it is widespread within this period, as if people were not yet ready to utter such strong statements about universal human equality without reservations and could only live with them if they were accompanied by qualifications. One thinks inevitably, for example, of the controversies surrounding the racism of the founders of the United States. To be sure, Jefferson and others belonged to a society where the racism was already embedded in the institutions of society, and it was the problem of making sense of those institutions at the same time that the country was declaring its independence in the name of liberty that led eventually to ever more extreme racial doctrines. The theory at the end of the nineteenth century that "the negro is a beast" arises precisely from the premise of human equality, as consistently as abolitionism. Racism is a set of practices first and foremost. Or, as Jean-Paul Sartre said, "racism is a passive constitution in things before being an ideology."[23]

Kant's racism was less tied to the practices of his day because Königsberg was relatively far removed from the slave trade for a major seaport. His racism, although in some respects conventional, was nevertheless confronted not only by the moral philosophy he was developing, but also by his philosophy of history. Kant's moral theory and his political theory would often come into conflict, as they would for many at that time. But whereas Hegel gives primacy to politics,

Kant put ethics first. It is our knowledge of that fact which provides the basis for thinking that Kant's moral theories should have trumped his convictions about what he saw as the fact of racial hierarchy, understood as inequality of capacities that he recorded in his lectures on physical geography and elsewhere. But he found it hard to reconcile these inequalities, which—in line with his racial theory—he believed to be permanent, with his hopes for humanity. He speculated about the extermination of all the races other than the White race, but particularly the extinction of the race of Native Americans.[24]

It is true that there were those in the late eighteenth century who saw that statements of moral universalism had clear implications for how society should be organized, but that does not mean that everybody saw it equally. In our own time we find that some of the people and nations, who are loudest in their proclamations about the value of human life and the equality of all, show a stunning indifference to the deaths of poor people and foreigners. Phrases like "collateral damage" will no doubt haunt historians for years to come as they try to make sense of the moral viewpoint prevalent at the dawn of the twenty-first century. Philosophers need to think less about saving the reputations of past philosophers and more about the ease with which moral theories are divorced from practice, precisely because so many of us fail in this regard.

Because this volume is devoted to Kant's *Physical Geography*, I will say a few words in the next section on the textual problems relative to its study. There is a danger that some scholars might attempt to make too much of Kant's virtual silence on race in this work and for this reason, it is important to rehearse some of the circumstances surrounding its publication. I will then proceed in a third section to a consideration of the relevant passages in the draft notes for *Toward Perpetual Peace* before turning to a consideration of the student notes from Kant's 1792 *Lectures on Physical Geography* in an effort to challenge Kleingeld's argument. In the fourth and final section I will leave Kleingeld's argument behind and focus on trying to learn what the 1792 *Lectures on Physical Geography* show about Kant's racism at the beginning of the final years of his philosophical activity.

Race in Kant's *Physical Geography*

Until Kleingeld's essay, Kant's lectures on *Physical Geography* played virtually no part in the current debate about Kant's theory of race and his racism. The main sources for an understanding of Kant's concept of race are—and certainly will remain—the published works: the 1775 essay, *Of the Different Races of Human Beings*, which was expanded in 1777;[25] the 1785 essay, *Determination of the Concept of a Human Race*;[26] and finally *On the Use of Teleological Principles in Philosophy* in 1788.[27] The 1775 version *Of the Different Races of Human Beings* was written by Kant specifically as an announcement of his lectures on physical geography,[28] which might have led one to suppose that Kant would develop his ideas further in lectures, but the best known edition of the *Lectures on Physical Geography*, Rink's edition from 1802, which was subsequently taken up to form part of the various editions of Kant's works, contains only two references to race. It was in the 1775 essay that Kant first introduced his new technical sense of human race, which he specified belonged to natural history rather than natural description.[29] But it is not primarily because physical geography was said to be confined to natural description that race is largely absent from Rink's edition of the *Physical Geography*, although in the context of saying that he distinguishes a description of dogs from a natural history of dogs that would show how the different races of dogs have arisen from the same stem through breeding and climate.[30] It is because, as Erich Adickes had already warned us in 1911, Rink drew his edition of the *Lectures on Physical Geography* from very early materials. Adickes judged that the first fifty-two sections of Rink's edition were chiefly drawn from a manuscript from the summer semester of 1775, although Stark has now reassigned it to 1774, thus predating the first essay on race. Rink also introduced many changes into this part of his text, including numerous additions, perhaps using another text from 1778, thereby compromising its integrity.[31] The rest of the lecture course, from section fifty-three to the end, was drawn from Kant's dictated text, which Adickes dated from before 1760.[32] So it is because the later sections were, with only a few exceptions, written long before Kant had arrived at his new scientific definition of race that one cannot expect to find it there, although it is true that Rink has Kant say toward the end of the book that "humanity attains its greatest perfection in the White race."[33] If there is any remaining doubt on this score it can be resolved with reference to Kant's lectures on anthropology from 1781–82. In the context of a discussion of the discovery of the role of the *Keime* he explained that the exploration of the difference between the races, each of which has a particular character, belongs to physical geography.[34]

Rink put together the edition hurriedly in response to an edition by Vollmer, who had begun publication of a rival and much larger unauthorized edition a year later. The speed with which Rink worked is reflected in some remarkable errors in the original Rink edition, which have since been corrected. Rink did what he could with the manuscript available to him, but he has Kant saying that the Pacific Ocean is between Europe and Asia and, a little later, that the rivers of Asia flow into the Atlantic.[35] Rink conceded in the Preface the need for another edition, but in fact it was never produced, and it was Vollmer's text, not Rink's, that first went into a second edition.[36] Rink's edition was not reprinted until 1839.[37] However, the defects of Rink's edition do nothing to make Vollmer's edition any more useful, palatable, or readable.[38] The point is rather that following Werner Stark's extraordinary labor in transcribing notes of

Kant's lectures, we should set aside everything we thought we knew and start again from scratch, commentating on the materials he has supplied.[39] From this point on both Vollmer's and Rink's editions are useful primarily for discussions of Kant's influence and are best avoided if we want to understand Kant's own thought on the subject.

In any event, the absence of a prolonged discussion of race in Kant's technical sense from Rink's version of these lectures has little or nothing to do with any alleged hesitation on Kant's part. It would be distinctly odd, not to mention a case of deception in advertising, to introduce a new concept of race in the announcement of a series of lectures only to ignore it during the course itself. Kant had from early on included in the *Lectures on Physical Geography* a detailed description of the human varieties. In a 1765–66 report that Kant submitted on his lectures, and using the word *Geschlecht* rather than *Rasse*, Kant presented the idea of a great map of the human races.[40] He highlighted the moral differences between the races, as well as those in physical properties, emphasizing that without knowledge of them one can scarcely pass universal judgments on man.[41] These remarks are in line with the treatment in Rink's edition of the *Physical Geography*, but their whole tenor runs counter to the treatment to the same themes addressed in 1775.

In §38 of Rink's edition Kant alluded to the kinds of considerations that belong to a reflection on human varieties, but he limited himself to giving the standard explanation for human differences in terms of climate: "Man is therefore made for the whole earth, and precisely because his body is formed by nature in such a way that he can become habituated to any climate—indeed can become accustomed to even the greatest differences of climate—perhaps this is partly the reason for the origin of differing national characters."[42] However, if the environment does not produce permanent characteristics, then it does not produce races in Kant's sense. With the introduction of his notion of race Kant effectively abandoned this explanation of human races in terms of environment, except insofar as he believed that climate was, along with diet, one of the main forces responsible for actualizing the germs or seeds (*Keime*) that he postulated and associated with one race rather than another.[43] The two most decisive factors in Kant's account of the formation of the four main races from the original stem are, firstly, migration into a different environment that subjects people to a different climate and diet and then, secondly, race mixing.[44] The problem is that if the different characteristics of the races are a result of environment, then one would expect that after a number of generations in a cold climate Blacks would become White, which was what Buffon anticipated.[45] Kant, however, insisted that there was no evidence for this. So he posited the existence of seeds and inclinations that were actualized only on contact with a given environment as they colonized the different parts of the world. On his account the actualization of the seeds inhibited any reversal and he explained that intermediary forms between the four races that he identified were largely the result of further migrations as the seeds were actualizing.[46] Kant's was the first theory to provide an account of how from the same stem such permanent diversity was possible. He established the racial differences as *permanent*, unless there was race mixing.[47] Over the years in his lectures on physical geography he frequently speculated on the effects of race mixing, wondering, for example, if the Chinese and Japanese were not produced by mixing the Indians with the Calmucks.

Race mixing is crucial to Kant's thinking about race in two ways. First one can only establish the existence of distinct races by mating across races. The production of fertile offspring across several generations shows, following Buffon's rule, that one is dealing with animals from the same species. However, if certain inherited characteristics from the two types produce on mating progeny in which those characteristics are midway between that of the parents, then one has identified two distinct races. The characteristics that are modified in this way can then be identified as racial characteristics.[48] Kant's favorite example of a racial characteristic among humans is skin color, but he is also clear that racial differences embraced not only physical characteristics but also mental or moral characteristics.[49] Kant seems to have changed his view about race mixing. In lectures on moral philosophy in the early 1770s Kant in a rare early use of the term *Racen* argued that nature intended the races to combine with each other.[50] However, when introducing his racial theory, Kant suggested that race mixing was a disturbance of nature.[51] This would seem to be because he now attributed to nature providential insight in creating the races.[52] In lectures delivered in 1791 Kant gave one of his clearest denunciations of race mixing.[53] And yet in his lectures on physical geography in 1792 he seems to suggest that Europeans should follow the instruction to mix with Indians (that is to say, Hindus, not Americans), although he says they do not do so as they consider it to be degrading and because they do not value the half-caste. Nevertheless, he says almost at once that nature seems to prevent half-castes from spreading out. However, one should not confuse Kant's concerns about race mixing with the later obsession about so-called miscegenation, which eventually gave rise to the one-drop rule. According to his lectures on physical geography from 1784, Kant believed that if Whites continued across five generations to mate with the offspring of a Black and a White parent, all trace of the Black color would disappear.

Kant on Slavery and the Slave Trade

Kleingeld entertains but does not embrace the idea that Kant might have given up what she calls somewhat tendentiously his "purely physiological concept of race."[54] To be sure, if Kant had maintained a "purely physiological concept of race," then his retention of the concept would not of itself implicate him in racism, but although Kant presented what we might call a largely physiological definition of race in 1785, three years later in *On the Use of Teleological Principles in Philosophy*, he reverted to one in which physical and moral characteristics were fully mixed. Other commentators take a more extreme view. Muthu declares that "In Kant's later years . . . the hierarchical and biological concept of race disappears in his published writings."[55] Fenves devotes several pages to expressing his puzzlement about Kant's "erasure of race," but he does not give a coherent account of why he considers this erasure to be a fact that the rest of us are blind to.[56] Part of the explanation for this failure of Muthu and Fenves to see the continuing role of the concept of race in Kant's writing lies in their inability to see its centrality to the argument of the *Critique of Teleological Judgment*, simply because the word is not mentioned there.[57] Even more significant, because it takes us as late as 1798, is their failure to give full weight to Kant's endorsement of Christoph Girtanner's *Über das Kantische Prinzip für die Naturgeschichte* in his *Anthropology from a Pragmatic Point of View*. In the section entitled "On the Character of Races" Kant offers the following ringing endorsement: "With regard to this subject I can refer to what Herr Privy Councilor Girtanner has presented so beautifully and thoroughly in explanation and further development (in accordance with my principles)."[58] Kant was not being disingenuous when he referred to Girtanner developing his ideas. Girtanner was, among other things, a respected scientist and he supplied a level of detail about the natural world that Kant could never have matched on his own.

If Kant had wanted to correct his earlier views on race or even if he had wanted to be silent about race, this would have been the time to do so: in 1798 when the *Anthropology* was published or two years later when it was republished in an improved version. Indeed, there is some question as to whether Muthu had actually seen Girtanner's book because he insists that "Girtenner's [sic] *Uber das Kantische Prinzip* . . . is an application of Kant's hereditary theories to *nonhuman* species," implying that it does not address the human races, which is to overlook the fact that over half the book is devoted specifically to the human races: some 230 pages of a 422-page book.[59] Finally, and perhaps even more telling is the fact that, as Mark Larrimore points out, Kant republished his 1785 essay on race in 1795, his 1785 and 1788 essays on race in 1793, and all three essays on race in 1797 and 1799.[60] This would be strange behavior for someone who had supposedly renounced the central category promoted in these essays.[61] By the same criterion, given that, as I shall show later, Kant promoted a hierarchical view of races particularly strongly in the 1788 essay, this seems to speak against Kleingeld's claim that he abandoned that view too. If one disregards the republication of these essays then it is true, as Kleingeld says, that after that time "he makes no mention of a racial hierarchy anywhere in his published writings."[62] But if silence is evidence of recantation, who knows how much of Kant's theoretical work survived the final decade of his life?[63]

Kleingeld's claim that Kant abandoned his racism is inevitably a much more complex issue, not least because attribution of "racism" to the distant past is always especially complicated. Racism is always shifting its character, and what is racist in one context is liberating in another. Indeed it is almost certainly not the kind of issue that will ever be resolved with the same level of finality that we might attain with regards to Kant's adherence to an idea of race. In fact a major reason why I am concerned with the question of Kant's racism is to throw light on the nature of racism and not because I believe we have at our disposal a ready-made concept of racism that we can all agree to apply in this or any other case.

The one unambiguous positive piece of evidence that Kleingeld brings to bear to support her thesis that Kant modified his racism concerns the shift in his position on slavery. Kleingeld is clear that Kant accepted the institution of slavery at least until the end of the 1780s. Most, if not all, of Kant's comments that recognize slavery without questioning the institution refer to the enslavement of Blacks. For example, in lectures on *Menschenkunde* given in the winter semester 1781–82 Kant declared that the only education Blacks could acquire was that of slaves: "they allow themselves to be trained."[64] There is also a manuscript in which Kant observed that neither Native Americans, nor Negroes, could govern themselves: "Thus they serve only as slaves."[65] Kleingeld quotes this passage and associates it with Kant's philosophy of history, but fails to mention that in this same context he writes that all the races will be extinguished except that of the Whites.[66] Kant also used here the phrase "born slaves" and he clearly associated this with the fact that whereas Negroes can be disciplined and cultivated, they can never be civilized.[67] Kant is absolutely clear that Negroes are fully human in the sense that they meet the criterion of Buffon's law of propagation. He also gives them legal status in the sense that they can enjoy the protection of cosmopolitan right, but as I speculated already in "Kant as an Unfamiliar Source of Racism" there is a question as to whether all races were capable of participating in the full development of humanity in the narrow and specific sense of participating fully in the cosmopolitan whole.[68] Mellin's *Encyclopädisches Wörterbuch der Kritischen Philosophie* has separate entries for the cosmopolitan whole and cosmopolitan right.[69] This appears to be justified, and, as I shall briefly suggest in the final section, Kant's attempt to bridge them is one source of the problems he faced in negotiating the implications of his concept of race.

However, as Kleingeld points out, Kant in the draft notes for *Toward Perpetual Peace* explicitly opposed the slave trade in Africans itself and not just the way it operated. The context of the discussion is an attack on certain seafaring civilized peoples who saw no limits on their conduct other than those that arise from their lack of power. Kant was here addressing the fact that social contract theory left nations in a state of war against each other. This was especially a problem on the high seas, because the sea belongs to nobody. Kant was concerned with the impact that this might have on commerce, which he sees as an agent of progress. The extent to which his concern with hospitality is also motivated by a concern for trade is often missed by commentators, but it is natural that Kant's mind should shift to a lucrative form of trade, the slave trade, that nevertheless violated the cosmopolitan law of hospitality that he wanted to promote for the sake of commerce. So Kant wrote: "The trade in Negroes that is in itself a violation of the hospitality of peoples among Blacks is even more of a violation for Europe because of its consequences."[70] Kant then proceeded to reflect on how sea power, which might be used for commerce, can be diverted to war, to pillaging, and to blockades that lead to people starving to death.

These sentences do not resolve the problem of Kant's attitude toward Africans. It is telling that Kant in his lectures in 1792 immediately after condemning the slave trade insisted that it would continue, even if the Europeans were not involved, because African kings engaged in it among themselves.[71] In his courses on physical geography Kant repeatedly insisted on this fact that Negroes sold Negroes, as if this somehow excused European involvement. To be sure, there is a slight shift from the 1792 lectures to the draft of the essay on perpetual peace in that Kant initially seems to suggest that the African slaves were no worse off simply because Europeans were involved, whereas in the later text he conceded that the violation of the hospitality of peoples through the slave trade was greater because of European involvement. Kant anticipated wars between the European navies resulting in a large number of men being buried at the bottom of the sea, the coastal regions being laid waste, and "entire peoples slowly starving to death as a result of restrictions in the circulation of foodstuffs."[72] So in the later text Kant seems to have been willing to acknowledge that Europe's involvement did alter the level of violence, but he did not focus on the treatment of the slaves so much as on the wars that follow from the greed of Europeans.

Kleingeld fails to reflect on two important features of Kant's discussion. The first and most striking is that Kant decided *not* to include this discussion in his published text. In *Toward Perpetual Peace*, as in the draft, Kant attacked the pirates of the Barbary Coast for enslaving stranded shipfarers, but when he came to criticize the inhospitable behavior of "the civilized, especially commercial, states," their use of slavery was not explicitly mentioned, and instead he referred only in general terms to the horrifying treatment of the inhabitants of the countries visited.[73] The overall criticism of the European states is as harsh in the published text as in the draft, but the criticism no longer highlights the slave trade as such. He limited himself to condemning "the cruelest and most calculated slavery," which he identified as taking place in the Sugar Islands.[74]

In other words, Kant was not so strongly committed to his opposition to the race-based slave trade that he was prepared to express his objections in print, even though in Europe at this time there were fewer and fewer defenders of it. The growing tendency of "defenders of slavery" in Europe was to accept its illegitimacy but argue that the time was not ripe for its abolition. If Kant had, as Pauline Kleingeld insists, "second thoughts," then it seems that he must have had "third thoughts" that led him not to come out and join those who at that time were opposing the slave trade.

I am not suggesting that Kant changed his mind and concluded that the slave trade as conducted was legitimate. We do not know the reason why he hesitated to express his opposition to it. Perhaps he lacked the moral courage to do this, although it is not clear that it called for moral courage in Germany by 1795. For all we know Kant might simply have feared that announcing his opposition to the African slave trade would distract from other causes to which he was more committed. We do not know his motivation, but we do know that he decided not to publish his direct opposition to the slave trade in Africans after having entertained the possibility of doing so. Nevertheless, even this comment directed against the cruelty of the slave trade marks a departure for Kant in contrast with his much earlier insistence that Black slaves had to be treated harshly: for example, in Rink's edition of the *Physical Geography* Kant records that all inhabitants of the hot zones have a thick skin and so must be hit not with sticks but whipped with split canes.[75]

The second relevant feature of Kant's draft on perpetual peace that Kleingeld fails to observe is that it is directed against the slave trade and not against the institution of slavery itself. It should be remembered that this was a crucial distinction at this time. The slave trade was abolished, for example, in Britain, in 1807, but Britain did not end slavery and emancipate the slaves in its dominions until 1838. It is because he was arguing specifically against the slave trade that Kant referenced the cosmopolitan law of hospitality as the basis for his objection, but he did not use the opportunity, even in these unpublished notes, to call directly for the emancipation of African slaves in the Americas. In other words, the shift in Kant's views after 1792 is much less dramatic than Kleingeld maintains.

It is telling that in the one place where Kant attacks the slave trade in Africans as an institution, he appeals to the cosmopolitan law of hospitality to condemn slavery and not to the categorical imperative. To be sure, Onora O'Neill among others has attempted to show that on the basis of the first formulation of the Categorical Imperative, the so-called Formula of Universal Law according to which one should act only on that maxim through which you can at the same time will that it should become a universal law,[76] Kant should have been against slavery.[77] But as even relatively sympathetic commentators have pointed out, there is a problem determining which is the appropriate maxim under which to describe one's maxims, particularly once slavery is in place.[78]

In any event, Kant's opposition to the slave trade does not underwrite the conclusions Kleingeld tries to draw from it. She concludes on this basis that Kant must have given up his hierarchal view of the races at this time, and yet she provides no reason to believe that the attribution of cosmopolitan right in this case implies equality of capacities between the races.[79] It would be perfectly possible to be against slavery and still maintain a racial hierarchy and indeed in his *Lectures on Physical Geography* in 1792 Kant did so. The most striking problem for Kleingeld's narrative, according to which Kant underwent some conversion experience after 1792, is that in the 1792 lecture course in which Kant reasserted racial hierarchy, he already called the trade in Negroes "morally reprehensible." So far as I am aware this lecture course is the only occasion where we know that Kant *publicly* condemned the slave trade, and he did so in the context of a reassertion of his hierarchical view of race. This breaks the link Kleingeld tries to establish between his opposition to the slave trade and his abandonment of a racial hierarchy.[80]

The fact that Kant declared the slave trade "morally reprehensible" in his lectures on physical geography in 1792 invites a reading of the discussion of slavery in *Metaphysics of Morals* from 1797 as an indirect comment on the Atlantic slave trade, and yet when we examine these pages more closely they appear for the most part to be largely conventional, reflecting opinions that had been offered by political philosophers for decades without reference to the slave trade in Africans. Given that Kant was so direct in his condemnation of colonialism with specific reference to events in North America and elsewhere,[81] the fact that in the same context he was, as I will show, less eloquent about the condition of slaves in North America and the extent of the Atlantic slave trade is telling. In *The Metaphysics of Morals* Kant was clear that all human beings are born free.[82] This of course, meant that the children of slaves could not legitimately be enslaved for the crimes of their parents. This was a widely held position already in the seventeenth century, by the likes of John Locke, for example.[83] If this idea had been taken seriously, the practice of slavery would have been very different, but it was rarely used to the advantage of African slaves.

The little that Kant does have to say in *The Metaphysics of Morals* directly about the enslavement of Africans is somewhat enigmatic. Kant allowed that one can be enslaved as the result of a crime and that in such a case one would be a tool of a state or of a citizen. This had been the main justification for the enslaving of Africans in the seventeenth and eighteenth centuries, since prisoners captured in a just war were thought of as criminals in this sense.[84] Furthermore, one would never agree to place oneself in such an arrangement by contract: "No one can bind himself to this kind of dependence."[85] At this point he explicitly referenced the enslavement of Africans: "For if the master is authorized to use the powers of his subject as he pleases, he can also exhaust them until his subject dies or is driven to despair (as with the Negroes on the Sugar Islands); his subject will in fact have given himself away, as property, to his master, which is impossible."[86] What does this mean? It would seem to mean among other things that if anyone believed that the African slaves had contracted to be slaves or bondsmen, then there was no basis for treating them as they were treated in the Sugar Islands. But, of course, nobody did believe this. Given that Kant did not seem to exclude criminals being so treated and, furthermore, given that this argument at least allows that those African slaves who were enslaved on that basis could be used up in this way, the defense of at least a certain form of slavery had been accomplished. What had been excluded was hereditary slavery, but this would be a great deal more impressive if it was not by Kant's time a familiar argument among those who were indifferent to the institutional form slavery had taken in the Americas. In other words, the weight to be placed on this point, if we want to discern Kant's intentions, is not yet clear.

It is helpful to compare Kant's presentation of slavery with that of the Marquis de Condorcet in his essay *Réflexions sur l'esclavage des Nègres* from 1791. Unlike Kant, Condorcet used to the full the abolitionist potential of the arguments that all human beings are born free and—given that he doubted that Africa could have produced so many criminals—the restriction of legitimate slavery to criminals. But he also raised a good question about whether, after due process, criminals could be enslaved, thereby leading to individuals and not just states owning slaves.[87] There is no reason to believe that Kant knew Condorcet's essay, but anybody wanting to read these statements by Kant, as intended by him to indicate his opposition to the institution of slavery in the New World, needs to explain why he explicitly allows for the possibility of individuals—and not just states—taking ownership of former criminals as slaves who can be used up until they are exhausted.[88]

Tempting though it is to read the discussion of slavery in *The Metaphysics of Morals* as an attack on the contemporary slave trade, what one finds for the most part is a highly stylized rehearsal of the standard European discussion of this topic that remains as indifferent as Locke was to the fact that reality did not correspond to theory. If Kant had wanted to rally support among his readers for abolitionism, he could have done so. Instead he fell into line with the common tendency of eighteenth-century moral philosophy in which opposition to slavery in general was commonplace, but where these philosophical assertions were rarely translated into calls for action to end the practice as it related to Africans.[89] The fact that Kant was ready to step out of the closed world of academic discussion to condemn the practice of colonialism is commendable, but it makes all the more striking his reluctance to do the same in the context of the slave trade. Even if one reads his reference to the Negroes who are used up by their owners in the Sugar Islands as an objection to the practice, which would bring it into line with his disapproval in *Toward Perpetual Peace* to the cruel slavery in the same place, the complaint highlights the treatment of the slaves rather than the very existence of slaves on the plantations.

Kant's Late Racism

Now that I have shown that we must abandon any suggestion that Kant eventually resolved his issues about race in the way we all wish he had done, I will examine the 1792 lectures in their own right to see what they show about Kant's racism. Werner Stark has found records of twenty-seven manuscripts or sets of lecture notes for Kant's *Physical Geography*, but only seventeen of them survive. Dating some of these texts is often difficult, but the 1792 text is very precisely dated to the summer of that year (April to September) and, furthermore, it is the only full set of lectures that we have which can be dated with confidence to the 1790s. In 1796 Kant lectured on physical geography for the last time but no record of these lectures is known to survive. In 1838 Fr. Chr. Starke published in a 20-page extract from the 1791 lectures. Eight of those pages are on human beings. What one finds is certainly significant as one discovers there that Kant had not abandoned the theory of *Keime* or germs on which his theory of race had been explicitly based, but which some commentators see him as having abandoned in 1790.[90] The account of race given in 1791 seems to be largely continuous with his first statements on the topic.

Even though these extracts from the 1791 course do not prepare us for what we find only a year later, we have no reason to doubt the accuracy of the notes from the 1792 course. They were made by Graf Heinrich Ludwig Adolph zu Dohna-Wundlacken, who also left notes for Kant's lectures on anthropology, which were edited by Arnold Kowalewski and published in 1924.[91] The 1792 course shows Kant talking about race much as he had done in 1775 when he first introduced the term: racial indifference is marked by inheritable characteristics and if two people of different races mate then those specific characteristics will, among the offspring, reflect both parents in equal measure. Kant also reiterated his view that crossbreeding was against nature's purpose because it led human beings to become increasingly similar, a possibility he clearly opposed. Nor is there any sign of Kant withdrawing or modifying his account of racial hierarchy, even if it is not presented in as much detail as it was, for example, in lectures on physical geography in 1784 where he insisted that the White race had the greatest perfection; that Indians have a little talent; that Negroes are below them; and that Americans are most deeply degraded, particularly those who lived in the hottest parts.

Nevertheless, in these 1792 *Lectures on Physical Geography*, in the context of recalling Hume's footnote attacking any claim Africans might have to equality, Kant integrated his racism into his philosophy of history in a particularly disturbing way. He attributed to Hume the idea that among the many thousand freed Negroes there is no example of any who had distinguished themselves with particular skill (*Geshiklichkeit*) (*sic*). Kant had already referred to this passage almost thirty years earlier in his *Observations on the Feeling of the Beautiful and Sublime*. Kant wrote: "The Negroes of Africa have by nature no feeling that rises above the ridiculous. Mr. Hume challenges anyone to adduce a single example where a Negro has demonstrated talents . . ." and so on.[92] This is what Hume wrote in a note added in 1753 to his essay of "National Character": "I am apt to suspect the negroes and in general all other species of men (for there are four or five different kinds) to be naturally inferior to the whites. There never was a civilized nation of any other complexion than white, nor even any individual eminent either in action or speculation. No ingenious manufactures amongst them, no arts, no sciences."[93] Kant developed this view in an extraordinary footnote to his 1788 essay *On the Use of Teleological Principles in Philosophy*. Although the major part of this essay is on race, and in particular a response to Georg Foster's attempt to challenge Kant's view on race, it also raises, as the title indicates, broad philosophical issues that lead directly to the third Critique only two years later.[94] In the footnote he cited an anonymous essay in Sprengel's *Beiträge zur Völker und Länderkunde*, a series that published translations of extracts or sometimes paraphrases of major essays from European travel literature.[95] According to this essay, among the many thousand of freed Negroes in Britain one can find none who actually works.[96] His exact words in the footnote are in part:

> In Hr. *Spengel's* Contributions . . . a knowledgeable man, adduces the following against Ramsay's wish to use all Negro slaves as free laborers: that among the many thousand freed Negroes which one encounters in America and England he knew no example of someone engaged in a business which one could properly called *labor*; rather that, when they are set free, they soon abandon an easy craft which previously as slaves they had been forced to carry out, and instead become hawkers, wretched innkeepers, lackeys, and people who go fishing and hunting, in a word, tramps.[97]

Kant then added on his own authority, again with that gratuitousness that is so alarming: "The same is to be found in the gypsies among us."[98]

These remarks may appear entirely gratuitous but in fact Kant was responding to Foster's challenge on the issue of slavery. Foster knew that polygenesis was associated with the defense of slavery. He questioned this connection: "When we differentiate the Negro as an original distinct stem from White men do we not cut through the last thread through which this mistreated people are connected with so and could still find some protection and some relief from european cruelty? Let me rather ask you whether the thought that Blacks are our brothers has anywhere even once meant that the raised whip of the slave-driver was put down?"[99] In other words, Foster challenged the monogenists to show how their views led to effective arguments in support of better treatment of the slaves and in response Kant produced an argument in favor of slavery.

We now know, although Kant almost certainly did not, that the text to which Kant referred in order to support his contention that free Blacks do not work, was by James Tobin. Tobin had written a critique of James Ramsay's *An Essay on the Treatment and Conversion of African Slaves in the British Sugar Colonies*. It is significant that the same issue of Sprengel's *Beiträge* includes a major extract from Ramsay's book that puts the alternative case,[100] but Kant in 1788 chose Tobin as his authority over Ramsay. The essay by Tobin published in *Beiträge* is an amalgam of extracts from Tobin's pamphlet, which in turn is at times a page by page response to Ramsay's book.[101] Sometimes described as proslavery, Tobin was careful to deny this. He presented himself as simply an opponent of Ramsay's arguments, even though he went beyond this in declaring himself against race mixing and in favor of the expulsion of Blacks from England.[102]

Tobin, after estimating the number of "able negro men" in England in 1785 at ten or twelve thousand, addressed to Ramsay the question of whether he had ever seen any of them in any laborious task.

> Did he ever meet with a black ploughman, hedger, ditcher, mower, or reaper, in the country; or a black porter, or chairman, in London? On the contrary, I will be free to affirm, that out of the whole of this number, those who are not in livery are in rags; and such as are not servants, are thieves or mendicants.[103]

Tobin extended the polemic to include Ignatius Sancho and, anticipating the response that Blacks were failing to work because of the climate, extended the argument to cover free Blacks in Jamaica.[104] The much condensed version of Tobin's essay omits some of the rhetoric as well as the reference to Sancho, but repeated all the essential points.[105]

Returning now to the 1792 lectures on physical geography, Kant did not mention Tobin's essay from *Beiträge* but his reference to Hume's observations about Negroes has as much to do with Tobin as with Hume. Kant completely ignored all the evidence that the abolitionists had amassed to refute Hume's view, which in the meantime Hume himself had backed away from to allow for exceptions. But the most disturbing feature of this comment is the use of this word *Geschicklichkeit*, which only two years earlier had played a prominent role in the account of the cosmopolitan whole in the *Critique of Judgment*. In the third critique Kant wrote that *Geschicklichkeit* is, together with discipline

(*Zucht*), the condition for developing that culture which is "the ultimate end that one has cause to ascribe to nature in regard to the human species."[106] Given this context in the *Critique of Judgment* it is impossible not to suspect sinister tones underlying Kant's claim that "Skill cannot very well developed in the human race except by means of inequality among people."[107] Kant did not say it is impossible, only that it is hard. Nor did he mention slavery specifically in this context. But could this give us a clue to what Kant thought about slavery? To the extent that an answer is possible it lies in his philosophy of history, which Kant seems to have closely associated with natural history, perhaps even viewing it as an extension of it.[108] In both his natural history and his philosophy of history he employed the same language of germs (*Keime*) and inclinations (*Anlagen*). Kant saw it as part of nature's purpose that the capacities of human beings are fully developed only in the species and not in the individual.[109] This is what he called human perfectibility. But does this apply to all races? It is striking that in his anthropology lectures for 1775–76, delivered immediately after Kant had formatted his racial theory for the first time, he appealed to the *Keime* to argue that the savage Indian and the Greenlander had the same germs as the French, but that they were not yet developed.[110] His claim at this time seems to have been that all human beings can thus be raised to perfection. Subsequently he would deny this on the basis of this same racial theory. This shows that Kant did not develop his racial theory to explain a preexisting belief in the permanent inequality of the races. Whether or not the racial theory, according to which one effect of the first migrations was to block the development of some of the germs in the original human type, was the cause of this shift is a matter of contexture. However, his conviction that racial characteristics are permanent points in this direction: it becomes at that point which characteristics are racial.

This shift toward advancing belief in a permanent racial hierarchy needs to be seen in a larger context, both with reference to Kant's own writings and the political issues of the day. I will have to restrict myself here to a few remarks on the former. Starting in 1784 with his *Idea for a Universal History with a Cosmopolitan Aim* Kant began raising serious questions about the purpose of the human race on the grounds that law was visible in nature but not in human affairs. Statistics gave Kant some hope for finding law in human affairs, but that was not enough for him.[111] He looked to history and the onward path of future generations toward a cosmopolitan whole for doing so. There had to be something more to human affairs than a succession of wars interrupted by moments of peace. Kant's solution involved giving a positive meaning to war. In the fifth proposition of the *Idea for a Universal History with a Cosmopolitan Aim* he proposed that antagonism be seen as "asocial sociability." That is, it was the way in which human affairs advanced, not according to a human plan, but a plan of nature.[112] Kant recognized that human beings want peace, but that nature operates differently. According to Kant's *Idea for a Universal History with a Cosmopolitanism Aim* "the crude natural capacity for moral discrimination" is transformed from a "*pathologically* enforced agreement into a society and finally, into *a moral whole*."[113] Nevertheless, the passage to this moral or cosmopolitan whole does not proceed morally but violently. Kant is clear in *Toward Perpetual Peace* that one should expect morality to arise from a good national constitution but that the order cannot be reversed.[114]

One could say that antagonism operates for Kant, like Adam Smith's "invisible hand,"[115] behind everyone's back. Through antagonism, cosmopolitanism in the operative sense was promoted. A similar—albeit more modest—version of this idea was proposed by Christoph Wieland, who in 1788 explained that cosmopolites believe that "the general laws of nature advance the perfection of the whole, while each in his own particular style and fashion busies himself in the pursuit of his own prosperity."[116] One does not need to pursue cosmopolitanism as such, one pursues one's own interest, just as one is cosmopolitan by being patriotic. And one should not fail to notice how *Toward Perpetual Peace* is dominated by commercial interests. Nevertheless, it is one thing to find the meaning of human existence not among the peacemakers but among those who pursue war, and it is another altogether when one oneself pursues peace through war and in the process invests the prospects of future humanity in one's own victory.[117] It is one thing to believe in an invisible hand, another entirely when one believes that that invisible hand is one's own, which is what Kant did when in his essay on cosmopolitan history he wrote that Europe gives law to the rest of the world.[118]

Because Kant believed that the lack of asocial qualities that would lead human beings to live like "Arcadian shepherds in perfect concord, contentment and mutual love"[119] at the same time excluded them from the end for which they were created, he was led in the context of his review of Herder to dismiss the Tahitians to the point of asking why they even existed.[120] In the *Critique of the Power of Judgment* Kant asked about the purpose of the Greenlanders, the Lapps, the Samoyeds, and the Yakuts, among others.[121] The initial puzzle for Kant had been that past generations must sacrifice themselves for future generations, but his attempt to resolve this problem created a problem for those races which lacked the capacities that White people allegedly had. In negotiating the question of human purposes Kant conceded that the question of why a people should have to exist was not always easy to answer.[122]

In his *Conjectural Beginning of Human History* Kant made clear that all human beings are equal recipients of nature's gifts of freedom and reason and that this is their right to demand they be treated as ends and not as means.[123] And yet he also in the same essay acknowledged the role of inequality as "that source of so many evils, but also of everything good" beginning with the first colonists.[124] If, as seems likely, this is a reference to the first migrations that led to the origin of the different races that Kant had outlined eleven years earlier in his essay on the human races, then it means that the essay provides an indispensable insight into how Kant thought the inequality of human races in the light of his other concerns. Inequality in one's rights as humans is "an inequality that cannot be separated from culture as long as it proceeds, as it were, without a plan (which is in any case, inevitable throughout much of time)."[125] Jennifer Mensch comments on this passage: "just as the philosopher can justify the mechanism of war as an ultimate force for good, so too must we accept and understand that inequality among the races is inseparable from the progress of culture and thus serves nature's purposes, that is, has the force of a natural necessity, in moving the species forward to greater perfection."[126] This strikes me as a position one can plausibly attribute to Kant and even extend to embrace his support of slavery at that time.

I am not claiming that this resolves the tension within Kant's philosophy between his racism and his universalism. I can only claim to have transferred the contradiction into a realm where we know that at that time Kant—and not only Kant—was conflicted: did morality or politics hold primacy? My approach can in this respect be contrasted with that of commentators who try to attribute to Kant the idea that he had doubts about whether Africans were fully human. In other words, what appears to us as a contradiction between Kant's racism and his universalism appears in his own work as the tension between a philosophy of history rooted in his account of natural history that authorized violence and a moral perspective that questioned much of this same violence. It is clear that Kant believed that the moral perspective trumped the political perspective favored by many of his successors, including Hegel.[127] One sees this, for example, in Kant's response to Anton Busching's critique of his essay on history where Kant insisted that although colonialism might be to the world's advantage because it would civilize crude peoples and at the same time cleanse one's own country of corrupt people, good intentions did not eradicate the injustice.[128] To this extent, in eventually condemning the slave trade, Kant was acting in conformity with the general tenor of his moral philosophy, even if we can now also begin to understand what was pulling Kant in the other direction: he came to see that he had to condemn the slave trade and the cruelty with which slaves were treated, but at the same time he also had an account of how slavery contributed to bringing about "a moral whole." There is also every reason to believe on the basis of *The Metaphysics of Morals* that had Kant bothered to think more about the practice of slavery in the Americas as a condition inherited across generations, he would have been as outspoken in condemning it as he was about some of the abuses associated with colonialism. This failure perhaps can be put down to a level of indifference on his part, but it is disturbing that his view of cosmopolitanism—the kind that moves outward by degrees—not only explains, but seems to explain away his lack of concern for African slaves, even though many of his contemporaries were by the 1790s deeply concerned about the basis on which these slaves were being held.

All of this helps to make a little more sense of Kant's conflicted treatment of Negroes in the 1792 lectures on physical geography. When Kant denied that Negroes possess skill or *Geschicklichkeit*, he was making a comment of an entirely different tenor from the one by Hume, which he was misremembering. Hume was raising doubts about the originality and excellence of the contributions of Negroes in various areas as a basis for denying their equality. Kant, by contrast, was raising a doubt about the capacity of a whole race to be raised to the level that would allow them to participate fully in the ultimate end nature had set for the human species. This question was imposed on him by his racial theory according to which racial characteristics were permanent. Kant may not have had a ready answer to the question of why the Tahitians or the Lapps existed, but he seemed to have believed that Negroes had a subservient role assigned to them. They could participate by serving those with the skill, the talents, and the discipline to contribute to human perfectibility. Beginning in the early 1780s the tension between Kant's new moral theory and his natural history broadly conceived slowly began to emerge and his lectures on physical geography were one of the places where this happened. Werner Stark's transcriptions of these lectures will prove a valuable resource for all future attempts to investigate this tension.

Notes

1. Pauline Kleingeld, "Kant's Second Thoughts on Race," *The Philosophical Quarterly* 57:229 (October 2007), 573–92.

2. See, for example, Susan Meld Shell, *The Embodiment of Reason* (Chicago: University of Chicago Press, 1996), 418 n. 53; Peter Fenves, *Late Kant* (London: Routledge, 2003), 103–5; Sankar Muthu, *Enlightenment against Empire* (Princeton: Princeton University Press, 2003), 183–4.

3. Kleingeld, "Kant's Second Thoughts," 574.

4. Pauline Kleingeld, "Six Varieties of Cosmopolitanism in Late Eighteenth-Century Germany," *Journal of the History of Ideas* (1999), 505–24.

5. Allen W. Wood, *Kant* (Oxford: Blackwell, 2005), 188.

6. For example, Thomas E. Hill and Bernard Boxill, "Kant and Race" in *Race and Racism*, ed. Bernard Boxill (Oxford: Oxford University Press, 2001), 448–71. I have responded in "Will the Real Kant Please Stand Up: The Challenge of Enlightenment Racism to the Study of the History of Philosophy," *Radical Philosophy* 117 (January/February, 2003), 10–19.

7. Todd Hendrick, "Race, Difference and Anthropology in Kant's Cosmopolitanism," *Journal of the History of Philosophy* 46:2 (2008), 246. He presents his own view at 262. I took up this question in a lecture in Denmark in October 2008, which will be published in *Acta Institutionis Philosophiae et Aestheticae* 27 (2009, forthcoming).

8. Kleingeld, "Kant's Second Thoughts," 586.

9. In her important and sensitive study of Kant's sexism, Pauline Kleingeld, "The Problematic Status of Gender-Neutral Language in the History of Philosophy: The Case of Kant," *The Philosophical Forum* 25 (1993), 134–50.

10. Kleingeld, "Kant's Second Thoughts," 586 n.

11. Charles W. Mills, "Kants *Untermenschen*," in *Race and Racism in Modern Philosophy*, ed. Andrew Valls (Ithaca: Cornell University Press, 2005), 184. There are other problems with Mills's analysis as when he denies that people of color are regarded by Kant as legal persons (185).

12. Kleingeld, "Kant's Second Thoughts," 583–4.

13. Kleingeld, "Kant's Second Thoughts," 575.

14. Kleingeld, "Kant's Second Thoughts," 586.

15. Kleingeld, "Kant's Second Thoughts," 582.

16. Kleingeld, "Kant's Second Thoughts," 584 and 586.

17. *Kant on the Metaphysics of Morals: Vigilantius's Lecture Notes*, in the Akademie edition of *Kants gesammelte Schriften* (hereafter cited as AK, followed by volume and page numers), AK 27: 674. Translated as *Lectures on Ethics*, trans. Peter Heath (Cambridge: Cambridge University Press, 1997), 406. Subsequent references will give the translation reference in parentheses.

18. *Lectures on Ethics*, AK 27: 673 (405–6).

19. *Lectures on Ethics*, AK 27: 673-4 (406).

20. *Lectures on Ethics*, AK 27: 673 (405).

21. Kuno Fischer, *Fichte*, 3rd ed. (1900), 627. Cited in Friedrich Meinecke, *Weltbürgertum und Nationalstaat* (Munich: R. Oldenbourg, 1908), 92; Robert B. Kimber, trans., *Cosmopolitanism and the National State* (Princeton: Princeton University Press, 1970), 73. Subsequent references will give the translation reference in parentheses.

22. Wilhelm Windelband, *Fichte's Idee des deutschen Staates* (Freiburg: J. C. B. Mohr, 1890), 11. Cited in Meinecke, *Welbürgertum*, 92 (73).

23. Jean-Paul Sartre, *Critique de la raison dialectique* (Paris: Gallimard, 1960), 692; Alan Sheridan Smith, trans., *Critique of Dialectical Reason* (London: Verso, 1976), 739.

24. *Reflexion 1520: Character der Race*, AK 15 (2): 878 and *Reflexion 1513*, AK 15 (2): 840. See Robert Bernasconi, "Why do the Happy Inhabitants of Tahiti Bother to Exist at All?" in *Genocide and Human Rights*, ed. John K. Roth (New York: Palgrave Macmillan, 2005), 142.

25. *Of the Different Races of Human Beings*, AK 2: 429–43, Holly Wilson and Günter Zöller, trans., in *Anthropology, History and Education*, eds. Robert B. Louden and Günter Zöller (Cambridge: Cambridge University Press), 2007, 84–97.

26. *Determination of the Concept of a Human Race*, AK 8: 91–106, Holly Wilson and Günter Zöller, trans., in *Anthropology, History and Education*, 145–59.

27. *On the Use of Teleological Principles in Philosophy*, AK 8: 157–84, Günter Zöller, trans., in *Anthropology, History and Education*, 195–218.

28. Kant had first announced his lectures on physical geography in 1757 with a consideration of the west winds in that region, AK 2: 1–12.

29. *Of the Different Races of Human Beings*, AK 2: 434 n. and 443 (89 and 97). On the novelty of Kant's concept of race, which I believe can be considered the first scientific definition of race, see Robert Bernasconi, "Who Invented the Concept of Race?"

in *Race*, ed. Robert Bernasconi (Oxford: Blackwell, 2001), 11–36. There has recently been new speculation on the fate of natural history in Kant's late work, but I believe it is misconceived. See Phillip R. Sloan, "Kant on the History of Nature: The Ambiguous Heritage of the Critical Philosophy for Natural History," *Studies in History and Philosophy of Biological and Biomedical Sciences* 37:4 (December 2006), 627–48. I intend to address this issue on another occasion. Meanwhile, see Marcel Quarfood, *Transcendental Idealism and the Organism* (Stockholm: Almqvist and Wiksell, 2004), 77–117.

30. *Physical Geography*, AK 9: 162.

31. Erich Adickes, *Untersuchungen zu Kants physischer Geographie* (Tübingen: J. C. B. Mohr, 1911), 278.

32. Adickes, *Untersuchungen zu Kants physischer Geographie*, 279.

33. *Physical Geography*, AK 9: 316. Parallel comments can be found in texts that can be dated to the early 1780s. See Robert Bernasconi, "Kant as an Unfamiliar Source of Racism," in *Philosophers on Race*, eds. Julie K. Ward and Tommy L. Lott (Oxford: Blackwell, 2002), 147–8.

34. *Anthropologie Menschenkunde*, AK 25 (2): 1195.

35. Immanuel Kant, *Physische Geographie*, ed. Friedrich Theodor Rink (Königsberg: Göbbels und Unzer, 1802), vol. 1, 62 and 68.

36. The first two parts of the second "thoroughly revised" editions are undated. The subsequent two volumes were improved and expanded under the editorship of Ferdinand Stiller: *Immanuel Kant's Physische Geographie*, 2nd ed., 2 vols. (Hamburg: Vollmer, 1816 and 1817).

37. It appeared under the editorship of Karl Rosenkranz and Friedrick Wilhelm Schubert in F. W. Schubert, ed. *Immanuel Kants Schriften zur physischen Geographie*, vol. 6 (Leipzig: Leopold Voss, 1839), 415–775. Rink's edition must have recommended itself to all subsequent editors of Kant's complete works because it is so much shorter than Vollmer's edition, whatever other merits it may or may not have.

38. It is theoretically possible that an expert might glean some information about Kant's thoughts on physical geography during this final year from Vollmer's text because he himself attended Kant's lectures in the 1790s, but whatever conclusions were drawn they would always be open to question.

39. See Werner Stark, ed. *Vorlesungen über der Physische Geographie*, AK 26 (1), forthcoming. In preparation for this edition, the editor, Werner Stark, generously prepared a website that gives scholars access to hundreds of pages of transcripts from Kant's lectures on this topic. Hitherto we have had to rely largely on Rink's text that was produced somewhat hurriedly in 1802 as a counter to an unauthorized edition that Vollmer had begun to publish. Neither edition is even remotely satisfactory to modern standards of scholarship, and the soon to be published English translation of the Rink edition in the works must be used with great caution. I am grateful to Werner Stark for giving me access to these unpublished notes.

40. *Nachricht von der Einrichtung seiner Vorlesungen in dem Winterhalbenjahre von 1765–1766*, AK 2: 313.

41. *Nachricht von der Einrichtung seiner Vorlesungen in dem Winterhalbenjahre von 1765–1766*, AK 2: 312.

42. *Physical Geography*, AK 9: 236.

43. *Of the Different Races of Human Beings*, AK 2: 434 (89).

44. *Of the Different Races of Human Beings*, AK 2: 432–3 (87–8).

45. George Louis le Clerc, Comte de Buffon, *Histoire Naturelle*, vol. 14 (Paris: Imprimerie royale, 1766), 314.

46. E.g. *Of the Different Races of Human Beings*, AK 2: 437 (91–2).

47. *Of the Different Races of Human Beings*, AK 2: 442 (96).

48. *Of the Different Races of Human Beings*, AK 2: 430 (85).

49. *On the Use of Teleological Principles in Philosophy*, AK 8: 175–6 (211).

50. Werner Stark, ed. *Vorlesung zur Moralphilosophie* (Berlin: Walter de Gruyter, 2004), 245.

51. *Of the Different Races of Human Beings*, AK 2: 431 (87).

52. *Of the Different Races of Human Beings*, AK 2: 434 (89).

53. Sabina Laetitia Kowalewski and Werner Stark, eds. *Königsberger Kantiana, Kant-Forschungen* 12 (Hamburg: Felix Meiner, 2000), 448.

54. Kleingeld, "Kant's Second Thoughts," 592.

55. Muthu, *Enlightenment against Empire*, 183.

56. Fenves, *Late Kant*, 101–5.

57. See Robert Bernasconi, "Kant and Blumenbach's Polyps: A Neglected Chapter in the History of the Concept of Race," in *The German Invention of Race*, eds. Sara Eigen and Mark Larrimore (Albany: SUNY Press, 2006), 73–90.

58. *Anthropology from a Pragmatic Point of View*, AK 7: 328, trans. Robert B. Louden in *Anthropology, History and Education*, 415.

59. Muthu, *Enlightenment Against Empire*, 184. The source of the error seems to be his misreading of Susan Shell's remark that "Girtenner's [sic] *Über das Kantische Prinzip für das Naturgeschichte* . . . applies Kant's concept of race to nonhuman species." Shell, *Embodiment of Reason*, 307 n. 23. Shell meant that Girtanner includes a consideration of the application of Kant's concept of race to nonhuman species but Muthu mistakenly reads this to say that that was Girtanner's exclusive concern. See Christoph Girtanner, "Erste Abtheilung. Von den Menschenrassen," in *Über das Kantische Prinzip für die Naturgeschichte* (Göttingen: Vandenhoeck und Ruprecht, 1796), 57–285.

60. Mark Larrimore, "Antinomies of Race: Diversity and Destiny in Kant," *Patterns of Prejudice* 42:4–5 (2008), 341–63.

61. Kant's views of race mixing across his philosophical career are more complex than I presented them in "Kant as an Unfamiliar Source of Racism." I am unable to explore that complexity here, but I underwrite Larrimore's treatment of this issue as it is presented in Kant's *Anthropology* and will not rehearse these issues here.

62. Kleingeld, "Kant's Second Thoughts," 586. She chooses the 1792 date because of the surviving lecture course. The term "race" is also prominent in Kant's anthropology lectures in 1790–91. See Immanuel Kant, *Anweisung zur Menschen- und Welterkenntnis*, ed. Fr. Ch. Starke (Leipzig: Die Expedition des europäischen Aufsehers, 1831), 109 and in 1791–92, see *Königsberger Kantiana*, 446–8.

63. The fact that Fenves suggests that Kant had already dropped his defense of the concept of race prior to the *Critique of Judgment* shows how arbitrary these arguments from silence can be (Fenves, *Late Kant*, 104).

64. *Anthropologie Menschenkunde*, AK 25 (2): 1187.

65. *Reflexion 1520*, AK 15 (2): 878.

66. Kleingeld, "Kant's Second Thoughts," 577.

67. *Reflexion 1520*, AK 15 (2): 878.

68. Bernasconi, "Kant as an Unfamiliar Source of Racism," 152–62.

69. Georg S. A. Mellin, *Encyclopädisches Wörterbuch der Kritischen Philosophie* (Jena: Friedrich Frommann, 1804), vol. 6, 147–51.

70. Loose Leaf F9 (draft of *Toward Perpetual Peace*), AK 23: 174.

71. Dohna-Wundlacken, *Lecture Notes on Physical Geography* (1792), courtesy Werner Stark.

72. Loose Leaf F9 (draft of *Toward Perpetual Peace*), AK 23: 174.

73. *Toward Perpetual Peace*, AK 8: 358, in *Practical Philosophy*, eds. Mary J. Gregor and Allen W. Wood, trans. Mary J. Gregor (Cambridge: Cambridge University Press, 1996), 329.

74. *Toward Perpetual Peace*, AK 8: 359 (330).

75. *Physical Geography*, AK 9: 313.

76. *Groundwork of the Metaphysics of Morals*, AK 4: 421, in *Practical Philosophy*, 73.

77. Onora O'Neill, *Constructions of Reason. Explorations of Kant's Practical Philosophy* (Cambridge: Cambridge University Press, 1989).

78. Richard F. Galvin, "Slavery and Universalizability," *Kant-Studien* 90 (1999), 191–203, especially 202.

79. Kleingeld, "Kant's Second Thoughts," 592.

80. On the basis of the draft to *Toward Perpetual Peace* Kleingeld takes me to task for writing in "Kant as Unfamiliar Source of Racism" that Kant "failed to speak out against slavery" (Bernasconi, "Kant as an Unfamiliar Source," 150), but given that the text she refers to refute me was never published by Kant, I suppose I could claim I was not proved wrong in that case (Kleingeld, "Kant's Second Thoughts," 587–8). But he did "speak out" in the 1792 lectures, so I was wrong after all.

81. *The Metaphysics of Morals*, AK 6: 353, in *Practical Philosophy*, 490.

82. *The Metaphysics of Morals*, AK 6: 283 (432).

83. On Locke, see Robert Bernasconi and Anika Maaza Mann, "The Contradictions of Racism," in *Race and Racism in Modern Philosophy*, 100.

84. See, for example, John Atkins, *A Voyage to Guinea, Brazil and the West Indies* (London: Caesar Ward and Richard Chandler, 1735), 176.

85. *The Metaphysics of Morals*, AK 6: 330 (472).

86. *The Metaphysics of Morals*, AK 6: 330 (471).

87. Condorcet, *Esquisse d'un tableau historique des progrès de l'esprit humain; suivi de Réflexions sur l'esclavage des Nègres* (Paris: Masson, 1822), 318–26.

88. *The Metaphysics of Morals*, AK 6: 330 (471).

89. Robert Forbes, "Slavery and the Evangelical Enlightenment," in *Religion and the Antebellum Debate Over Slavery*, eds. John R. McKivigan and Mitchell Snay (Athens: University of Georgia Press, 73). Many of these discussions took their lead from seventeenth-century discussions belonging to the natural right tradition that predate or are contemporary with the great expansion of slavery. Kant was no exception.

90. Fr. Chr. Starke, ed. "Betrachtungen über die Erde und den Menschen," in *Immanuel Kant's vorzügliche kleine Schriften und Aufsätze* (Leipzig: Ernst, 1838), vol. II, 276. This calls into question certain claims by Phillip Sloan in "Preforming the Categories: Eighteenth-Century Generation Theory and the Biological Roots of Kant's A Priori," *Journal of the History of Philosophy* 402 (2002), 250.

91. Ed. Arnold Kowalewski, *Die philosophischen Hauptvorlesungen Immanuel Kants nach den neu aufgefundenen Kollegheften des Grafen Heinrich zu Dohna-Wundlacken* (Munich: Rösl and Cie, 1924). The similarities between these notes on anthropology and a fuller set of notes prepared by Daniel Matuszewski show that they are reliable and give us reason to believe that the notes on physical geography are also of good quality. See *Königsberger Kantiana*, 182–452.

92. *Observations on the Feeling of the Beautiful and Sublime*, AK 2: 253, trans. Paul Guyer in *Anthropology, History and Education*, 59.

93. David Hume, *Essays, Moral, Political, and Literary*, ed. Eugene F. Miller (Indianapolis: Liberty Fund, 1987), 629. For a history of the note see John Immerwahr, "Hume's Revised Racism," *Journal of the History of Ideas* 53:3 (1992), 481–6. The footnote was revised for the posthumous 1777 edition.

94. See for example, John D. McFarland, *Kant's Concept of Teleology* (Edinburgh: Edinburgh University Press, 1970), 56–68 and John H. Zammito, *The Genesis of Kant's Critique of Judgment* (Chicago: University of Chicago Press, 1992), 207–13.

95. Matthias C. Sprengel, ed. *Beiträge zur Völker und Länderkunde* 5 (1786).

96. *On the Use of Teleological Principles in Philosophy*, AK 8: 174 (209). See further, Bernasconi, "Kant as an Unfamiliar Source of Racism," and Monika Firla, "Philosophie und Ethnographie. Kants Verhältnis zu Kultur und Geschichte Afrikas," *Zeitschrift der Deutschen Morgenländischen Gesellschaft* Supplement X (1994), 439.

97. *On the Use of Teleological Principles in Philosophy*, AK 8: 174 (209).

98. Kurt Röttgers, "Kants Zigeuner," *Kant-Studien* 88 (1997), 60–86.

99. Georg Foster, "Noch etwas über die Menschenrassen," *Kleine Schriften zu Philosophie und Zeitgeschichte*, ed. Siegfried Scheibe (Berlin: Akadamie, 1991), 154.

100. Sprengel, ed., *Beiträge*, 1–72.

101. It is theoretically possible that there was another essay that formed the basis of the translation, but it seems most likely that the German translator performed the work of altering the original to construct a shorter piece of work.

102. James Tobin, *Cursory Remarks Upon the Reverend Mr. Ramsay's Essay on the Treatment and Conversion of African Slaves in the Sugar Colonies* (London: Wilkie, 1785), 5 and 118 n.

103. Tobin, *Cursory Remarks*, 117.

104. Tobin, *Cursory Remarks*, 119.

105. In Sprengel, ed., *Beiträge*, 287–8.

106. *Critique of the Power of Judgment*, AK 5: 431, trans. Paul Guyer and Eric Matthews (Cambridge: Cambridge University Press, 2000), 299. On *Geschicklichkeit*, see Mellin, *Encyclopädisches Wörterbuch der Kritischen Philosophie*, vol. 2, 879–86.

107. *Critique of the Power of Judgment*, AK 5: 432 (299). But this is only one aspect of what Kant is denying when he denies skill to Negroes. In the *Groundwork of the Metaphysics of Morals* Kant talks of imperatives of skill, where the issue is not whether an end is rational and good, but what one must do in order to attain it. (AK 4: 415 [68]). Kant talks about how parents try to teach children skill in the use of money. The notion of skill plays a more prominent role in the *Critique of Practical Reason*: "But the concept of perfection in the *practical* sense is the fitness or adequacy of a thing for all sorts of ends. This perfection, as a characteristic of the human being and so as internal is nothing other than *talent* and what strengthens or completes this, *skill*" (AK 5: 41 [172–3]).

108. For a general account of the relationship of natural history and philosophy of history, see Friedrich Kaulbach, "Der Zusammenhang zwischen Naturphilosophie und Geschichtsphilosophie bei Kant," *Kant-Studien* 56 (1966), 430–51.

109. *Idea for a Universal History with a Cosmopolitan Aim*, AK 8: 18, trans. Allen W. Wood in *Anthropology, History and Education*, 109.

110. *Anthropologie Friedländer*, AK 25 (1): 694.

111. *Idea for a Universal History with a Cosmopolitan Aim* AK 8: 17 (108).

112. *Idea for a Universal History with a Cosmopolitan Aim* AK 8: 18 (109).

113. *Idea for a Universal History with a Cosmopolitan Aim* AK 8: 21 (111).

114. *Toward Perpetual Peace*, AK 8: 366 (335).

115. Adam Smith, *An Inquiry into the Nature and Causes of the Wealth of Nations* (Oxford: Oxford University Press, 1976), vol. 1, 456.

116. Christoph Martin Wieland, "Das Geheimnis des Kosmopolitenorders," in *Ausgewählte Werke* (Essen: Phaidon, nd), 736.

117. I have argued elsewhere that it is when one's enemies come to be seen as enemies of humanity that one is most likely to engage in genocide, civilian bombing, torture, and other crimes against humanity. See Robert Bernasconi, "Ewige Friede und totaler Krieg," trans. Christina Schües, in *Denkwege des Friedens*, eds. Alfred Hirsch and Pascal Delhom (Freiburg: Alber, 2007), 22–42.

118. *Idea for a Universal History with a Cosmopolitan Aim* AK 8: 29-30 (119).

119. *Idea for a Universal History with a Cosmopolitan Aim* AK 8: 21 (111–2).

120. *Reviews of Herder's Ideas on the Philosophy of the History of Mankind*, AK 8: 65, trans. in *Political Writings*, ed. Hans Reiss (Cambridge: Cambridge University Press, 1991), 219–20.

121. *Critique of the Power of Judgment*, AK 5: 369 (241).

122. *Critique of the Power of Judgment*, AK 5: 378 (250).

123. *Conjectural Beginning of Human History*, AK 8: 114, trans. Allen W. Wood in *Anthropology, History and Education*, 167–8.

124. *Conjectural Beginning of Human History*, AK 8: 119 (172).

125. *Conjectural Beginning of Human History*, AK 8: 117–8 (170).

126. Jennifer Mensch, "Morality and Politics in Kant's Philosophy of History," in *Toward Greater Human Solidarity*, ed. Anindita N. Balslev (Calcutta: Dasgupta, 2005), 80.

127. For example, G. W. F. Hegel, *Vorlesungen über Naturrecht und Staatswissenschaft* (Hamburg: Meiner, 1983), 248–54; trans. J. Michael Stewart and Peter C. Hodgson, *Lectures on Natural Right and Political Science* (Berkeley: University of California Press, 1995), 298–304.

128. *The Metaphysics of Morals*, AK 6: 353 (490).

17

The Darker Side of the Enlightenment

A De-Colonial Reading of Kant's *Geography*

Walter D. Mignolo

Introduction

In the proposal for their volume, Stuart Elden and Eduardo Mendieta, reminded the participants that

> . . . in his essay on "The Conflict of the Faculties," Kant divides the philosophy faculty into two parts—the one that deals with "pure rational knowledge" and one that deals with "*historical knowledge.*" The former contains metaphysics of nature and morals, along with pure philosophy and mathematics; the latter includes history, geography, philology, the humanities and the empirical knowledge of the natural sciences.[1]

My interest in Immanuel Kant, beyond reading him as an educated man majoring in philosophy and literature, in Córdoba, Argentina, is indirect. Kant himself—and his *Geography*, are not the target of this article. I am first and foremost interested in certain issues and, secondly, to what Kant has to say about them. Or if you wish, Kant himself, through his *Geography* is part of the problem not of the solution. His work triggers a wealth of issues for someone who is interested—like myself—in the darker side of the Renaissance and, to be oxymoronic, the darker side of the Enlightenment—which means that I am more interested in what Kant hides than in what he reveals. I will be addressing the links (and complicities), in Kant, between historical and rational knowledges, on the one hand, and the silences that his *Geography* contributes to create. The global totality Kant searches in his *Geography*, parallels to other fields of his inquiry, is an anxious will to control knowledge and a blind sensitivity toward what he overrules by means of what he appropriates.

A Geo-Historical Detour on the Way to Kant's *Geography*

If you imagine the world around 1500, it was a polycentric world. China and Japan were standing in their millenarian histories. South of China, the millenarian histories of India and the wide geographic stretch, intellectual sophistication, and influence of Sanskrit can only be compared to Latin under the Roman Empire and Arabic—from the seventh century on—under the Islamic Caliphate. By 1500, Western Christians were still a marginal society, about one thousand years after the collapse of the Western Roman Empire and about eight hundred years after the Crusades (1095–c. 2130) and the loss of Jerusalem as the hub of Christianity. By 1500 vernacular languages derived from Latin (Italian, Spanish, Portuguese, French, German, English), were just unfolding and Latin was still a language of scholarship and theology. Parallel to the origination and unfolding of modern European imperial languages from Latin, Hindi (in the area which is today India) and Urdu (in Central Asia and today in Pakistan) originated and unfolded from Sanskrit, through Hindustani.

By 1500 also, the Ottoman Sultanate was entering into its prime period, with Suleiman the Magnificent leading it between 1520 and 1566; the Mughal (Persian for "Mongol") Sultanate began to take form around 1526 and lasted until the second half of the eighteenth century when British merchants began to contribute to their decay. Between 1501 and 1732, the Safavid Dynasty constituted the third Islamic Sultanate. The three Islamic Sultanates originated and spread out from the foundation of the original Islamic Caliphate in the seventh century and the Arabic language became prominent next to the Hebrew, Greek and Latin. Last but not least, when Moscow was declared the "Third Rome" (by 1520, that is coeval with the rise of Suleiman the Magnificent and Charles V of the Holy Roman Empire), a polycentric world was in its splendor. But that was also the beginning of its demise.

By 1500, the territory that is today Europe was the land of marginal Western Christians, or Christendom, that began to consolidate itself by the coincidence of the "discovery of America," the expulsion of the Moors (and Jews) from the Iberian Peninsula, and Charles I of Spain being also Charles V of the Holy Roman Empire of the German nations. Venice, Genoa, and Florence, the wealthiest cities of Italy, which was composed of city-states after the collapse of the Roman Empire, were looking toward the East and benefiting from commercial trade with the South and East of the Mediterranean and from there to China and India. If the fall of Constantinople (1453) gave Italian merchants a pause, the "discovery of America" definitely made them turn their heads toward the West and to contribute (through banking and lending money) to the emergence of a type of economy known today as "capitalist economy." While it is impossible to describe in two lines "the spirit of capitalism," I would stress, in view of my argument, two fundamental aspects:

a. *The massive appropriation of land and resources (gold, silver).* This circumstance, made possible by the "discovery and conquest of America" and the massive exploitation of labor, Indians, and the trade of enslaved Africans, allowed for a qualitative jump in the use of "capital" already accumulated in the banks of Florence, Venice, and Genoa. It was thanks to Genovese lending to the Spanish Monarch that facilitated trans-Atlantic explorations and the emergence of capitalism. The combination of capital, massive appropriation of land and resources, and massive exploitation of labor made possible the production of commodities for a global market for the first in the history of the human species.
b. *The expendability or dispensability of human lives.* Again, for the first time in history, human lives became dispensable and irrelevant to the primary goal of increased production and accumulated benefits. Enslaved Africans were not only an exploited labor force, they became just one more commodity. Christian ethics condemned slavery, but could not stop it. By the second half of the eighteenth century, when Immanuel Kant was delivering his lectures on geography, Adam Smith was providing the theoretical frame for the wealth of nations, looking toward the future and overlooking the recent past history that allowed him to celebrate free trade and be silenced on the ethical consequences of such an economic breakthrough.

Seen from the perspective of Islamic history, historian Karen Armstrong dated the "arrival of the West" to the Islamic world toward 1750. In fact, the first "targets" were the Mughal, the Safavid, and the Ottoman sultanates. What made this possible was, on the one hand, the new type of economy and associated mentality (e.g., what it takes to increase gains), and on the other, the blossoming of knowledge-making. The Atlantic economy (the historical foundation of capitalism) made possible, in Armstrong's view, the change from reliance on "a surplus of agricultural produce (as was the case in the economies of the polycentric world until 1500)," to the rising Atlantic economy that "was founded on a technology and investment of capital that enabled the West to reproduce its resources indefinitely, so that Western society was no longer subject to the same constraints as an agrarian culture."[2] She adds that, although not planned in advance, the transformation of economic practices resulted in a second Axial Age "which demanded a revolution of the established mores on several fronts at the same time: political, social and intellectual."[3] Indeed there were impressive advances in astronomy, medicine, navigation, industry, and agriculture. These are the headings provided by Armstrong. And she adds:

> None of these was in itself decisive, but their cumulative effect was radical. By 1600 innovations were occurring on such a scale that progress seemed irreversible: a discovery in one field would often lead to fresh insights in another. Instead of seeing the world as governed by immutable laws, Europeans had found that they could alter the course of nature. They were now prepared to invest and reinvest capital in the firm expectation of continuing progress and the continuous improvement of trade.
>
> By the industrial revolution of the nineteenth century, Westerners felt such assurance that they no longer looked *back to the past* for inspiration, as in the agrarian cultures and religions, but *looked forward to the future.*[4]

Economy and knowledge joined forces with navigation through that Atlantic and from there through the world. Cartography underwent a radical transformation in the sixteenth century that culminated, in 1570, with Abraham Ortelius's *Theatrum Orbis Terrarum,* first published in 1570 and repeatedly reproduced until 1612.

The historical and theoretical consequences of *Theatrum,* particularly for understanding Kant's lectures on geography taught about two centuries after Ortelius, is twofold. On one side, there is Ortelius's atlas mapping the world and showing its configuration in land and water masses. Ortelius's depiction of the world overruled all existing territorial conceptions and descriptions prior to 1500. On the other side, we find that polycentric territorialities were subsumed in the *Orbis Universalis.* The end result and the consequences were formidable for European political, economic, and gnoseologic projects. They were devastating for the rest of the world, as was becoming clear by the end of the eighteenth century: every civilization or culture began to be seen as stagnated and falling behind "modernity," receding toward the past. At the time of Ortelius's *Orbis Terrarun,* the idea that pagans, infidels, and the "new" barbarians of "America" were in the space beyond Europe; *humanitas* was being defined in contrast to those who missed the norm, either because excess or because of lack. Cartography, interestingly enough, was crucial to leaving about 80 percent of the world out of eighteenth-century history. By the end of the eighteenth century, "primitives" in time replaced "barbarians" in space.

The significance of the second aspect (e.g., subsuming all existing territorialities on the world map) comes from the word chosen by Ortelius: *Theatrum.* "Theatrum" belongs, in ancient Greek, to the family of "theorein, theorem, theory." From the *Online Etymology Dictionary* we find, for "theory:"

> 1592, "conception, mental scheme," from L.L. theoria (Jerome), from Gk. theoria "contemplation, speculation, a looking at, things looked at," from theorein "to consider, speculate, look at," from theoros "spectator," from thea "a view" + horan "to see." Sense of "principles or methods of a science or art (rather than its practice)" is first recorded 1613. That of "an explanation based on observation and reasoning" is from 1638. The verb theorize is recorded from 1638.

And for "theatre" we find:

> c.1374, "open air place in ancient times for viewing spectacles," from O.Fr. theatre (12c.), from L. theatrum, from Gk. theatron "theater," lit.[5] "place for viewing," from theasthai "to behold" (cf. thea "a view," theater "spectator") + -tron, suffix denoting place. Meaning "building where plays are shown" (1577) was transferred to that of "plays, writing, production, the stage" (1668). Spelling with -re prevailed in Britain after c.1700, but Amer. Eng. retained or revived the older spelling in -er. Generic sense of "place of action" is from 1581; especially "region where war is being fought" (1914). Theatrical "pertaining to the theater" is recorded from 1558; in the sense of "stagy, histrionic" it is attested from 1709.

"Contemplation, a place of viewing, spectator (theoros)," are some of the key words. I have argued, in *The Darker Side of the Renaissance*,[6] that the cartographic breakthrough of the sixteenth century was to displace and replace, on the one hand, the ethnic center by the geometric center. Coexisting territorialities beyond Western intellectual history were relegated to the past. It appears as if only Western cartography continued its historical march). On the other hand, the very *ethnic center of Ortelius* himself, and all the Dutch geographers and cartographers around him, assumed (and I believe that this assumption was honest blindness, not perversity), that their maps where not the maps that *projected their ethnic view of the world but that it was,* indeed the world as is and should be for everybody.

When Kant was delivering his lectures on geography, the epistemic foundation of this particular field (mapping and describing the earth) was not only already "mapped" (to be redundant) but it was above all, epistemologically deeply grounded already in the belief that knowledge-making about the world was detached from the knower. Although Kant insisted that knowledge starts from senses and experience, he assumed that there was a universal formula and therefore all human senses and experience lead to the same reasoning and conception of the world. Kant's philosophy, his lecture on geography and anthropology, as well as the anthropological observation surrounding his *Observations on the Feeling of the Beautiful and Sublime* (1764), are all ashore on the sixteenth-century theological and cartographic assumptions according to which not only knowledge was universal, but the knower was also equally a universally endowed epistemic subject who embodied the universality of sensing and experiencing. Hence, a subject that was beyond the racial and patriarchal hierarchies that the system of knowledge Kant himself was embracing had already been established.

The splendors and miseries in Abraham Ortelius and Immanuel Kant are both in what they have achieved and in what they have ignored and dismissed. In the internal history of Europe, both where part of the cause that, from the European Renaissance on, originated a frame of mind that we now describe as "modern/modernity." The Middle Age and Antiquity (Greece, Rome) were invented as both the foundation of Christian European history and as difference: the past from the present. The European Renaissance founded itself by colonizing time.[7] That was one of their recognized achievements and splendors.

However, both happened to live (Ortelius was born in 1527, at the time that Charles V and Suleiman the Magnificent were establishing themselves in world history) at crucial moments of European imperial expansion: Ortelius during the first wave, the Spanish sixteenth century; and Kant during the second wave, when the British Empire and France were setting their sights on Asia and ready to go to Africa. Both Ortelius and Kant gathered their information thanks to wide-ranging European navigations around the world. Both dismissed the rest of the world to the benefit of their own "universal surface": the map and alphabetically written descriptions allowed them "to eat" the world—metaphorically and epistemically—and to posses it. The accumulation of meaning was an avid enterprise, parallel to the anxiety to accumulate money and wealth. In this era of "irrational epistemic exuberance" both Orthelius and Kant were blinded by the excesses of information, by the enthusiasm of the *theatrum mundi* filling the eyes, the senses, and the imagination of Europeans for whom information from around the world was available and enticing. It was a messy and splendid world in front of them, which they had the pleasure of arranging, organizing, and making it "understood." As such, they missed that while they were "seeing and conceiving the world,"—the enunciated—they were doing so within the "limits of their own subjectivities and places,"—the enunciation.

Thus, the modern-imperial epistemic subject was in its infancy with Orthelius and in its full maturity in Kant.[8] Colombian philosopher Santiago Castro-Gómez saw the links between cartography and theology-philosophy and how we find in both and epistemology that assumes and celebrates the "hubris of the zero point."[9] The "hubris of the zero point" is not only what signals European "modernity," but also what legitimizes Europe imperial/coloniality.

What Questions Could be Asked About the Knowing Subject?

Stuart Elden and Eduardo Mendieta reminded us also, in the proposal for this book that, in his *Logic*, Kant suggests that there are four fundamental questions.

1. "What can I know?"
2. "What ought I to do?"
3. "What may I hope?"
4. "What is the human being?"

Kant suggests that "*Metaphysics* answers the first question, *morals* the second, *religion* the third, and *anthropology* the fourth. Fundamentally, however, we could reckon all of this to anthropology, because the first three questions refer to the last one."[10]

It is a very helpful quote to juxtapose next to a version of what I would call the "de-colonial epistemic platform." From a de-colonial viewpoint (and sensing, shall I say, not to put all the weight on the eyes) Kant's are not the fundamental questions. The fundamental de-colonial perspective would look more like this:

1. Who is the knowing subject and what is his or her material apparatus of enunciation?
2. What kind of knowledge/understanding is he or she engaged in generating and why?
3. Who is benefiting or taking advantage of such and such knowledge or understanding?
4. What institutions (universities, media, foundations, corporations) are supporting and encouraging such and such knowledge and understanding?

These questions cannot be properly answered from the established disciplines, which are themselves part of the problem and they have been under trial. At the same time, the formulation of the questions, as well as the answers we can provide, cannot avoid what is common to all Western disciplines: a certain way of gathering information, of reasoning, of interpreting. Simultaneously, the questions and the answer have to be epistemically disobedient,[11] that is, teasing out and betraying certain principles of "epistemically correct" reasoning and interpretation. Otherwise, de-colonial thinking cannot take place.

A de-colonial platform is transdisciplinary and originates at the moment that standard conceptions and practices of knowledge (e.g., all disciplinary formations) have been both, "advancing" modernity and "contributing" to coloniality of knowledge and of being. Coloniality of knowledge does not mean that knowledge was colonialized, but that hegemonic ways of knowing and disciplinary world-making, since the European Renaissance, were instruments of colonization and, as a consequence, of colonialization of non-European knowledge. By the same token, the modern subject, and therefore the modern-knowing subject, became the model of all knowing subjects, whether European or not. Coloniality of being refers to the enforcement of a being in the world, a subjectivity, to which W. E. B. DuBois responded with his "double consciousness" and Frantz Fanon with the schism between "black skin/white masks."

Addressing the type of questions formulated above—and starting and following up on DuBois's and Fanon's epistemic and hermeneutic foundations—allows the reader to de-link or disengage from disciplinary formations while, of course, addressing them in their imperial complicity. In this precise sense one can say that de-colonial thinking is transdisciplinary; not interdisciplinary.

Consequently, instead of assuming a universal "human nature" as the starting point, de-colonial thinkers start by assuming that, since the European Renaissance, and particularly during and after the Enlightenment, human kind were divided between *humanitas* (who control knowledge) and *anthropos* (pagans, infidels, Indians, Orientals, Africans, women, etc.), who were not quite to the level of *humanitas* and, therefore, their knowledge was not "sustainable." For the Christian, the civilizing mission to development in the twentieth century, *humanitas* controlled knowledge maintaining, at the same time, the idea of a "human nature." We are born equal, as the dictum goes.[12]

But it so happens that shortly after we are born, we may be in trouble and discover that we are not equal. Or, we may not be born socially and economically privileged and do not realize that we are not all equal; or we may choose to ignore that we are not equal. The inequalities start from a patriarchal and racial matrix in which knowledge is universal and beings are measured on an ideal of humanity that is basically constructed on the model of a white, European, and Christian male. This may be rude, but a glance at the canon of Western knowledge since the Renaissance will show that 98 percent of canonical figures, Christian and secular, fall under those categories. The categories of knowledge are reinforced by institutions and institutions are supported by capital.

Changing the questions means changing the terms of the conversation. Therefore, instead of engaging Kant with his own rules and to question the content, I propose to change the rules of the game, de-link from his presuppositions, and change the terms of the conversation.

The key is the first question, from which the other three logically follow. French linguist and philosopher of language Emile Benveniste, described several decades ago the "formal apparatus of enunciation."[13] For Benveniste, the formal apparatus of enunciation is structured by the pronominal system of any language and by the spatial/temporal deictic found in any language.

The speaker, by assuming the first person pronoun (generally singular, but it could be plural), mobilizes the pronominal system and locates the listener (plural or singular) in the second person and the referent (person, event, things) in the third person, plural or singular, as the absent referent. Now, not every language has a pronominal system governed by the same logic. In Tojolabal, for example, the third person does not exist as such.[14]

Temporally, the speaker takes for granted that he is speaking "now" in the present, although he or she can speak about the past, the present itself, or the future. Spatially, the speaker locates itself "here," in the very place where he or she is enunciating. The possibility of registering a conversation and listening to it someplace else in the future, doesn't change the basic formal apparatus of enunciation.

Now, thinking de-colonially, the formal apparatus is hardly sufficient if we are to change the terms of the conversation and ask the questions I listed above. It is not enough to say that "knowledge is situated" or that "experience is the source of knowledge" if the options and possibilities in which knowledge is situated and experiences experienced are not spelled out. Otherwise, it has to be accepted that there is a universal organization of the world: a universal ontological dimension of "human beings" in which knowledge is situated and experiences are experienced, it doesn't matter what. If we do not accept that there no universal common ground of experiences and situated knowledge has to be spelled out in the colonial matrix, or that knowledge is situated in the colonial matrix, we shall then spell out in what sense, de-colonially speaking, knowledge is situated and experiences experienced. To put it blatantly, Kant's and Fanon's experiences and situatedness are not the same, and there is neither an epistemic privilege from where to claim an universal truth or the relevance of the problems to be addressed.

The first set of assumptions (which I believe are the assumptions upon which Kant was able to state that "there are four fundamental questions," presupposes that the rules of the game and the terms of the conversation are those of "objectivity without parenthesis" (in terms of Chilean neurophysiologist and philosopher, Humberto Maturana).[15] The second set of assumptions presupposes that the rules of the game and the terms of the conversation are those of "objectivity within parentheses."

Kant operates under the assumptions that knowledge is objective without parenthesis and that the knower (or observer) can map objectively (that is, there is a correspondence between the description, in words or in cartography, and the world). The knower occupies a place—the place of knowing. And the place of knowing is beyond geo-political histories and beyond body-political subjectivities: the hubris of the zero point. The knower operates in the domain of the mind: beyond the body and beyond history. De-colonial thinking starts from the assumption that imperial epistemology racialized bodies and places: bodies out of rationality and places out of history.

Take, for example, the section of *Geography* in which Kant maps the four continents. He begins by setting up six fundamental principles valid for all the continents:

> On dry land, however, we find:
>
> 1. Lands whose extent and interior we know.
> 2. Lands we only partially know.
> 3. Lands of which we know only the coasts.
> 4. Lands that we have actually seen, but not relocated.
> 5. Such of which the Ancients were aware, but are now as lost.
> 6. Finally, lands whose existence we only suppose.[16]

In the following paragraph Kant observes:

> Europe belongs to the first named. To the lands of the second sort, however, belongs Asia.[17]

Kant concedes that a better knowledge of Africa is necessary:

> The cause, for which the interior of Africa is as unknown to us as the lands of the moon, lies more with us Europeans than with the Africans, in that we have let ourselves be made so shy, through the Negro trade.[18]

And when he gets to America he points toward the North "whose northern part, situated towards to Russia, is still so good as undiscovered, and in whose southern half likewise, particularly on the Brazilian coasts, there still exist many unknown regions."[19]

And he continues by acknowledging that "we don not know any better" and that "the northern part . . . is still so good as undiscovered." Kant erases in less than one line the knowledge that people dwelling "in the interior of

Africa" and in "the northern part" have of those places. Europeans are those who do not know the place and have not explored yet. People living there have situated knowledge and knowledge grounded in their experiences, like Kant. Operating under the hubris of the zero point blinds you to the fact that other people (their existence and knowledges) do not have the same problems you have and, therefore, could not care less about your knowledge until the moment that you impose it upon them and tell them they are inferior and ignorant, their reasoning is defective, and their sense of the beautiful does not exist. You do not stop to think that they are as ignorant about you, your interests and values, as you are about them. Except you presume to "know" them because you describe them and put them in your *system* and in your *architectonic*.

The point here is to underscore the differential of power between the "institutional" sphere in which Kant was operating and the sphere of the "institute" in which Africans and people living in "the northern part" had of themselves and of the world.[20] The power differential between the institutional and the institute, which is not of course a privilege of eighteenth-century enlightened thinkers, was put in place in the sixteenth century. In Europe, Latin language and knowledge (theology) set itself up against and above Hebrew and Arabic. From the European Renaissance to the Enlightenment, from theology and theo-politique of knowledge to egology and ego-politics of knowledge, the control and "institutionalization" of meaning makes invisible the epistemic production of other parts of the world, either by producing the effect that the world, except Europe, stopped in 1500 or by watering down knowledge-making in African inlands and in the North extreme of the Americas.

Operating on the hubris of the zero point resulted in conjoining epistemic and ontological differences.[21] Now, as far as an ontological difference in the modern/colonial world is the philosophical foundation of racism, then racism is not only to be looked at ontologically but also epistemically.[22] Mentioning "epistemic racism" is upsetting for many, and not just for right-wing thinkers. But if we do not mention epistemic racism, we run the risk of maintaining the terms of the conversation and trying to be "radical thinkers" by operating only at the level of the enunciated.

Kant's *Geography* is not only one side of his *Anthropology*; they are interconnected.[23] Kant in the *Geography* refers to and describes places, and in the *Anthropology* he talks about and describes people. But what is more important to my argument are the connections between both places and people and his philosophical reflections on the beautiful, the sublime, and pure reason. Kant was a human being, brilliant and committed, but as every brilliant human being, he was just human, not god. It would be a sin to dismiss Kant's contribution to the history of European thought; and it would be a crime to hide or undermine his epistemic racism.

Kant's Conceptual Matrix

Beyond what Kant enunciated in different fields of inquiry (philosophy, ethics, religion, anthropology, geography, aesthetics, and education—e.g., *The Conflict of the Faculties*), there is a recurrent "matrix" of the enunciation that I will attempt to unveil by looking at his *Geography* in tandem with his *Anthropology* (as many have already remarked) and his *Philosophy* (e.g., *Critique of Pure Reason)*, which has been less analyzed. What I am suggesting through these three fields is applicable to the Kantian corpus.

Four Partite Division of the Earth

Part III of Kant's *Geography* is organized—not surprisingly—into four sections and in a very revealing order: Asia, Africa, Europe, and America. Asia is the place of Ancient civilizations where, a few decades later, Hegel will locate the materialization—to be oxymoronic—of the Spirit. But from where did Kant get the four partite divisions of the earth masses? Not from the land masses themselves, unless the Spirit underneath was whispering the land masses' names to Kant or putting a badge on them. To be sure, the four partite divisions may have been known and adapted, in Chinese viewpoint—with the Pacific Ocean in the center and not the Atlantic. Jesuit Father Mathew Ricci introduced Ortelius to the Ming Dynasty in 1582, and since then there is a history of the adaptation and an inversion of what continent appears on the East and the West side of the Map. But it is unlikely that the image of the world with which Kant was operating was meaningful in the Arabo-Islamic world, which was divided into seven regions, as reported by Ibn Khaldun and other sources. Perhaps, at the time Kant was writing, an Ortelius-type of map was being introduced by the British in South Asia, among the elites of the Mughal Sultanate. The cosmology in the Incanate in Cuzco and the Tlatoanate in the valley was based on a four-partite division of the world, but certainly not in such a continental divide.

I have told the story both in chapter 5 of *The Darker Side of the Renaissance* (1995) and in *The Idea of Latin America* (2005) that the four-partite division of the earth prompted by the European invention and appropriation of "America" and the erasure of what for Incas was "Tawantinsuyu," for the Aztecs "Anahuac," and for Kuna (Indians) dwelling in the area of today Panama, was "Abya Yala." "America" was included in the metaphorical Christian tri-partite division of the earth on Asia, Africa, and Europe (in the well-known T-in-O maps of Western Christendom, before the Renaissance).[24]

Now, the distribution of the earth in four parts is not merely descriptive. In the T-in-O map, a hierarchy was clearly established. Since Christendom was located in Europe, and Christians were the "enunciators," Europe was attributed to Japheth, Asia to Shem, and Africa to Ham. Whomever knows a little about

Noah's three sons will immediately remember that Japheth was the hope for the future and the preferred son; Shem was not bad, after all; and Ham, well, he was willfulness. When America came into the picture, it was too late for Noah to have a fourth son. In a way, America was first conceived as "Indias Occidentales," that is, Japheth extending to the West, as it was predicted in biblical narratives. On the other hand, this part of Occident was "Indian." When the name "America" began to be accepted, "Indians" were already one of the trademarks of the fourth continent.

Four Types of Human Beings

But that is not all. As is well known, Kant linked continents with people's skin color. The racist issue in Kant's work that surfaced in the past decade was brought about, to my knowledge, by Emmanuelle Chukwidi Eze in his landmark article "The Color of Reason: The Idea of Race in Kant's Anthropology" (1997). And it is here where his *Geography* and his *Anthropology* come together through Carl Linnaeus's classification of "four types of human beings." Linnaeus's four types are *Homo sapiens Asiaticus, Homo sapiens Africanus, Homo sapines Europeaus,* and *Homo sapiens Americanus.*[25]

It is possible, then, to ask the same question about Linnaeus that I have asked about Kant: where did he get the idea that there were four types of homo sapiens (with variations, of course, and edges), and the idea that each type corresponded to one of the four continents?

I am aware that Kant's *Geography* has Parts I and II before this last one (Part III), where he deals with the four continents. But allow me to push Part III a little further before attending to Parts I and II.

The final section (Part IV) of Kant's *Observations on the Feeling of the Beautiful and Sublime* (1764) is devoted to "national characteristics" and the distinct "feelings" that national characters have of the beautiful and the sublime. He returns to the topic of "national characters" in his *Anthropology*, but it is less developed. In *Observations* it is quite extended and it covers the globe. However, the basic structure is the same in both works. In *Observation* Kant opens Part IV with this striking observation (which is geographic, anthropological, and philosophical):

> Of the peoples *of our part* of the world, in my opinion those who distinguish themselves among all other by the feeling for the *beautiful* are the Italians and the French, but by the feeling of the *sublime,* the Germans, English and Spanish. Holland can be considered as that land where the *finer taste* becomes largely *unnoticeable.*[26]

Let's parse this paragraph, thinking of the assumed (not analytical) conceptual structure of the *Geography.* "Our part of the world" doesn't need comment: it is the part devoted to Europe in the *Geography.* Secondly, in this part of the world there are six countries named first: Italian (the Renaissance), France, England, and Germany (forget about the difference that the first excel on the beautiful and the second on the sublime—but think of the European Union now)—these are the heart of Europe in Hegel's formulation a few decades after Kant, that is, the imperial power since the Enlightenment. Spain is the last of the imperial countries mentioned. Holland barely made it, and Portugal was forgotten. Interestingly enough, Portugal and Holland—contrary to the other four countries—where marginal in the imperial epistemic "race," was concerned more with commerce than with conversion and civilization. Consequently, the Dutch and Portuguese are at the margins of the beautiful and the sublime.

Basically, we have here the part of *Anthropology* dealing with national characteristics of "our part of the world," which will serve as the "standard model" to judge the rest of "national grouping" in other continents. It is not just the beautiful and the sublime that are being tested around the world, but also religion connected with the beautiful and the sublime: "The religion of our part of the world is not a matter of an arbitrary taste, but is of more estimable origin."[27] "Religion" will allow Kant to ridicule Asian religions (China, Japan, and India): Japanese, said Kant, could be regarded as the Englishmen of "this part of the world" (here referring to Asia). "Being regarded as the Englishmen" is flattering but at the same time stamping the racial difference coded in national characters: "being regarded as" is not like "being the model." They are like Englishmen in some aspects, said Kant, but "for the rest, they display few signs of finer feeling."[28] Then come the Chinese and Indians of which he said, "their religions consist in grotesqueries." The "despotic sacrifice of wives in a very small funeral pyre that consumes the corpse of the husband is a hideous excess."[29] And for the Chinese Kant expressed his surprise at the "verbose and studied compliments of the Chinese" and marvels at the grotesque of their painting as well as the strange and unnatural figures "such as are encountered nowhere in the world."[30]

Kant was writing shortly after the Battle of Plassey (1757), a date that in most histories inaugurates the British takeover of Mughal Sultanate. At that point India was not yet on the list of colonies of European empires and Kant was looking at its millenarian civilization. Consequently, I would say that Kant here is redrawing the "imperial difference" that was inaugurated in the sixteenth century when Spanish intellectuals (men of letters) established it in relation to the Ottoman Sultanate. The "colonial difference" Kant will just remap: Indians of the Spanish colonies will "vanish" from Kant's view and instead he will highlight the "savages of North America" as those who display the most sublime mental character, among all the savages. Native North Americans (including Canada certainly) won the consolation prize. And for the Negros, Kant is famous for his observation about them, quoting Hume who "challenges to cite a single example in which a Negro has shown talents. . . ."[31]

Let me give you one more interesting example. The section on Europe, in *Geography*, begins by "European Turkey,"—but there is no text under it. Then is the next title: Bulgaria. If we now go to *Anthropology*, this is what Kant has to say about the Turks:

> Since *Russia* has not yet developed definite characteristics from its natural potential; since *Poland* has no longer any characteristics; and since the nationals of *European Turkey* never have had a character, nor will even attain what is necessary for a definite national character, the description of these nations' characters may properly be passed over here.[32]

At the time Kant was writing and lecturing, Catherine the Great was becoming the Emperatrice (not longer the Tsarina) (1762–1796). She followed the trend of Peter the Great in extending Russian domain into the Black Sea of central Europe; and promoted modernization and Westernization. But it was too early for Kant to predict the future. The Poland-Lithuanian Commonwealth, which flourished in the sixteenth century, was falling apart in Kant's time. And the Ottoman Sultanate had already lost his proper name, and had become European Turkey. In this move, Kant's reaffirms the imperial reference to Russia, which was still in its formation in the sixteenth century (Ivan the Terrible became Tsar about 1558, in the same decade than Philip II of Spain and Elizabeth of England inaugurated their ascension to the throne). By the same token, Kant contributes to the translation of the imperial difference that was established in the sixteenth century in relation to the Ottoman Sultanate into the colonial difference that generated Orientalism. The Ottoman Sultanate, like the Incanate and the Aztec Tlatoanate in the sixteenth century, was dismantled (the Ottoman Sultanate would fall apart in 1922, but by the end of the eighteenth century was already on its way down), and dethroned, so to speak, and the inhabitants devalued as Indians and Turks.

Briefly, the fourth part of Kant's *Geography*, organized around the four-partite continental divide, is where we can see the racialization of people and regions: not only Chinese, Japanese and Indians, but also Asians; not only the South of Europe (and what he has to say about Spaniards) but also the North Eastern part of Europe (what he and then Hegel have to say about Poland and Russia); not only Africa, but Africans, that is, Negros; and not only America, but also the "savages" and their red skin. Since during Kant's time neither the steamboat nor the railroad were in place (both were technological by-products of the Industrial Revolution, which were in full force from 1850 on), the idea of the correlation between people and territory was stronger than today, even as Kant himself acknowledged migrations.

Let us now attend the first two parts of Kant's *Geography.*

The Accomplished Among the Races

The second part, "On Human Beings" (*Vom Menschen*) is the part that most clearly connects his *Geography* to his *Anthropology.* The difference however is that *Anthropology* is devoted to cognitive faculties, desires, national characteristics and races (genus), and species, in *Geography* the section devoted to Man is centered on skin color. A telling paragraph is the following:

> In hot regions, people mature earlier in every sense, but do not reach the perfection of the temperate zones. Humanity is in its greatest perfection in the race of the whites. Yellow Indians have somewhat less talent. Negros are far lower, and at the bottom lies a portion of the American peoples.[33]

In 1590, José de Acosta's *Historia Natural y Moral de las Indias*[34] was published in Europe and translated in several languages (Latin, French, German) and was widely read. In the mid-nineteenth century Alexander von Humboldt highly prized the book. Apparently Kant did not know it. Otherwise he would have not been so dismissive and ignorant on the *Geography's* section devoted to America. The "white race" that attained such perfection also had its internal hierarchies. The "Black Legend"[35] was, by Kant's time, a fait accompli. There is no other explanation for Kant's dismissive description of the Spaniards not only in his *Anthropology* but also in the *Geography*. Racism, and not just in Kant, is both ontologic (obvious in the previously quoted paragraph) and epistemic. Epistemic racism consists in devaluing ways of thinking as well as just ignoring them, not taking them into account.

In retrospect, looking at both Kant's and Acosta's enunciation, and delinking from both, we can see that Kant's *Geography* and *Anthropology* would not have been possible without the work done by Spanish men of letters in the sixteenth century. Acosta is a paramount example. Acosta's book, particularly in the first two chapters originally written in Latin, set the record straight and corrected the views that Aristotle, Saint Thomas, and the Bible had about the antipodes. He also set the record straight vis-à-vis the natural history of Pliny the Elder and contributed to the debates about the humanities of the Indians in Europe and among Europeans, since Indians did not participate in that debate about their own humanity. When Kant, and before him Carl Linnaeus, were in a position to receive information from around the world and create a system of classification and a systemic apparatus (as Kant does in the introduction to *Geography*), the foundation had been already established. In other words, although Kant ignores it, it is not possible to go from Aristotle and Pliny to the eighteenth century without tumbling into the massive corpus of Spanish intellectuals who (a) changed the directions in which natural history was being told and conceived, and (b) placed the debate on human nature on a different terrain. It is only through Kant and other enlightened philosophers who, through the Black Legend, created the division between the South of Europe and its Heart (England, France, and Germany) that the peninsular sixteenth century was obliterated from their own history in a manner similar to the way Spanish intellectuals totally obliterated Arabo-Muslim's contribution to their own field of knowledge. The first established the external epistemic difference; the second, the internal epistemic difference.

Linnaeus and Kant were living and acting during the historical period in which Western thinking was moving from "barbarians in space" to "primitives in time." Joseph Francois Lafiteau, *Moeurs de sauvages Americaines* (1723–1724), is one of the references in the conversion from barbarians into primitives and for starting the primacy of time over space in closing the borders of Western history and civilization. Natural history entered into this frame as natural history ended up in the present, with the earth divided into four continents. In the eighteenth century, the entire debate of the New Worldwas centered around the idea that even nature was still young in the New World. So, if we take Man, for instance, there would be a linear history of Homo sapiens, from primates to sapiens, that accounts for Man in the present and Men in the four continents.[36] And when we get to the present, the linear history is translated into the spatial distributions of Homo sapiens around the globe. In third place, Homo sapiens were organized in a hierarchy that, as Kant said, places the "White Race" as the one that reached the most perfection. But, it so happened that the perfection of the "White Race" was such that it was the only race that was allowed to classify all the races.

The epistemic trick to which Kant so much contributed is the following:

a. In the sphere of the enunciated, it was stated that there are four continents and four races;

b. In the enunciation, it was only within the white race that knowledge-making took place and that the four continents and four races could be stated as reality.

Consequently, while "the white race" was one among others, it was the most accomplished. People who belonged to the white race and lived in Europe (the temperate zone) had the privilege of generating and institutionalizing such knowledge. The privilege of the "white race" consisted in being at the same time one unit of the classification and the only instance in which the classification was made: the hubris of the zero point was, and still is, an imperial epistemology. It is not dialogic in the specific sense that does not allow room for any other type of knowledge to enter into dialogue. And this is one of the key issues of the twenty-first century, as we are already witnessing among Arabo-Islamic scholars, intellectuals, politicians, and religious leaders, as well as in the wide spread of "Indians" from the Americas to Australia and New Zealand.[37]

Shifting the Geography of Reason

Let us close the argument by going back to the basic distinction between Kant's four fundamental questions and the four fundamental questions asked within de-colonial projects, by de-colonial thinkers. Let us look at Kant, in other words, from the sidewalk across the street from where he is walking.

Kant is well known for his use and meaning of the term *architectonic*. The problem with the *architectonic* is not the term in itself. The problem is that it positions itself as *the architectonic*. The concept did not exist among theologians in the sixteenth century. However, the assumption that a Jesuit, and the Jesuit order, can provide to Europeans the knowledge of what the Indies and the Indians were, naturally and morally, and totally ignore, dismiss, or not even think that Incas and Aztecs and Mayas had built through centuries a sophisticated system of knowledge, did not cross theologians' minds.[38]

Kant's secular philosophy is unthinkable without the apparatus of Western Christian theology.[39] Kant's critique of theology is a critique of the enunciated, while he remained within the logic already established by Christian theology. The point I am trying to underscore was clearly stated by Carl Schmitt, not in relation to Kant, but uncovering the common assumptions and foundation of Christian theology and secular philosophy. "It should have been noticed that any elaboration of political theology [Schmitt stated] are not grounded in a diffuse metaphysics. They bring to light the classical case of a transposition of distinct concepts which has occurred within the systematic thought of the two—historically and discursively—most developed constellations of "western rationalism:" the Catholic *church* with its entire juridical rationality and *the state of the ius publicum Europaeum*, which was supposed to be Christian in even Thomas Hobbes's system." [40] It is my claim that Kant's remains within the horizon of theological rationality, which Schmitt has here circumscribed, even as he critiqued its content. In other words, Kant doesn't shift the geography of reasoning. He only secularizes the content of Western rationality.

It is the same order of knowledge, the same imperial epistemic assumptions. The only difference is that secular knowledge affirmed itself and its right by dismissing theology in the same way Renaissance men of letters created and dismissed the Middle Ages. But all of this is like a family that feuds within the same set of assumptions, structure of enunciation, and knowledge-making. I am proposing, and hopefully enacting a way to de-link from the system set by the four Kantian questions and reading Kant from the four de-colonial questions. Here, then, is my last move.

Kant begins the introduction by reminding us that the fountain and origin of all knowledge is pure reason and experience. The introduction to *Geography* is congruent with the four questions asked in his *Logic*.[41] Thus, the introduction lays out the *architectonic of systemic knowledge*: the system goes from the whole to the parts. *Empirical knowledge* is an procedural aggregate that from the parts mounts to the whole. Empirical knowledge comes to us through our senses. Sensing would be then an activity parallel to reasoning.

For de-colonial thinkers, the senses are also foundational. Agreeing with Kant and with the Aztec's *tlamatini* and Inca's *amauta* ("wise man" in Náhuatl and Quechua-Aymara, respectively)—for whom thinking was located in the heart rather than in the brain—[42] knowledge begins with the senses. Now, the "senses" are not equally and universally affected in the same way. The modern/colonial world grounded on a type of economy described by liberals and Marxists as "capitalist" went hand in hand with a type of knowledge mounted on theology and then secularized to philosophy and science. It is not just the modern State that is unthinkable without Political Theology behind it, as Carl Schmitt argued. It is also the entire epistemic secular apparatus of Reason that is mounted on Epistemic and Hermeneutic Theology, to parallel Schmitt. Economic and epistemic/hermeneutical practices complemented and still complement each other.

So, then, if knowledge starts with the senses and our (human beings) mark our experience, the senses and experiences of Quobna Ottobah Cugoano, an enslaved African shipped to the Caribbean and then taken by his Master to London, where he was liberated, cannot be the same as that of Kant. And they both were coetaneous. Cugoano was born circa 1757. He published his classical book, *Thoughts and Sentiments on the Evil of Slavery*, in 1786. The date of his death is unknown, but suppose that it was some time between 1807 and 1817. Kant was born in 1724 and died in 1804. Although two generations apart, Kant should have been aware of and read *Thoughts and Sentiments* . . . published in London when Kant was in his prime. However, this is not the point; the point is what Kant names "sense and experience," which is the most basic body-reaction to the world around us (seasons of the years as well as people and institutions) and is connected to the geo-historical configuration of *that world around us*. Cugoano's body was colored Black. Kant's was White. Kant was born and raised in the Kingdom of Prussia at a time when Europe was benefiting from the Caribbean colonies as well as from the existing commercial routes with regions situated to the East of Europe. Ottobah Cugoano was born in Africa, captured and detached from his community, and worked as a slave in the Caribbean. The world must look different from each of these perspectives. So, while Hume and Kant thought that it would be impossible to "cite a single example in which a Negro has shown talents," we (now, here) can say that they are entitled to their own opinion but there is no obligation to take their opinion *ad pedem litterae.* It is time to de-link and to look at Kant from Ottobah Cugoano's point of view instead of maintaining Hume's and Kant's. In other words, it is not sufficient to say, "Oh, well, yes Kant was racist" and remain within Kant's epistemic dwelling. It is necessary to shift the geography of reason and start reasoning from the senses and experiences of people like Ottobah Cugoano. If we can do that, we can begin to think de-colonially.

Why de-colonially? Because one of the basic hypotheses of de-colonial thinking is that knowledge in the modern world was and is a fundamental aspect of coloniality. In other words, knowledge is not just something that accounts for (describes, narrates, explains, interprets) and allows the knower to sit outside the observed domain and, from above, be able to observe imperial domination and colonial societies ignoring or disguising the fact that *knowledge itself is an integral part of imperial processes of appropriation.* Colonialism has been basically analyzed in terms of economic and political control of territory and population, as if the knowledge being generated was outside Western imperial/colonial history (by which I mean, since 1500). Coloniality (I and others have argued in the past decade) is constitutive of modernity. There is no modernity without coloniality: hence modernity/coloniality. Modernity is constituted by rhetoric: the rhetoric of salvation by conversion, civilization, progress, and development. In order to implement what the rhetoric of modernity preaches, it is necessary to drive the juggernaut over every single difference, resistance, or opposition to modernity salvation projects. Knowledge is of the essence. Thus, coloniality of knowledge means not that modern knowledge is colonized but that modern knowledge is epistemically imperial and, as we have seen in Kant, we must devalue and dismiss epistemic differences. And epistemic differences go hand in hand with ontologically ones: "show me a Negro that has shown talent" or "what trifling grotesqueries do the verbose and studied compliments of Chinese contain."

Thus, thinking de-colonially and reading Kant de-colonially means to operate on the basis of geo- and body-politics of knowledge[43] and to shift the geography of reason.[44] And that is, I hope, what I have done here. I disengaged from Kant's categories of thought and assumptions. Disengaging or de-linking does not mean that it is possible to "get out" of modern epistemology (or Kant himself if you wish), as one walks out from the summer resort and goes home, just like that. De-linking means not to operate under the same assumptions although acknowledging that modern categories of thought are dominant if not hegemonic, and in many, if not in all, of us. This is one of the problems that Muslim thinkers have been dealing with for a long time, and more so recently.

De-linking means to think from the silences and absences produced by imperial modern epistemology and epistemic practices—like Kant's. It means to read Kant from the silences he himself has produced.

The foundational acts of de-colonial thinking are (a) de-linking and (b) border gnosis (epistemology or hermeneutics). Imperial epistemology rolls over epistemic differences. "Tradition," "folklore," and "myth" are some of the terms invented to dismiss differential knowledge. *Reversing* the terms of the conversation will not work, mainly because it will remain within the same rules of the game and play under inferior conditions. *Changing the terms* of the conversation is, instead, the de-colonial project through border gnosis and border/de-colonial thinking. De-colonial thinking consists then of deploying their co-evalness forms of knowledges and ways of being pushed aside or buried in the past, in order to make room for the triumphal march of modernity. *Pushing aside* and *burying in the past* are two operations of the logic of coloniality; they are invisible, the darker side of modernity—the darker side of Kant's *Geography*, for that matter.

Second, de-colonial thinking runs away from the aim of building a system and putting together an architectonic. Indeed, de-colonial thinking privileges hermeneutic over epistemology. The expression "border gnosis and border epistemology" already indicates it. For, even if the word "epistemology" is maintained, epistemology proper, in the Western tradition, cannot be in the border between diverse languages and systems of thought (cf. Sanskrit, Latin, and Arabic). Of course, the borders between French and German will not really count as border epistemology since both are differentials within sameness. De-colonial thinking presupposes pluri-topic, not mono-topic, hermeneutics.[45]

Kant distinguished between external and internal dimensions of knowledge. By the external we know the world, and by the internal we know man. Thus a knowing subject is being formed in Kant's epistemology, "through the experience we have of *nature* and of *man*." What law of enunciation legitimizes this distinction? An assumption of the universal nature of man that is no other than the invention of Man and Humanity that runs from Renaissance Humanism to Enlightenment Rationalism. In that lineal drive, within a limited geo-historical part of the world—North of the Mediterranean—an idea of Humanity and of Knowledge was created as if it was good for the rest of the world. Kant's geography is on the one hand founded in the system of knowledge (as shown in the introduction) and deployed in the structure of *Geography* in which the anguish to "totalize," to get and control knowledge is, from a de-colonial point of view, frankly surprising. However, around the world he totalized, there are several non-European languages, concepts, epistemologies, and hermeneutics (to use the Western world in the same way the Zapatistas use "democracy"—that is, de-colonially already), that are not in the past or no-longer sustainable, although there is no way to erase Kant's *Geography* in the same way Kant erased all other territorial conceptualizations.

Thus, if Kant's *Geography* aimed at the universal, de-colonial thinking aims at the global. Modernity, in its internal diversity—Martin Luther is not Karl Marx but neither are Ibn Rush or Mahatma Gandhi or Ottobah Cugoano—spread globally a system of thought; and from its foundation in Western languages (Greek, Latin, Italian, Spanish, Portuguese, French, English, and German), disqualifying any other language, borders emerge in its pluri-versality. Thus, the way to the future is the way toward pluri-versality as a universal project.[46] Pluri-versality as a universal project is, in Enrique Dussel's words, the route to a transmodern, not a postmodern, world.[47]

Once we are in the borders where imperial Western epistemology meets with its global differences, we are also facing the needs of what Raymundo Pannikar described as "imperative method."[48] Once we start thinking de-colonially, it is not possible to stand in the borders *comparing* both sides. Because if you do that, you are placing yourself in *the hubris of the zero point*. "Comparative methology," put in place in the nineteenth century, was precisely that: a method to insure that the observer remains uncontaminated and that Western epistemology remains on top, controling all other forms of knowledge. *Imperative* method, Pannikar stated is "the effort at learning from the other and the attitude of allowing our own convictions to be fecundated by insight of the other." In contradistinction to the *comparative* method that privileges dialectics and argumentative reasoning (system and architectonic), the imperative method (for Pannikar, diatopical hermeneutics) focuses on dialogue, praxis, and existential encounters. De-colonial thinking then is one type of imperative practice that aims to de-link from the colonial matrix of power and to engage in border de-colonial thinking. In reading Kant de-colonially, I have aimed at entering in a dialogue with Kant's (and modern Western epistemology) de-linking from the rules of the game by being epistemically disobedient.[49] I have tried to show that a different kind of game is possible and, in so arguing, entering into it with an independent set of rules.

Notes

1. "The Conflict of the Faculties," in *Religion and Rational Theology*, ed. and trans. Allen W. Wood and George Di Giovanni (Cambridge: Cambridge University Press, 1996), 256.

2. Karen Armstrong, *Islam: A Short History* (New York: Modern Library Chronicles, 2002), 142.

3. Armstrong, *Islam*, 142.

4. Armstrong, *Islam*, 142 (my italics).

5. Walter Mignolo, "Coloniality at Large: Time and the Colonial Difference," in *Time in the Makings*. (Rio de Janeiro: Universidad Cándido Mendes, 2000), 106–49;

"Introduction," in *Decolonizing the Middle Ages*, eds. John Dagenais and Margaret Greer, special issue, *Journal of Medieval and Early Modern Studies* 30:3 (2000): 431–48.

6. *The Darker Side of the Renaissance* (Ann Arbor: University of Michigan Press, 1997), ch. 5.

7. Walter Mignolo, "Introduction," *Decolonizing the Middle Ages*, eds. John Dagenais and Margaret Greer, special issue, *Journal of Medieval and Early Modern Studies* 30:3 (2000): 431–48.

8. Walter Mignolo, *De-linking: Globalization, Don Quixote and the Colonies* (Institute for Global Citizenship: Macalaster College Publications, 2005), 19–56.

9. *De-linking*, 19–56. Santiago Castro-Gómez makes this argument in the introduction of his book *La hibris del punto cero. Ciencia, raza e ilustración en la Nueva Granada (1750–1816)* (Bogotá: Pontificia Universidad Javeriana, 2005). A short version in English can be found in his article "The Missing Chapter of Empire. Postmodern Reorganization of Coloniality and Post-Fordist Capitalism," *Cultural Studies* 21:2–3 (2007): 428–48.

10. Immanuel Kant, *Lectures on Logic* (Cambridge: Cambridge University Press, 1992), 538.

11. For this concept see my "Epistemic disobedience, independent thought, decolonial freedom." In *Theory, Culture and Society*, 26(7), 2009, 1–23.

12. I find it telling that a Japanese scholar will be sensitive to a time-period where the racial logica was articulated in what he describe as *humanitas and anthropos*: those dwelling in the *humanitias* command the enunciation while *anthropos* are always enunciated. Nishitani Osamu, "Anthropos and Humanitas: Two Western Concepts of 'Human Being,' " trans. Trent Maxey, in *Translation, Biopolitics, Colonial Difference*, ed. Naoki Sakai and Jon Solomon (Hong Kong: Hong Kong University Press, 2006), 259–74.

13. Emile Benveniste, "L'appareil formel de l'enunciation," in *Problémes de Linguistique Générale II* (Paris: Gallimard, 1974), 9–88.

14. Carlos Lenkersforf, *Los Hombres Verdaderos. Voces y testimonios tojolabales* (México: Siglo Veintiuno, 1996), 27-150.

15. Humberto Maturana, "Reality: The Search for Objectivity or the Quest for a Compelling Argument," *The Irish Journal of Psychology* 9:1 (1988): 25–82.

16. *Physische Geographie*, cited according to the Akademie edition of *Kants gesammelte Schriften* (hereafter AK, with volume and page number), AK 9: 228.

17. *Physische Geographie*, AK 9: 228.

18. *Physische Geographie*, AK 9: 229.

19. *Physische Geographie*, AK 9: 229.

20. Siba N'Zatioula Grovogui, "Mind, Body and Gut!: Elements of a Postcolonial Human Rights Discourse," in *Decolonizing International Relations*, ed. Branwen Gruffydd Jones (London: Rowman and Littlefield, 2006), 179–96.

21. Walter Mignolo, "Dussel's Philosophy of Liberation. Ethics and the Geopolitics of Knowledge," in *Thinking from the Underside of History*, eds. Linda Alcoff and Eduardo Mendieta (London: Rowman and Littefield, 2000), 27–50; Walter Mignolo, "The Geopolitics of Knowledge and the Colonial Difference," *South Atlantic Quarterly* 101 (2002): 57–96; Nelson Maldonado-Torres, "On the Coloniality of Being: Contribution to the History of a Concept," *Cultural Studies* 21:2–3 (2007): 240–70.

22. Emmanuel Chukwudi Eze, "The Color of Reason: The Idea of 'Race' in Kant's Anthropology," in *Postcolonial African Philosophy: A Critical Reader* (London: Blackwell, 1997), 103–40.

23. See the introduction to this volume for the interconnections between Kant's teaching and the editing of both works.

24. Marshall G. S. Hodgson, *Rethinking World History. Essays on Europe, Islam and World History* (London: Cambridge University Press, 1993), 29–35.

25. "Philosophical Anthropology," *Encyclopædia Britannica Online*, http://www.britannica.com/EBchecked/topic/456743/philosophical-anthropology.

26. *Observations on the Feeling of the Beautiful and Sublime*, AK 2: 243 (97).

27. *Observations on the Feeling of the Beautiful and Sublime*, AK 2: 250 (107).

28. *Observations on the Feeling of the Beautiful and Sublime*, AK 2: 252 (110).

29. *Observations on the Feeling of the Beautiful and Sublime*, AK 2: 252 (110).

30. *Observations on the Feeling of the Beautiful and Sublime*, AK 2: 252 (110).

31. *Observations on the Feeling of the Beautiful and Sublime*, AK 2: 253 (111).

32. *Anthropology with a Pragmatic Intent*, AK 7: 319 (245).

33. *Physical Geography*, AK 9: 316 (G, 223).

34. A relevant article, for the argument developed here, is Thayne R. Rod, *Stranger in Foreign Land: José de Acosta's Scientific Realization in Sixteenth-Century Peru," Sixteenth Century Journal* 29 (1998), 19–33. See also my "Commentary" to the edition of José de Acosta, *Natural and Moral History of the Indies*, ed. Jane E. Mangan, trans. Frances López-Morillas (Durham: Duke University Press, 2002), 451–518.

35. Mignolo, "Introduction," *Decolonizing the Middle Ages*.

36. "Homo Sapiens: Classification Within Primates," *Encyclopeadia Britannica Online*, http://www.britannica.com/EBchecked/topic/1350865/Homo-sapiens.

37. For example, Maori anthropologist Linda Tiwai Smith who assumes both, being anthropologist as a disciplinarian identity and being Maori as an imperial identity attributed by attributing non-European indigenous identities by European indigenous actors. See her *Decolonizing Methodologies, Research and Indigenous People* (London: Zed Books, 1999).

38. I analyze this issue in "Commentary," *Natural and Moral History of the Indies*, as well as in *The Darker Side of the Renaissance*.

39. The myth that secularism was a racial break with theology lasted for a while, perhaps a century or so. But it is clear today that theology provided the frame for secular intellectual to cut the Gordian knot with God. For an early argument, see Carl Schmitt, *Theologie Politique*, trans. Jean-Louis Schlegel (Paris: Gallimard, 1988), 168. Schmitt, like Kant, operates in a pure secularized frame of reference where the sacred in the domain of religion or the metaphysical in the domain of philosophy are out of consideration. For Schmitt, "political theology" refers to the juridical rationality of the Catholic Church translated into the *ius publicum Europeum*, still accepted as Christian in Thomas Hobbes's political system.

40. Carl Schmitt, "Postscript: On the Current Situation Problem: The Legitimacy of Modernity." In *Political Theology II. The Myth of the Closure of Any Political Theology*, translated and introduced by Michael Hoelzl and Graham Ward (Malden: Polity Press, 2008), 117.

41. See section titled *A Geo-Historical Detour on the Way to Kant's Geography* above.

42. At this point neuro-philosophy, for example, which operates under Cartesian principles (of which Kant was not excent), becomes questionable. Of course, given the ideological and economic weight today of "science," the argument cannot be refuted in a footnote. But it is worse leaving a record of things to come. I am thinking here in neuron-philosophical arguments advanced by Patricia Smith Churchland, *Brain-Wise. Studies in Neurophilosophy* (Cambridge: MIT Press, 2002) and Paul Churchland, *Neurophilosophy* at *Work* (Cambridge: Cambridge University Press, 2007).

43. See Mignolo, "Geopolitics of Knowledge and the Colonial Difference."

44. See Lewis Gordon, *Disciplinary Decadence. Living Thought in Trying Times* (Boulder: Paradigm, 2006). See particularly chapter 6, "Próspero's Words, Caliban's Reason," 107–32.

45. Mignolo, *The Darker Side of Renaissance*, 219–58.

46. I developed this elsewhere. See "The Zapatistas's Theoretical Revolution (Its historical, ethical, and political consequences)" available at: http://translated.bythe-zapatistas-s-theoretical-revolution-its-historical-ethical-and-political-consequences/original/. See also the "introduction" to Walter Mignolo, compilador, *Capitalismo y geopolítica del conocimiento* (Durham and Buenos Aires: Ediciones del Signo, 2001), 9–54.

47. See Enrique Dussel, *Hacia una Filosofia Politica Critica* (Bilbao, Desclée de Brouwer, 2001).

48. Raymundo Pannikar, "Aporias in the Contemporary Philosophy of Religion," *Man and World* 13:3–4 (1980): 370ff.; Pannikar, "What is Comparative Philosophy Comparing?," in *Interpreting Across Boundaries*, eds. Gerald J. Larson and Eliot Deutsch (Princeton: Princeton University Press, 1988), 127ff. A preliminary version of these ideas, following Pannikar's work, is in the introduction to Mignolo, *The Darker Side of the Renaissance.*

49. Walter Mignolo, "Epistemic Disobedience and the De-colonial Option: A Manifesto," in *The De-colonial Turn*, ed. Nelson Maldonado-Torres (2008). The Spanish version can be found in *Tabula Rasa* [online]. 2008, n.8, pp. 243–282. ISSN 1794-2489.

18

Geography Is to History as Woman Is to Man

Kant on Sex, Race, and Geography[1]

On the way to an Epilogue

Eduardo Mendieta

> In 1797—Kant was still alive—Lichtenberg wrote in his notebook: "It might be possible that some tenets of Kantian philosophy would be fully understood by no one, and that each would believe someone else understood it better and therefore would content himself with an indistinct insight or would even believe at times that it was his own inability that hindered him from seeing as clearly as others." In fact the second part of the notation missed the situation in the fraternity entirely: none ever believes the other has understood better. But the suggestion that possibly no one could ever understand everything in Kant is wonderfully perceptive of the continent relationship in the reception of the great work. Whether and when it is understood is every bit as contingent as how, when, and that it originated and appeared. Even if there is something to the idea that the extraneous occasion of the bicentennial of Kant's birth helped bring to light the first precise understanding of the fundamental ideas of his major work—or even just a confirmation of the inadequacy of all previous pronouncements about it—then also implicitly confirmed is how futile it was that this work had already lain on the table and in our hands for a century and a half. Why so needlessly early? Why so scandalously long?[2]

In the following I want to build on discussions and analysis of Kant's views on race and sex by bringing in a third element, namely geography. Additionally, while discussions of race and gender in Kant's work have increased, they remain peripheral to Kant scholarship to the detriment of our better understanding of how Kant's corpus as a whole was held together. As has been argued so far by the outstanding contributors gathered between these book covers, Kant's lectures on anthropology arose out of the lectures on physical geography, although theoretically and philosophically they originated in the section on empirical psychology in the lectures on metaphysics.[3] Kant scholars have argued about the context of origin of Kant's anthropology lectures in terms of a distinction nicely articulated by Holly Wilson as the distinction between "origin" (*Ursprung*) and "arise" (*Entstehen*).[4] Whereas the lectures on anthropology are a branching out of the physical geography lectures, the motivation for a more thorough and separate treatment of humanity as a natural creature originated in Kant's lectures concerning the possibility of knowledge, i.e., empirical psychology.[5] Kant began to lecture on *Physical Geography* in 1756 and continued to do so for the next forty years. Sixteen years later, in 1772, he began his course on Anthropology. In 1775 Kant published an essay that served as an advertisement for his Physical Geography and Anthropology lectures, namely the famous *Of the Different Races of Human Beings*. It is evident that this text, and I underscore that it was meant to advertise Kant's lectures, is a direct outgrowth of Kant's work on both physical geography and anthropology. In this famous text we have Kant's articulation of his "acclimatization" and "epigenesis" theories of human racial differences.[6] Kant published his *Anthropology from a Pragmatic Point of View* in 1798, which he saw into press himself. In the preface to this book Kant states in a note appended at the very end that the book arose out of lectures that were "popular" and were attended by people of different classes or standings. In this note, he also excuses himself for not being able to publish the other set of lectures, those on physical geography. All of this is fairly incontrovertible. Now, I want to complicate this chronology by foregrounding the following. Kant deals with human sexual differentiation in the second part of the *Anthropology*, entitled "Anthropological Characteristics." Those pages are very concentrated and relatively brief, but they are also extremely important. What is uniquely distinctive about these pages is that they resemble closely Kant's writings from 1764, specifically *Observations on the Feeling of the Beautiful and Sublime*. Kant returned to the question of sexual difference in 1786 in his "Conjectural Beginning of Human History." Additionally, Kant dealt with sexual difference in terms of marriage in the *Metaphysics of Morals* of 1797. Evidently, there are also scattered remarks on women and sex in his *Lectures on Ethics*, in particular those gathered by Herder and Vigilantius, as well as throughout the notes gathered in volume 15 of the Prussian Academy's edition of Kant's collected works, entitled *Reflexionen zur Anthropologie*. When we read these texts together, we begin to develop a very different picture than that handed down to us by Kant scholars. Kant scholars have divided Kant's intellectual itinerary into two periods, the precritical and the critical, taking the publication of the *Critique of Pure Reason* in 1781 as the dividing point.[7] My argument will be, however, that when we read Kant on woman and sex in general, from the 1764 to 1798, we are able to discern a striking stability and endurance of certain views about one of the most central issues of humanity: sexual difference. Additionally, when we relate Kant's views on sexual difference to those on racial difference, which is an expression of geographical difference, we begin to see that Kant held on to a dualistic anthropology that correlated woman to geography as it correlated man to history. In this inter-textual reading we also see how sexual difference is akin or similar to racial difference. It will be argued that sexual and racial difference, qua modification of human propensities due to the challenges of climate and geography, are essential devices of what Kant called in his *Anthropology* "nature's economy."[8] The organizing principle of this economy is what Kant called the "unsociable sociability" of humans.[9] Both sexual and racial difference are expressions of the essential antagonism, if not enmity, that leads humanity to rise above nature to rationally self-regarding moral dignity. In other words, sexual difference and racial distinction are tools, elements of the design of nature, that instigate a discord, disharmony, conflict, antagonism, and quarrelsomeness that leads humans to rise from the state of nature to the state of civil society.[10] One could speak along with Hans Saner of nature being a "theater of war"[11] that is the very indispensable condition, the design by means of which nature engineers the departure of humans from that condition of animalistic egotism, to rise to the level of rational self-legislation. In tracking the durability and endurance of certain tropes in Kant's work, I am thus concerned with the stability of what Michèle Le Doeuff called a "philosophical imaginary,"[12] or what Hans Blumenberg called "metaphorology."

Kantian Metaphorology

In 1960 Hans Blumenberg published in the *Archiv für Begriffsgeschichte*, then edited by Erich Rothacker, a well-known article of philosophical anthropology, his programmatic essay "Paradigms for Metaphorology."[13] At the time, Blumenberg conceived metaphorology as an aid or handmaiden of *Begriffsgeschichte*, the history of concepts, or what in the English-speaking world is sometimes called "history of ideas." The assumption was that a study of the key metaphors that have guided the development of intellectual paradigms, or worldviews, would allow us to discern the patterns, processes, and logics of concept formation. Thus, the study of metaphors was conceived as a kind of archeology of the preconceptual, or what Blumenberg called much later, "a special case of nonconceptuality."[14] The task of metaphorology, then, was to "approach the substructure of thinking, its underground, where systematic crystallizations are unleashed."[15] But in an essay appended to the 1979 volume *Shipwreck with*

Spectator, entitled "Prospect for a Theory of Nonconceptuality," Blumenberg expands his understanding of the tasks of the metaphorology. Now, it is no longer a mere prolepsis, or prolegomena, to the work of the concept, or conceptuality, but rather a discipline that excavates the metaphors that have guided the human imagination in order to trace the interaction between what he calls following Husserl, the "life-world" and the humanity's approximation to it. As he put it, "metaphors are fossils that indicate an archaic stratum of the trial of theoretical curiosity."[16] These fossils, however, are traces of what he called "nonconceptuality," that which resists clear and distinct theoretical articulation, even if the later makes use of it. The metaphor is indispensable for theoretical reflection. Theoretical reflection and philosophical articulation are elaborated in terms of metaphors that it would like to dispense of once its goals have been accomplished. Yet, these metaphors remain indispensable, irreducible, and more enduring that some theoretical or philosophical articulation. Even as theoretical reflection and philosophical articulation seek to dispose of their form of ascent, by way of a metaphor, it can only do so by means of a new metaphor. Thus, "Demythicization is in large measure nothing more than remetaphorization."[17]

This metaphorology, which has been called by Anselm Haverkamp "The technique of rhetoric,"[18] is what guided the production of Blumenberg's magisterial studies from the seventies: *The Genesis of the Copernican World, The Legitimacy of the Modern World, Work on Myth*, and in the eighties *Die Lesbarkeit der Welt* [The Legibility of the World], *Lebenszeit and Weltzeit* [Living Time and World Time], and *Höhlenausgänge* [Cave Exits].[19] In *The Genesis of the Copernican World*, Blumenberg has a fascinating chapter that illustrates how metaphors both enable, but also remain inassimilable; they remain irritants in the theoretical work that tries to make use of them. The chapter is entitled "What is 'Copernican' In Kant's Turning?"[20] and in it Blumenberg excavates the use and misuse by Kant of the reference to Copernicus' scientific revolution and demonstrates how Kant, in trying to find "an adequate expression, but above all an intuitive, a graphic expression for the radicalness and the special character of this philosophy,"[21] reached out to an analogy and metaphor that went contrary to the very spirit of it. In turn, Kant's analogizing of the transcendental method with Copernicus's scientific theory of the motions of the earth and the planets unleashed a history of reception that has left layers of misunderstanding of Kant's critical project.

Metaphorology is to the philosophical imaginary as metaphor is to imagery. If the former was excavated by Blumenberg, the latter has been archived by Michèle Le Doeuff. In a work that should be read in tandem with many of Blumenberg's, Le Doeuff has focused more exclusively on the way in which images, tropes, allegories, and similes operate in philosophical texts. The philosophical imaginary is one that both enables and hinders, it is one that thinks with and against images. In Le Doeuff's view, then, a history of philosophy should also be a history of those images, and most specifically of the ways in which these images work both for and against the system that deploys and marshals them. That images function within philosophical texts and systems can be expressed in a maximalist and minimalist form. In the minimalist version, these images point to tensions, irritations, and difficulties in an "intellectual venture."[22] Once such disturbances are located, then they may be removed. The maximalist, or broader, view of the operation of images in texts argues that "the meaning conveyed by images works both for and against the system that deploys them. *For*, because they sustain something which the system cannot itself justify, but which is nevertheless needed for its proper working. *Against*, for the same reason—or almost: their meaning is incompatible with the system's possibilities."[23] Le Doeuff proceeds to illustrate how these images, metaphor, allegories function both *for* and against a philosophical system or text by analysis the image of the island in the *Critique of Pure Reason*; specifically, in chapter III of the "Transcendental Doctrine of the Power of Judgment." The passage reads:

> We have now not only traveled throughout the land of pure understanding and carefully inspected its every part, but have also surveyed it throughout, determining for each thing in this land its proper place. This land, however, is an island, and is enclosed by nature itself within unchangeable bounds. It is the land of truth (a charming name), and is surrounded by a vast and stormy ocean, where illusion properly resides and many fog banks and much fast-melting ice feign new-found lands. This sea incessantly deludes the seafarer with empty hopes as he roves through his discoveries, and thus entangles him in adventures that he can never relinquish, nor ever bring to an end. But before we venture upon this sea, to search its latitudes for certainty as to whether there is in them anything to be hoped, it will be useful to being by casting another glance on the map of the land that we are about to leave, and to ask two questions. We should ask, first, whether we might not perhaps be content with what this land contains, or even content with it from necessity if there is no other territory at all on which we should settle. And we should ask, second, by what title we possess even this land and can keep ourselves secured against all hostile claims. Although we have already answered these questions sufficiently in the course of the Analytic, a summary account of the Analytic's solutions may still reinforce one's conviction by uniting their moments in one point.[24]

Le Doeuff's reading is brilliant, suggestive, and too lengthy for analysis here. I want to focus, however, on two key elements of her exegesis of the imagery at work in this passage.

The text is articulated as a mere repetition, and thus it deflates or derogates itself, while also censoring itself. It seduces while denouncing the seduction: The "land of truth"—what a charming name! (Kemp Smith translated this expression as "seductive name!"). But here, the land is not a continent, but an island. This image, or iconography, Le Doeuff tells us, has been copied by Kant from Francis Bacon's *In Temporis Partus Maximus.*[25] Moreover, this island has its symmetrical antithesis, the island of the South Seas, that holds out the "shadowy image" of a "*golden age,*" about which so many "Robinsonades" have been written, and of which Kant writes in his "Conjectural Beginning of Human History."[26] If the former is the island of strife, hazard, and danger, which is only to be assailed by stringent methods, the latter is the island of repose, satiation, happiness, and utopian satisfaction. Le Doeuff could have added that the island of truth has another antithetical island, it is the island that cannot be outside history, and beyond the reach of the civilizing violence that raises humans to their human dignity. This is the island that Kant refers in his review of Herder's *Ideas for the Philosophy of the History of Humanity*, Tahiti.[27] Still, as Le Doeuff puts it ". . . this island [the island of truth] of the *Critique of Pure Reason* is thus the emblem of the Kantian enterprise (an emblem unthinkable in the logic of the Great Instauration), the self-justification of a project which is, for good measure, conducted on the very terrain where that project could least readily be defended: that of the pleasure principle."[28] Indeed, through the image of the island, how we land on it, how we are to distinguish it from other false islands, how we can acquire the proper map of it, Kant has mobilized what Le Doeuff calls most felicitously a "calculus." In Kant's philosophical imagery, the island is an emblem of an affective calculus that refers us to a math of pleasure and pains: "Playing on a calculus of pains, a distribution of affective values, it also works a seduction, produces and structures a fantasy."[29]

Hans Saner has offered us a catalogue of the metaphors and images that structure Kant's corpus in his book *Kants Weg vom Krieg zum Frieden, Vol. 1: Widerstreit und Einheit. Wege zu Kants politischen Denken*, (which could be translated as "Kant's Path from War to Peace. Vol. 1: Contention and Unity. Paths to Kant's Political Thinking"[30]) that echoes both Blumenberg and Le Doeuff. It would be helpful to quote at length:

> Among the metaphors we meet in these functions [of clarity and interpretability], in the field of theoretical philosophy, are some from navigation, representing the relation of the knowable to the unknowable (of a continent to the vast ocean) and the course of thought that remains possible on the borderline (along the shore). We come upon metaphors from astronomy (30: B xvi), which help (by way of the zenith of the sun) to understand the possibility of taking our bearings in what we cannot survey; we encounter

> metaphors from architecture (30: A 707/B 735 and others), which confront the proud pretensions of metaphysics (the tower it seeks to build) with the modest result (the solid foundation); we meet metaphors of road clearance, not infrequently carried by a chiliasm of reason and opening paths that lead to assuredly scientific metaphysics. More frequent and important than these, however, are the metaphors from politics: the images of combat and war, the thought model of peacemaking, and, above all, the image of legal process that encompasses all of *Critique of Pure Reason.*[31]

Along with Blumenberg, Le Doeuff, and Saner,[32] I want to claim that gender and race are caught in the logic of a Kantian metaphorology and philosophical imaginary that is more than the simple concatenation of specific affirmations that may or may not have been disavowed sometime between 1792 and 1795, to pick two dates as Pauline Kleingeld has done in order to defend the position that Kant gave up on his racism towards the end of his career,[33] or between the first lectures on *physical geography* and the publication of the *Anthropology* in 1798.

"Women" Is to the Beautiful as "Men" Is to the Sublime

Kant's *Observations on the Feeling of the Beautiful and Sublime* was one of Kant's most successful works.[34] It was the most popular, and stylistically, the most accomplished. The work gravitates around the distinction between two feelings that are indispensable to human appreciation of art, but, most importantly, these two feelings anticipate Kant's views about the human predisposition for morality. Very succinctly, the feeling of the sublime is associated with horror, with what is awe inspiring, and moves us to unexpected levels of friendship, and also inspires us to disdain eternity and even the world itself. The sublime elevates us beyond or above our human frailty. The feeling of the beautiful is associated with what is pleasant, small, cheerful, sweet, and is accompanied by mirth and joy. The beautiful is orderly, calculable, and foreseeable; like hedges and flowerbeds, it is a product of human industriousness and fulfills what is customary and expectable.[35] In section two of this book, Kant expands his qualifications of the difference between the sublime and the beautiful. There he states, "Understanding is sublime, wit is beautiful. Boldness is sublime and grand, cunning is petty, but beautiful."[36] Not only is rational insight linked to the feeling of that which exalts human nature, but it is the means by which humans make themselves into the admirable moral creatures that are the paragon of humanity which alone makes them deserving of the feeling of the sublime. In other words, "true virtue alone is sublime."[37] In fact, *genuine virtues* alone are sublime, whereas those we *adopt*, are merely beautiful. In this way, the moral character of humans is linked to the attributes of either the sublime or the beautiful. While sociability may be beautiful, pleasant, and joyful, morality alone is sublime and worthy of our reverence.

In section three of *Observations*, Kant addresses the way in which sexual difference correlates with either the beautiful or the sublime. While Kant states at the very outset that persons of either sex should bring together both feelings, it become clear that Kant elaborates a division of labor among the sexes: ". . . one expects that each sex will unite both [noble qualities and beauty], but in such a way that in a woman all other merits should only be united so as to emphasize the character of the beautiful, which is the proper point of reference, while by contrast among the male qualities the sublime should clearly standout as the criterion of his kind."[38] In fact, all "education and instruction" as well as all efforts to "advance the moral perfection" of either sex, must keep before its eyes these proper points of reference: beauty in and for woman, the sublime and noble in and for man. Kant is fairly unambiguous: "Women have a stronger innate feeling for everything that is beautiful, decorative, and adorned."[39] Even as children, they take delight in dressing up, and love what is pleasant and can easily be entertained by trivialities. In contrast to men, who are clumsy, unruly and confused, women are composed, elegant, and display a modest "demeanor." In general, woman has many "sympathetic sentiments, good-heartedness, and compassion. . . ."[40] Women prefer what is beautiful to what is useful. As Kant puts it, "they contain the chief ground for the contrast between the beautiful and the noble qualities in human nature, and even refine the male sex."[41] Thus, our ability to distinguish between beauty and the sublime is due to woman's inborn concern with what is beautiful, and in this way, they educate men and make them more sociable. Yet, even if women refine us, it is men who raise us to the heights of rational insight and moral self-legislation, for while women's understanding is beautiful that of men is deeper, or rather, sublime. Again, even if women "refine" men, their task is as admirable as that which is assigned to men: "For the beauty of all actions it is requisite above all that they display facility and that they seem to be accomplished without painful effort; by contrast, efforts and difficulties that have been overcome arouse admiration and belong to the sublime."[42] We should read these sentences with what Kant is going to claim in 1795 with respect to the moral worth of human actions, namely that those done out of duty alone have moral worth. Duty, as Kant famously argued, is never easy, or to be accomplished without painful toil, nor should it be done even to secure one's happiness. Duty is difficult and demands strenuous labor, a labor that fashions us into the kind of sublime creature that, like the "starry heavens above," "alone fills the mind with admiration and reverence."[43]

What if a woman were to undertake the painful, laborious, punishing, and challenging tasks involve in deep understanding? According to Kant she may do so but only at the expense of her own femininity. "Laborious learning or painful grubbing, even if a woman could get very far with them, destroy the merits that are proper to her sex, and on account of their rarity may well make her into an object of cold admiration, but at the same time they will weaken the charms by means of which she exercises her great power over the opposite sex."[44] In fact, women who may decide to fill their heads with learning, and seek to ape men, may as well have beards, "for that might perhaps better express the mien of depth for which they strive."[45] For this reason, the abstract sciences remain outside their reach. Yet, this does not mean that women are merely relegated to the realm of hedges, flowerbeds, and French gardens. Women do have access to a science, or rather they have their own science: "The content of the great science of woman is rather the human being, and, among human beings, the man. Her philosophical wisdom is not reasoning but sentiment."[46] Her science is that of seduction, slyness, coyness, politeness, and refined sociability. Women hae the task of civilizing male humans to sociability. Women, however, aspire to educate humanity, and man in particular, not because it is a noble duty, but because male uncouth and coarse ways are ugly, not morally reprehensible. Women turn away from what is wicked, evil, immoral, not because it transgresses the moral order, but because it is ugly. In fact, women do not know anything of duty, compulsion, or obligation.[47] Kant, however, manages to praise them in this way: "It is difficult for me to believe that the fair sex is capable of principles, and I hope not to give offense by this, for these are also extremely rare among the male sex."[48] Yet, that women may be beautiful, that they may captivating and coquettishly seduce us into refined behavior is enough.[49]

Notwithstanding the popularity of Kant's *Observations* during his time, it is a work that is now little read, much less seriously considered as part of Kant's true and lasting philosophical contribution. The contemporary point of departure for our assessment of Kant's contributions to aesthetics and the philosophy of art in general is the *Critique of Judgment* (1790). This third critique completes a system, which had as its other two pillars the *Critique of Pure Reason* (1781–87), and the *Critique of Practical Reason* (1798). This decade-long project of the formulations of the critique of theoretical, practical, and teleological reason crystallized what has been called the critical Kant. It is therefore difficult to approach the *third critique* without some reference to the other critiques. Yet, once we acknowledge the hazards and evidently, as Blumenberg would say, layers of reception, we could justifiably point to the "continuity" between the *Observations* and the *Critique of Judgment*. True, all the references to the gendered division of labor between beauty and the sublime have disappeared, as well as the references to national character, and evidently, to the relationship between race and the appreciation of beauty and the approximation to the sublime. As the architectonic of Kant's system unfolded, gender differences or differentiations were now properly assigned to anthropology, as were racial differences. Yet the dynamic, the very logic, of the distinction between the beautiful and the sublime remains the same, or, given now the proper geometry of reason, each experience is properly assigned to the determination of the imagination or reason. While beauty refers to the determination of the imagination by understanding, the sublime refers to reason that grasps itself in the act of overcoming something that is terrifying, abysmal, and incommensurable. In beauty we encounter others, for the judgment about beauty is a judgment that must be communicable. In the sublime, we encounter ourselves alone in our rational natures, for the sublime is about reason affirming its ability to delimit what nature offers us as boundless. If beauty refers us back to our animal, corporeal, sentient natures, the sublime refers us back to our rational natures. It is thus not by happenstance that Jean-François Lyotard personalizes Kant's dyad of the beautiful and the sublime in terms of a scene in which the former appears as a mother and the latter as a husband.[50] As Cornelia Klinger put it, "[t]he gendered implications of the beautiful and the sublime that were implicit in Kant's *Observations* will become invisible in the *Critique of Judgment*. But the implicit gendered connotations of immersion in nature versus distance from nature will play an even more important role. It is under this guise that a gender subtext is present in the *Critique*."[51]

Woman's Refusal and Favor

As was noted above, Kant returns to the question of sexual difference in the second part of his *Anthropology*, namely in the section entitled "Anthropological Characteristic." Characteristic can mean two things. On the one hand, it refers to the physical and physiological character of the human being. On the other, it refers to the moral character of a person. The former marks out the human being as a natural being; the latter as a "rational being with freedom." In the treatment of what is characteristic of the human being we can make the distinctions among what is humanity's *natural aptitude* or natural predisposition,[52] his *temperament* or sensibility, and his *character* or way of thinking. Whereas the first refers to what nature has made of humans (their physiological characteristics, so to say), the latter refers to what humans can make of themselves. Another way we can talk about characteristic is in terms of what is the natural or animal nature that constitutes the human being and therefore what we have to work with when engaging in the process of the moral improvement of the human species. Characteristic also contains an analysis of what individuals make of themselves, in terms of their submitting themselves to the rule of reason in accordance with freedom. Thus, characteristic makes simultaneous reference to the dual character of the human being within Kant's moral and physical anthropology. We are sensible, natural, animal beings, who are also rational and free. It is under this general discussion that Kant then addresses the "character of the sexes." Thus, Kant will discuss simultaneously what both woman and man are as natural and physical entities, and also what they can make of themselves, i.e., what their respective characters are.

Sexual difference is an expression of *providence*, i.e., the design of natural difference for an end.[53] Sexual difference expresses a natural economy that seeks, first, to preserve the species, and second, to secure the cultivation of society among humans. Yet, at the heart of this natural plan, or providence, there lies a tension, or struggle, one we may call the "war of the sexes." Inasmuch as nature invested more "art" in woman, and less in man, nature endowed man with strength, and woman with her own "science." Inasmuch as man is stronger, woman must yield to the stronger. It is this yielding, this submission that is the foundation of the preservation of the species. Yet, it is this very submission that is also the foundation for the union that becomes the family. Equality and equal claims leads to self-love, which in turns leads to disruptive and fragmenting squabbling. The *progress of culture* and the ascension of humanity from the state of nature to the realm of freedom, requires that each member of the union be "superior" in a different way. While man may be superior in strength, woman is superior in guile and cunning. "Feminine ways are called weakness,"[54] but this is a relative weakness, for in her weakness, she is also strong in terms of her own skills and strengths. Man may abrogate for himself the prerogative to give orders on the assumption of his physical strength; yet woman may reproach him for his lack of "generosity," and may even "disarm" him with her tears. So, to the male brute physical strength, Kant juxtaposes female emotional recrimination and censure. Women hold the upper hand over him, for they return us to our ability to make of ourselves what we will, which includes of course, making ourselves either into tyrants, or creatures of self-imposed regulations that constrain our inclinations.

Kant provides a very telling narrative of humanity's emergence from the state of nature in the following way, and it merits lengthy citation:

> There [in the *crude state of nature* where woman can't recriminate and disarm man with her tears and accusations] the woman is a domestic animal. The man leads the way with weapons in his hand, and the woman follows him loaded down with his household belongings. But even where a barbaric civil constitution makes polygamy legal, the most favored woman in his kennel (called harem) knows how to achieve dominion over the man, and he has no end of trouble creating a tolerable peace amid the quarrel of many women to be the one (who is to rule over him). In civil society the woman does not give herself up to the man's desire without marriage, and indeed *monogamous marriage*. Where civilization has not yet ascended to

> the feminine freedom of *gallantry* (where a woman openly has lovers other than her husband), the man punishes his wife if she threatens him with a rival. But when gallantry has become the fashion and jealousy ridiculous (as never fails to happen in a time of luxury), the feminine character reveals itself: by extending favors towards men, woman lays claim to freedom and, at the same time, to the conquest of the entire male sex.[55]

At the dawn of human history, thus, women were treated as beasts of burden. Later, as humans developed social networks and social institutions, they were treated better, but remained property, a property that had the uncanny power to rule over the house lord. In the teleological progression towards civil society, however, woman achieves equality only in monogamous marriage: "woman become free by marriage; man loses his freedom by it."[56] The highest state in the evolution of the war of the sexes, one that establishes a peaceful coexistence when both surrender in their own way, is attained when we are in the situation where "woman *refuses*," but man has to *woo*. Her "surrender is a favor."[57]

Nature's economy, according to Kant, has entrusted to woman's womb the "dearest pledge, namely the species." It is for this reason that nature has also implanted in woman fear of physical injury and in particular of sexual injury. By the same token, this fear has given rise to female timidity. And for this reason, woman must rightly claim the protection of the male. Woman's role as the preserver of the species requires of a protector of the womb of the species. In this way, we have a division of labor that attains its ultimate harmony in the monogamous marriage, where the strength of physical weakness and the weakness of physical strength find balance in male wooing and female surrender.

Woman is also the carrier of the "finer feelings," namely sociability and propriety that belong to culture. In fact, woman is the carrier of culture as such. Culture expands and is firmly implanted in the human soul by the artifice of woman. Man is ruled by woman's modesty, refinement, loquaciousness, and expressiveness. Nature, opines Kant echoing his earlier views about woman, "made her clever while still young in claiming gentle and courteous treatment by the male, so that he would find himself imperceptibly fettered by a child through his own magnanimity, and led by her, if not to morality itself, to that which is its cloak, moral decency, which is the preparation for morality and its recommendation."[58]

This last group of sentences is particularly important and remarkable. Females evoke and awaken in males their sense of power, but a power that must be restrained and controled. Like the sublime, which in his earlier *Observations*, was assigned to males only, woman's guile and cunning awakens in man the sense of his own ability to overcome and go beyond himself. Magnanimity is the ability to be generous even towards those undeserving of our favor. The conquering and triumphant lord grants to the vanquished the favor of not taking their lives, nor perhaps even shaming them for their weakness. This greatness of generosity and forgiveness is what woman awakens in man with her gentleness, courteousness, and vulnerability. Woman guides man if not directly to morality, then to that realm where we subject ourselves to our laws and subjugate our inclination to animality, at the very least to its semblance, namely moral decency. Moral decency is part and parcel of sociability, a realm over which woman reigns sovereign.[59]

Kant summarizes his *Anthropology* in the following way: "The sum total of pragmatic anthropology, in respect to the vocation of the human being and the Characteristic of his formation, is the following. The human being is destined by his reason to live in a society with human beings and in it to *cultivate* himself, to *civilize* himself, and to *moralize* himself by means of the arts and sciences."[60] If we bring together what Kant has said about the role of woman in "civilizing" man in the *Observations*, as well as numerous remarks through his *Lectures on Ethics*,[61] and what we just glossed from the *Anthropology*, it is not difficult to come to the conclusion that women occupy a secondary, even if indispensable, role in advancing the species. Woman's contribution to the cultivation and civilizing of humanity may be beautiful, but it is man's contribution to his moralization that is truly sublime. In short, and as Kneller put it eloquently, "sex represents the constant threat of moral devolution."[62]

Geography Is the Foundation of History

In the fourth section of the introduction to the *Physical Geography*, Kant addresses the related questions of the dependence of history on geography, and the possibility of natural history.[63] With respect to the later, Kant notes that even if it exists in name, it does not exist as an actual science yet. Kant then proceeds briefly to illustrate what a genuine natural history may include by making reference to the history of dogs. Such a history would demonstrate that they descended from one line, and that the different "breeds" (Kant uses here *Racen der Hunde*) developed due to differences in country, climate, and breeding (*Fortflanzung*). This description would alone provide us a natural history of dogs. This can and should be done for every plant and animal on the surface of the planet, and when it all is put together then we would have a genuine natural history. Kant, however, is quick to point out that we cannot rely on written history to compile such a history, for what is written is relatively miniscule, and does not reach very far back into the history of nature, which is not shorter than the history of the world (*Welt*) itself. Only if we were able to "journey through nature in such a way that one noted the changes it has undergone through the whole of time (*alle Zeiten*), then this would yield a genuine natural history."[64]

One must note here that in contrast to animals and plants, a natural history of humanity, or *Homo sapiens,* would not be sufficient. The *Anthropology* is not just a description of a natural entity; it is also a science that seeks to give an account of what humans have made of what nature has granted them. It is for this reason that Kant used the qualifier of *pragmatic* in the title of the lectures that accompanied the ones on *Physical Geography*. Nonetheless, a natural history of humans as a biological entity would be possible, and that history, like the natural history of dogs, horses, goats, and sheep would have to give an account of territories, climates, and different mating practices that gave given rise to the diversity of humans in the different parts of the earth.

This discussion is then followed by a reflection on the relationship between geography and history, in which Kant subordinates history to geography, for there is no history without a space or stage on which it can take place. Even if history is a continuous succession of events, as Kant also notes, things in nature do not cease to change, and thus give rise to different geographies. Nonetheless, there is no history without geography, and the latter is the foundation of the former. For this reason, if we speak of ancient history, then we must also speak of ancient geography. In fact, *Physical Geography* is full of interesting notes and remarks about the changes in rivers, mountains, sea coasts, etc., which demonstrate that Kant had arrived at this conclusion not through induction, but from empirical evidence, gathered from reports and travelogues.[65] At other points, Kant also remarks on the consequences of human actions on natural history. Geography and history, in fact, will be demonstrated to be linked in more than one way.

Physical geography, however, is the most general description of nature, and in this way, it is not only the foundation for history, but also of all other forms of geography. Thus, Kant lists at least five other forms of geography: mathematical, moral, political, mercantile, and theological.[66] Yet, it is most peculiar that Kant does not list here another geography that in fact he did speak of and talk about in the context of the *Geography* and *Anthropology* lectures, namely *Racial Geography*. Perhaps more accurately, Kant does not deem it necessary to distinguish a separate "racial geography," because *physical geography* in as much as it is about a description of all that is the earth is per definition one that assigns certain races to certain regions. In this, Kant stands within a long established tradition that wavers between racial geography and racist geography, i.e., between explaining racial differences with references to geography and assigning certain moral worth and excellence to specific races vis-à-vis those same geographical locations. We know, for instance, that Charles-Louis de Secondat Montesquieu, in his *Spirit of the Laws*, and Jean Jacques Rousseau, *On the Social Contract,* correlated political freedom, moral temperament, and geographical location. In book III, chapter 8 of *On the Social Contract*, Rousseau writes: "Freedom, not being a fruit of every Clime, is not within the reach of every people. The more one meditates upon this principle established by Montesquieu, the more one challenges it, the more opportunities one provides to establish it with new proofs."[67] Moral excellence and human dignity, adds Kant, also have their specific geography.[68]

In fact, Kant does address the "different forms (*Bildung*) and colors" of humans in the different parts of the globe in volume two, second part, first section, of the *Physical Geography*.[69] This particular section is entitled "Vom Menschen," i.e., concerning human beings. The first paragraph of this section begins by mapping different human colors and physiognomy with different regions, starting with the polar regions and moving towards the equator. According to Kant, extreme cold and extreme heat seem to have the same effect on humans, namely their complexion tends to be dark, their hair black. Humans from these regions tend to have scant facial hair, to be small in stature, with wide and flat faces, and large mouths. These observations are then followed by remarks on some "peculiarities" of black humans. I quote: "Negroes were born white, except for their *reproductive* parts and the circle around the navel, which are black. From these parts, the black color spreads over the entire body in the first months [after birth]."[70] This is an important remark because here color is linked to both birth and sexual reproduction. Then Kant proceeds to remark on the fact that while color is related to geography, it is not reversible. That a black person cannot become white even if he or she moves to the template zones of the earth, or conversely, even if a white person moves to the equator he or she does not become black, even as the body becomes extremely tanned. Still, notwithstanding the irreversibility of racial coloring, Kant reaffirms that "climate is the cause" of the diversity of human coloring. Kant is quick to point out that degrees of coloring are related to altitude, in such a way that people who live in valleys and at low altitudes tend to be darker than those who live at the higher ones.

Climate, however, is not solely responsible for the color of the skin and the overall physiognomy of the human being. It is also related to the speed and degree of overall physical development of the human being as such. Different regions, in fact, affect the overall developmental speed and quality of human maturation. Thus, Kant writes:

> In the torrid zones, humans mature more quickly in all aspects than in the temperate zones, but they fail to reach the same degree of perfection. Humanity has its highest degree of perfection in the white race [*Die Menschheit ist in ihrer größten Vollkommenheit in der Race der Weißen*]. The yellow Indians have a somewhat lesser talent. The Negroes are much lower, and lowest of all is part of the American races.[71]

When we turn to the 1775 essay *Of the Different Races of Human Beings* we can easily recognize some of the similar arguments from volume two, part two, of the *Physical Geography*. The difference, evidently, is that the essay extends Kant's examples and argumentation. Essentially, however, the structure is the same. In the essay we find an affirmation of the homogenesis and epigenesis of the different races, whether they be of plants, animals, or humans. After a long evolutionary history, which involved thousands of years of migrations, humans settled on four major regions of the planet, giving rise to four fundamental or basic races of humanity. They are, according to Kant: "1) the race of the *whites*, 2) the *Negro Race*, 3) The Hunish (Mongolian or Kalmuckian) race, 4) the Hindu or Hindustani race."[72] Most interestingly, Kant then claims that he includes among the white race Moors, Arabs, Turkish-Tatars, and Persians, and even all those people from Asia who are not included among the remaining three basic races. The white race, Kant notes, is mostly to be found in Europe, even if these other peoples he lists are to be counted as white. Later on towards the end of this section where Kant discusses the different races, he concludes with a seemingly inconsistent passage: "The reason for assuming that Negroes and whites to be basic races, is in itself clear."[73] It is self-evident because all other racial variations can be explained in terms of different degrees of mixing of white and black. Still, later on Kant returns to the four basic races and lays them out in terms of climatic factors thusly:

> Phyletic Species
> Whites of brunette color.
> First race, High blondes (Northern Europeans) from humid cold.
> Second race, Copper-reds (Americans) from dry cold.
> Third race, Blacks (Senegambia) from humid heat.
> Fourth race, Olive-Yellow (Indians) from dry heat.[74]

As in the *Physical Geography*, in the essay "Of the Different Races of Human Beings" Kant links both disposition and the achievement of the fullest expression of human perfection to climate. Thus, Negroes, who are "amply supplied by their motherland [Africa?]" are "lazy, soft, and thrifling."[75] Similarly, the Native Americans also exhibit a "half extinguished life power,"[76] one that can also be exhibited by those living in the cold regions of the world. But if the New World, the torrid regions, and the Poles predispose humans towards indolence, lassitude, and weakness, there is nonetheless one region that is the most propitious for humans, and that is the region that lay between the "31st and 52nd degree of latitude in the ancient world."[77] As Kant explains it, this latitude "is rightly taken for that region of the earth in which the most fortunate mixture of the influences of the colder and hotter regions are found and also the greatest riches in creatures of the earth are found; and where also the human being must have diverged the least from his original formation, given that he is equally well prepared for all transplantings from there. Now here we do indeed find inhabitants that are white, however they are *brunette*, which shape we thus want to assume to be the one closest to that of the phylectic species."[78] Or, in other words, humanity descends from the original inhabitants of the old world, who were originally white. And, while the other races are manifestations of geographical accommodations of the original species seed, the European *brunettes* remain the most perfect expression of the original human species and the superior adaptation to the most generative region of the planet. Reinhard Brandt has summarized succinctly Kant's view on race:

> Kant's remarks . . . leave no doubt that [he, namely Kant, believes that] the white race is intellectually and morally superior to the remaining three not only in degree but qualitatively; only whites are capable of progress; only whites can act from moral principles, and, as a result, do justice to the demand of the categorical imperative. Whites, that is to say clearly and exclusively: the white man and not also the white woman.[79]

As with woman, so with races, a putative monogenesis gives way to a hierarchical chain of being, in which just as woman is subject to man, the non-white races are subject to the white race, as the hotter or colder regions are subordinate to the temperate regions of the planet. The economy of nature is what compels humans to ascend from animal egotism to self-legislation through reason. As Yovel put it, ". . . nature itself, even without the rational will, is working according to a hidden design, bringing about political progress by means of violence and passion. It is through wars, exploitation and calculated self-interest that new political institutions, domestic and international, are created, which in effect serve the goals of reason and freedom."[80] This hidden plan, which Yovel calls rightly "the cunning of nature," uses "antagonism" or what Kant calls the "unsociable sociability"[81] as a means to rise humans above the animal status to the status of rational dignity. This "cunning of nature," however, enacts a price, and it is the price of what is left over by nature as adaptations, or momentary manifestations. The detritus of the cunning of nature is nothing else than the "sublime waste"[82] of the human races and the sex that is never capable of overcoming its naturally determined "immaturity." In as much as the telos of nature's violence is the elevation of humans to rational dignity and moral worth, then what we have in Kant's anthropology is an *anthropodicy*,[83] one that just as theodicy justifies a violent and discriminating *Heilsgeschichte*, also justifies the violence of human history as indispensable to the very evolution of humans out of the state of nature. Individuals, the fair sex, and the lower races "might be sacrificed on the pyre that illuminates reason's way on its path to the whole [cosmopolitan history]."[84] As Kant had put it in 1755 in his *Universal Natural History and Theory of the Heavens*, such destruction should not be lamented for it is by means of that destruction that nature shows its prodigality. "What an innumerable multitude of flowers and insects are destroyed by a single cold day! And how little are they missed, although they are glorious products of the art of nature and demonstrations of Divine Omnipotence! In another place, however, this lost is again compensated for the superabundance. Man who seems to be masterpiece of the creation, is himself not excepted from this law."[85] In *Reflection* 1520, from manuscripts from the 1770s and 80s, Kant writes: "All the races will be wiped out (Americans and Negroes can not rule themselves. They are only good as slaves), except the Whites . . . All the revolutions in the world have come from the white race. Our (ancient) history of humanity goes dependably on the white race . . . All culture has often began in the north."[86] Race, like sex, is part and parcel of nature's creative destruction that gives birth to its masterpiece: man, in its fullest perfection, in the white race.

Conclusion

The economy of nature, or what we have called along with Yovel the "cunning of nature," includes not only sexual differentiation, with its respective division of labor, but also racial differentiation, with its respective degrees of exhibition of accomplishment of human perfection. If sexual differentiation is the artful device through which nature raises humans to the level of moral dignity, racial difference registers the way in which human have been "despotically willed"[87] by nature that they lived everywhere on the planet, so that they can be challenged by its different climates and environments. If sexual difference works to civilize and moralize humanity so that it can ascend to its rational *self-esteem*, racial difference analogously works to provide exemplars of what are the highest physical and cultural expression of humanity, so that some may be judged as either failed experiments or successes.[88]

In his "Idea for a Universal History from a Cosmopolitan Point of View" from 1784, Kant comes closest to expressing the organizing principle of nature's economy, when in the fourth proposition he writes:

> *The means that nature employs in order to bring about the development of all of its predispositions is their* ***antagonism*** *in society, insofar as the latter is in the end the cause of their lawful order.* Here I understand by "antagonism" the unsociable sociability of human beings, i.e. their propensity to enter into society, which, however, is combined with a thoroughgoing resistance that constantly threatens to break up this society. This predisposition for this obviously lies in human nature.[89]

This *unsociable sociability* expresses itself in the war of the sexes, as well as in the war of the races and cultures, and finally, in the war of man against his environment. Ultimately, however, the end is good, for the ultimate goal is the emergence of a cosmopolitan world order under the rule of law. In fact, nature's economy will lead towards the "progressive organization of citizens of the earth into and towards the species as a system that is cosmopolitically united."[90]

In this chapter I have aimed to show that Kant's work is not just animated, oriented, guided, and informed by a philosophical imaginary that has at its core conflict and war, but also that this metaphorology informs Kant's way of making sense of sexual, racial, and geographical difference. At the very least, I have made the case that we cannot read Kant's views on sex without considering what he has to say about both race and geographical difference. What is before us, thus, is the task of engaging Kant's work and his tradition on a larger canvas, the canvas of almost half a century of teaching on geography and anthropology. In his recent book *Kantian Ethics*, Allen W. Wood opens with a chapter entitled "Reason" in which he immediately takes the challenge that Kant's views on race and sex invalidate or cast a long shadow over Kant's views on reason and morality. In particular, Wood is at pains to dissociate Kant's more "disturbing" remarks—as Louden calls them[91]—from the way Kant got some basic principles of ethics and cosmopolitan law correct. Wood writes: "The correct standard for an ethical theory is whether it gets things right at the level of basic principles and values, not whether it contains some magical property that protects us, in the application of the theory, from every perversion or abuse through the influence of tradition and prejudice or the infinite human ingenuity of rationalization. All theories are equally subject to such abuse, and no theory is immune to it."[92] All of the essays gathered in this book have shown that it is not simply a matter of keeping the formulation of the basic principles immune from the abuses of their application. Each contributor has in his or her own way shown that Kant's aim to provide citizens of Königsberg with cosmopolitan knowledge with a pragmatic intent suffused all of his pedagogy as well as system building. The Kantian distinction between grounding (*begrunden*) and application (*anwendung*) is one that is challenged by all the contributors here gathered. The task, thus, remains to begin to disentangle what is to be saved and what is to be disposed off, to uncouple Kant's own prejudices from what he glimpsed as humanly invariant notwithstanding his own eighteenth-century chauvinist and racist views. At the very least, the chapters here gathered are an invitation not to dispose, bury, brand as useless and anachronistic, Kant. Rather, it is evident that all the authors gathered in this book share a profound respect and appreciation of Kant's work, and how much work there remains to be done to do justice to all the spheres of human knowledge to which he contributed in such an unparalleled fashion.

I began this chapter with a citation from Hans Blumenberg that ends with two questions: "Why so needlessly early? Why so scandalously long?" In this volume we have come too early, as the actual translation of Kant's *Physical Geography* has yet to appear, and when that translation appears it will be already obsolete, as some of our contributors have so eloquently shown. Even then, we will still have to wait for the translation of the notes from volume 15 of the Prussian Academy edition of Kant's Collected Writings, which include his *Reflections on Anthropology*. But, we have also come "scandalously" late. Why has it taken so long for scholars in the English speaking world—evidently with the exception of those here included—to engage what is surely one of Kant's longest, richest, most diverse contributions: his teaching on geography and anthropology. Why, beyond the obvious absence of reliable translations, has it taken this long for us to engage this witty, chatty, engaging, committed, but also prejudiced and provincial Kant? When we look into a book it is as if we are looking into a mirror: we see what we want to see, what we let ourselves see. When we look into Kant's works, why do we refuse to see what we in fact can see?

Notes

1. A very early version of this paper was presented at the workshop Stuart Elden organized in Durham, England. I benefited greatly from the comments and criticisms there articulated by Robert Louden, Robert Bernasconi, Emily Brady, Werner Stark, and Charlie Withers. A longer version was presented at the New School, where it was energetically debated. I also want to express my thanks to Stuart Elden, Holly Wilson, Roy Scranton, and Martin Woessner who read the manuscript and made detailed comments.

2. Hans Blumenberg, "Does it Matter When? On Time Indifference," trans. David Adams, *Philosophy and Literature* 22:1 (1998): 212–8.

3. See the excellent collection of essays on the *Anthropology*, Brian Jacobs and Patrick Kain, eds., *Essays on Kant's Anthropology* (Cambridge: Cambridge University Press, 2003), especially the essay by Werner Stark, which details the evolution of the lectures and how they related to the published version of the lectures. A useful, though highly polemical, work is Frederick P. van de Pitte, *Kant as Philosophical Anthropologist* (The Hague: Martinus Nijhoff, 1971). See also his essay "Kant as Philosophical Anthropologist," in *Proceedings of the Third International Kant Congress*, ed. Lewis White Beck (Dordrecht: D. Reidel, 1970), 574–81.

4. Holly L. Wilson, *Kant's Pragmatic Anthropology: Its Origin, Meaning, and Critical Significance* (Albany: SUNY Press, 2006), 7–26.

5. Roubert B. Louden, *Kant's Impure Ethics: From Rational Beings to Human Beings* (Oxford: Oxford University Press, 2000), 62–4.

6. See Alix A. Cohen, "Kant on Epigenesis, Monogenesis and Human Nature: The Biological Premises of Anthropology," *Studies in History and Philosophy of Biological and Biomedical Sciences* 37 (2006): 675–93, as well as Philip R. Sloan, "Kant on the History of Nature: The Ambiguous Heritage of the Critical Philosophy for Natural History," *Studies in History and Philosophy of Biological and Biomedical Sciences* 37

(2006): 627–48, and Phillip R. Sloan, "Preforming the Categories: Eighteenth-Century Generation Theory and the Biological Roots of Kant's Apriori," *Journal of the History of Philosophy* 40:2 (2002): 229–53.

7. For a discussion of this division see Paul A. Schilpp, *Kant's Pre-Critical Ethics* (Bristol: Thoemmes, 1989).

8. *Anthropology from a Pragmatic Point of View*, in the Academy edition of Kant's collected works (hereafter referred to as AK, followed by volume and page number), AK 7: 310.

9. See Allen W. Wood, "Unsociable Sociability: The Anthropological Basis of Kantian Ethics," *Philosophical Topics* 19:1 (Spring 1991): 325–51. See also Robert B. Louden, *Kant's Impure Ethics*, especially Part II, ch. 3, 62–106.

10. If space had permitted it, I would have also included discussions of the role of discord and conflict in Kant's *The Conflict of the Faculties*. Holly Wilson has generously already identified the relevant passages. I thank her for bringing this to my attention.

11. Hans Saner, *Kant's Political Thought: Its Origin and Development*, trans. E. B. Ashton (Chicago: University of Chicago Press, 1973), 230.

12. Michèle Le Doeuff, *The Philosophical Imaginary*, trans. Colin Gordon (Stanford: Stanford University Press, 1989).

13. Hans Blumenberg, "Paradigmen zu einer Metaphorologie," *Archiv fur Begriffsgeschichte* 6 (1960), 7-142. I will be quoting from the Suhrkamp re-edition. *Paradigmen zu einer Metaphorologie* (Frankfurt a.M.: Suhrkamp, 1998). For a brief discussion of "metaphorology" see Translator's Introduction in Hans Blumenberg, *Shipwreck with Spectator: Paradigm of a Metaphor for Existence*, trans. Steven Rendall (Cambridge: MIT Press, 1997), 1–5.

14. Blumenberg, "Prospect for a Theory of Nonconceptuality" in *Shipwreck with Spectator*, 81–102, quote at 81.

15. Blumenberg, *Paradigmen zur einer Metaphorologie*, 13.

16. Blumenberg, "Prospect for a Theory of Nonceptuality," 82.

17. Blumenberg, "Prospect for a Theory of Nonceptuality," 94.

18. See Hans Blumenberg, *Ästhetische und metaphorologische Schriften* (Frankfurt a.M.: Suhrkamp, 2001), 435–54.

19. For a bibliography and chronology of Blumenberg's writings see Franz Jose Wetz, *Hans Blumenberg zur Einführung* (Hamburg: Junius, 1993).

20. Hans Blumenberg, *The Genesis of the Copernican World*, trans. Robert M. Wallace (Cambridge: MIT Press, 1987), 595–614.

21. Blumenberg, *Genesis of the Copernican World*, 595.

22. Le Doeuff, *The Philosophical Imaginary*, 3.

23. Le Doeuff, *The Philosophical Imaginary*, 3.

24. *Critique of Pure Reason*, trans. Werner S. Pluhar (Indianapolis: Hackett, 1996), A236/B295 (303–4).

25. Le Doeuff, *The Philosophical Imaginary*, 9.

26. "Conjectural Beginning of Human History," AK 8: 122.

27. "Review of J. G. Herder's *Ideas for the Philosophy of the History of Humanity*, Parts 1–2," AK 8: 65.

28. Le Doeuff, *The Philosophical Imaginary*, 11.

29. Le Doeuff, *The Philosophical Imaginary*, 12.

30. Saner, *Kant's Political Thought.*

31. Saner, *Kant's Political Thought*, 217–8.

32. I should add Yirmiahu Yovel to this list, for he has also has discussed the central function that metaphors play in Kant's work, to the degree that Yovel argues that "reason" itself should be thought of as being conceived by Kant as a metaphor. See Yirmiahu Yovel, *Kant and the Philosophy of History* (Princeton: Princeton University Press, 1980), 16.

33. Pauline Kleingeld, "Kant's Second Thoughts on Race," *The Philosophical Quarterly* 57:229 (October 2007), 573–92.

34. John H. Zammito, *Kant, Herder, The Birth of Anthropology* (Chicago: University of Chicago Press, 2002), 106.

35. *Observations on the Feeling of the Beautiful and Sublime*, AK 2: 209–10.

36. *Observations on the Feeling of the Beautiful and Sublime*, AK 2: 211.

37. *Observations on the Feeling of the Beautiful and Sublime*, AK 2: 215.

38. *Observations on the Feeling of the Beautiful and Sublime*, AK 2: 228. Bold in the original.

39. *Observations on the Feeling of the Beautiful and Sublime*, AK 2: 229.

40. *Observations on the Feeling of the Beautiful and Sublime*, AK 2: 229.

41. *Observations on the Feeling of the Beautiful and Sublime*, AK 2: 229.

42. *Observations on the Feeling of the Beautiful and Sublime*, AK 2: 229.

43. *Critique of Practical Reason*, AK 5: 161.

44. *Observations on the Feeling of the Beautiful and Sublime*, AK 2: 229.

45. *Observations on the Feeling of the Beautiful and Sublime*, AK 2: 230.

46. *Observations on the Feeling of the Beautiful and Sublime*, AK 2: 230.

47. *Observations on the Feeling of the Beautiful and Sublime*, AK 2: 231.

48. *Observations on the Feeling of the Beautiful and Sublime*, AK 2: 232.

49. *Observations on the Feeling of the Beautiful and Sublime*, AK 2: 240.

50. See Cornelia Klinger, "The Concept of the Sublime and the Beautiful in Kant and Lyotard," in *Feminist Interpretations of Immanuel Kant*, ed. Robin May Schott (University Park: The Pennsylvania State University Press, 1997), 192–211.

51. Klinger, "The Concept of the Sublime and the Beautiful in Kant and Lyotard," 196.

52. *Anthropology from a Pragmatic Point of View*, AK 7: 285–6.

53. Kant also discusses sexual differences in the *Critique of Judgment*, §82 (AK 5: 425).

54. *Anthropology from a Pragmatic Point of View*, AK 7: 303.

55. *Anthropology from a Pragmatic Point of View*, AK 7: 304–5.

56. *Anthropology from a Pragmatic Point of View*, AK 7: 309.

57. *Anthropology from a Pragmatic Point of View*, AK 7: 306. As extreme as these formulations may sound to our ears, they express in pedestrian language what Kant himself formalizes as a theory of right within marriage. Thus, his views about gender carry over into his practical philosophy. See the excellent articles by Allegra de Laurentiis, "Kant's Shameful Proposition: A Hegel-Inspired Criticism of Kant's Theory of Domestic Right," *International Philosophical Quarterly* 40:3 (September 2000), 297–312; Jane Kneller, "Kant on Sex and Marriage Right" in *Kant and Modern Philosophy*, ed. Paul Guyer (Cambridge: Cambridge University Press, 2006), 447–76; Holly L. Wilson,

"Kant's Evolutionary Theory of Marriage" in *Autonomy and Community: Readings in Contemporary Kantian Social Philosophy*, eds. Jane Kneller and Sidney Axinn (Albany: SUNY Press, 1998), 283–306. See also Hannelore Schröder, "Kant's Patriarchal Order" in *Feminist Interpretations of Immanuel Kant*, 275–96.

58. *Anthropology from a Pragmatic Point of View*, AK 7: 306.

59. For the role of shame, humiliation, and decency in Kant see the provocative book by Paul Saurette, *The Kantian Imperative: Humiliation, Common Sense, Politics* (Toronto: University of Toronto Press, 2005).

60. *Anthropology from a Pragmatic Point of View*, AK 7: 324–5.

61. See *Lectures on Ethics*, ed. Peter Heath and J. B. Schneewind (Cambridge: Cambridge University Press, 1996), 22–3, 155–60.

62. Jane Kneller, "Kant on sex and marriage right," 465.

63. *Naturgeschichte*, 9: 162.

64. AK 9: 162.

65. See Charles W. J. Withers, *Placing the Enlightenment: Thinking Geographically about the Age of Reason* (Chicago: University of Chicago Press, 2007); H. Glenn Penny and Matti Bunzl, eds., *Worldly Provincialism: German Anthropology in the Age of Empire* (Ann Arbor: University of Michigan Press, 2003), especially the chapters by Harry Liebersohn and Pascal Grosse; John Gascoigne, "The German Enlightenment and the Pacific" in *The Anthropology of the Enlightenment*, eds. Larry Wolff and Marco Cipolloni, (Palo Alto: Stanford University Press, 2007), 141–71.

66. *Physical Geography*, AK 9: 164–5.

67. Jean Jacques Rousseau, *The Social Contract and Other Later Political Writings*, ed. and trans. Victor Gourevitch (Cambridge: Cambridge University Press, 1997), 100.

68. On Kant's relationship to Rousseau, see the still indispensable Ernst Cassirer, *Rousseau, Kant, Goethe*, trans. James Gutmann, Paul Oskar Kristeller, and John Herman Randall, Jr. (Princeton: Princeton University Press, 1945). See also Emmanuel Chukwudi Eze, "The Color of Reason: The Idea of 'Race' in Kant" in *Postcolonial African Philosophy: A Critical Reader*, ed. Emmanuel Chukwudi Eze (Malden: Blackwell, 1997), 103–40.

69. See Eze, ed., *Race and the Enlightenment*, for a generous selection of writing by Kant where he discusses race.

70. *Physical Geography*, AK 9: 312.

71. *Physical Geography*, AK 9: 316.

72. *Of the Different Races of Human Beings*, AK 2: 432.

73. *Of the Different Races of Human Beings*, AK 2: 433.

74. *Of the Different Races of Human Beings*, AK 2: 441.

75. *Of the Different Races of Human Beings*, AK 2: 438.

76. *Of the Different Races of Human Beings*, AK 2: 438.

77. *Of the Different Races of Human Beings*, AK 2: 440.

78. *Of the Different Races of Human Beings*, AK 2: 441–2.

79. Reinhard Brandt, *D'Artagnan und die Urteilstafel: Über ein Ordnungsprinzip der europaischen Kulturgeschichte* (Stuttgart: Steiner, 1991), 115, quoted in Louden, *Kant's Impure Ethics*, 98.

80. Yovel, *Kant and the Philosophy of History*, 8.

81. "Idea for a Universal History from a Cosmopolitan Point of View," AK 8: 20.

82. I am appropriating Mark Larrimore's most apt expression. See Mark Larrimore, "Sublime Waste: Kant on the Destiny of the 'Races' " in *Civilization and Oppression*, ed. Catherine Wilson (Calgary: University of Calgary Press, 1999), 99–126.

83. Saner, *Kant's Political Thought*, 4. See also Reinhard Brandt, "The Vocation of the Human Being" in *Essays on Kant's Anthropology*, 85–104.

84. Brant, "The Vocation of the Human Being," 99.

85. *Universal Natural History and Theory of the Heavens*, trans. William Hastie (Ann Arbor: University of Michigan Press, 1969), AK 1: 318 (150).

86. *Reflection 1520: Character of Race*, AK 15: 878–80.

87. *Perpetual Peace: A Philosophical Essay*, AK 8: 364.

88. In fact, Kant contemplated the thought that the cunning of nature included the possibility that all the other races would be destroyed and that only the white race would survive. See Robert Bernasconi's gloss on *Reflection 1520* in Robert Bernasconi, "Kant as an Unfamiliar Source of Racism," in *Philosophers on Race: Critical Essays*, eds. Julie K. Ward and Tommy L. Lott (Malden: Blackwell Publishers, 2002), 145–66, quote at 159.

89. "Idea for a Universal History from a Cosmopolitan Point of View," AK 8: 20.

90. *Anthropology from a Pragmatic Point of View*, AK 7: 333.

91. Louden, *Kant's Impure Ethics*, 98.

92. Allen W. Wood, *Kantian Ethics* (Cambridge: Cambridge University Press, 2008), 12.

Contributors

Robert Bernasconi is the Edwin Erie Sparks Professor of Philosophy at Pennsylvania State University. He has published widely in recent continental philosophy, and social and political philosophy, on Heidegger, Hegel, and race theory. He is the author of *How to Read Sartre* (W.W. Norton & Co., 2007), editor of *Race* (Blackwell, 2001), and the editor, with Sybol Cook, of *Race and Racism in Continental Philosophy* (Indiana University Press, 2003)

Samuel A. Butler earned a Ph.D. in Philosophy at the State University of New York at Stony Brook. He has translated articles from French and German from a variety of contemporary social and political philosophers. His original research, focusing on work, workers, and the working classes, has appeared in *International Studies in Philosophy*, *Regional Labor Review*, and *Socialist Studies.*

Edward S. Casey is Distinguished Professor of Philosophy at the State University of New York at Stony Brook. Among his books are *Remembering: A Phenomenological Study* (Indiana University Press, 1987); *Getting Back into Place* (Indiana University Press, 1993); and *The Fate of Place: A Philosophical History* (University of California Press, 1996). His most recent book is *The World at a Glance* (Indiana University Press, 2007)

Michael Church is Professor Emeritus in the Department of Geography at the University of British Columbia, and a fluvial geomorphologist. He is the current editor of the *Journal of Geophysical Research—Earth Surface.* He has a long-term interest in the history of geomorphology, and it is from this perspective that he comes to the consideration of Kant and his influence on the development of geography.

Jeffrey Edwards is Associate Professor of Philosophy at the State University of New York at Stony Brook. He works primarily in Kant and the history of modern philosophy. His *Autonomy, Right and Anthropology in Kant's Practical Philosophy* is forthcoming from the University of California Press, and his translation of the *Essay on Living Forces* appears in Volume 8 of the Cambridge *Works of Kant in English* (2007).

Stuart Elden is Professor of Political Geography at Durham University. He is the author of *Speaking Against Number: Heidegger, Language and the Politics of Calculation* (Edinburgh University Press, 2006) and *Terror and Territory: The Spatial Extent of Sovereignty* (University of Minnesota Press, 2009), as well as books on Lefebvre and Foucault. He is currently completing a history of the concept of territory.

David Harvey is Distinguished Professor of Anthropology at the City University of New York Graduate Center. One of the foremost theorists of geography and of geographical social theory, he has published extensively on the political economy of globalization, urbanization, and cultural change. A new edition of his classic work *Limits to Capital* was published by Verso Books in 2007, and among his recent books is *Cosmopolitanism and the Geographies of Freedom* (Columbia University Press, 2009).

Robert Louden is a Professor of Philosophy at the University of Southern Maine. He is the author of *Kant's Impure Ethics: From Rational Beings to Human Beings* (Oxford University Press, 2002) and the translator of a new edition of Kant's *Anthropology from a Pragmatic Point of View* (Cambridge University Press, 2006). He is the co-editor and translator of *Anthropology, History and Education* (2008) in the Cambridge Edition of the Works of Immanuel Kant in English Translation.

Jeff Malpas is Professor of Philosophy at University of Tasmania. He is the author of *Place and Experience* (Cambridge University Press, 1999), *Heidegger's Topology: Being, Place, World* (MIT Press, 2008), and the editor of volumes on the work of Donald Davidson, Hubert Dreyfus, Hans-Georg Gadamer, and Martin Heidegger.

Max Marcuzzi is maître de conférences at the Université d'Aix-en-Provence, and one of the French translators of Kant's lectures on geography. He has recently translated and edited a critical edition of Kant's *Vers la paix perpétuelle* (Librairie Philosophique J. Vrin, 2007) and *Fichte: la philosophie pratique* (Publications de l'Université de Provence, 2008).

Eduardo Mendieta is Professor of Philosophy at the State University of New York at Stony Brook. His most recent book, *Global Fragments* (SUNY Press, 2007) ties together many diverse areas of his research, including global ethics and globalization, Latin American philosophy, race theory, critical theory, and pragmatism.

Walter Mignolo is William H. Wanamaker Professor and Director of the Global Studies and the Humanities at the John Hope Franklin Center for Internatonal and Interdisciplinary Studies at Duke University. His recent publications include *The Idea of Latin America* (Blackwell Publishing, 2005), *Local Histories/Global Designs: Coloniality, Subaltern Knowledges and Border Thinking* (Princeton University Press, 2000), and *The Darker Side of the Renaissance: Literacy, Territoriality & Colonization* (University of Michigan Press, 2003).

David Morris is Associate Professor of Philosophy at Concordia University. He does work on Hegel, with a focus on the philosophy of the body, mind, and nature in relation to current biology and cognitive science. He is currently studying the problem of the genesis of meaning and sense in relation to biological and perceptual phenomena. His most recent work is *The Sense of Space* (SUNY Press, 2004). His forthcoming book is tentatively titled *Time, Life and Meaning in Hegel's Science of Logic.*

Onora O'Neill is Professor of Philosophy at the University of Cambridge, President of the British Academy, and a crossbench member of the House of Lords. She has published extensively on Kant and in the field of bioethics, including *Constructions of Reason: Explorations of Kant's Practical Philosophy* (Cambridge University Press, 1989) and *Rethinking Informed Consent in Bioethics* (Cambridge University Press, 2007).

Olaf Reinhardt is Visiting Associate Professor of German Studies at the University of New South Wales. He has recently completed a large government-funded project called *What Myth is That?*, a prototype of a multimedia CD-ROM enabling users to find any classical myth that occurs in modern films or works of literature. He has translated Kant's *Physical Geography* for the Cambridge Edition of the Works of Immanuel Kant in Translation.

Werner Stark is Honorarprofessor at Philipps-Universität, Marburg, and a researcher at the Berlin-Brandenburgische Akademie der Wissenschaften, working on the Akademie edition of Kant's works. He is the co-editor of volumes 25 (1997) and 26 (forthcoming) of *Kants Gesammelte Schriften*, which comprise the lectures on anthropology and geography.

Karsten M. Thiel is lecturer at the Philosophy Department of the Ludwig-Maximilians-Universität Munich and was ARC Linkage International Fellow with Jeff Malpas at University of Tasmania in 2008. He authored a volume on Kant's doctrine of transcendental illusion *Kant und die "Eigentliche Methode der Metaphysik"* (Olms, 2008)), and also works on Nietzsche, philosophy of history, and theoretical issues in business ethics.

Holly L. Wilson is Associate Professor of Philosophy at the University of Louisiana at Monroe. She is the author of *Kant's Pragmatic Anthropology: Its Origin, Meaning and Critical Significance* (SUNY Press, 2006).

Charles W. J. Withers is Professor of Historical Geography at the University of Edinburgh. A Fellow of the British Academy, he is the author of, most recently, *Placing the Enlightenment: Thinking Geographically about the Age of Reason* (University of Chicago Press, 2007) and co-editor of *Georgian Geographies: Essays on Space, Place and Landscape in the Eighteenth Century* (University of Manchester Press, 2004) and *Geography and Revolution* (University of Chicago Press, 2005).

Index